International Series in Operations Research & Management Science

Volume 378

The book series **International Series in Operations Research and Management Science** encompasses the various areas of operations research and management science. Both theoretical and applied books are included. It describes current advances anywhere in the world that are at the cutting edge of the field. The series is aimed especially at researchers, advanced graduate students, and sophisticated practitioners.

The series features three types of books:

• Advanced expository books that extend and unify our understanding of particular areas.

• Research monographs that make substantial contributions to knowledge.

• Handbooks that define the new state of the art in particular areas. Each handbook will be edited by a leading authority in the area who will organize a team of experts on various aspects of the topic to write individual chapters. A handbook may emphasize expository surveys or completely new advances (either research or applications) or a combination of both.

The series emphasizes the following four areas:

Mathematical Programming: Including linear programming, integer programming, nonlinear programming, interior point methods, game theory, network optimization models, combinatorics, equilibrium programming, complementarity theory, multiobjective optimization, dynamic programming, stochastic programming, complexity theory, etc.

Applied Probability: Including queuing theory, simulation, renewal theory, Brownian motion and diffusion processes, decision analysis, Markov decision processes, reliability theory, forecasting, other stochastic processes motivated by applications, etc.

Production and Operations Management: Including inventory theory, production scheduling, capacity planning, facility location, supply chain management, distribution systems, materials requirements planning, just-in-time systems, flexible manufacturing systems, design of production lines, logistical planning, strategic issues, etc.

Applications of Operations Research and Management Science: Including telecommunications, health care, capital budgeting and finance, economics, marketing, public policy, military operations research, humanitarian relief and disaster mitigation, service operations, transportation systems, etc.

This book series is indexed in Scopus.

Daniel P. McGibney

Applied Linear Regression for Business Analytics with Python

A Practical Guide Using Ravix with Case Studies

Springer

Daniel P. McGibney
Management Science
University of Miami
Coral Gables, FL, USA

ISSN 0884-8289 ISSN 2214-7934 (electronic)
International Series in Operations Research & Management Science
ISBN 978-3-032-23805-4 ISBN 978-3-032-23806-1 (eBook)
https://doi.org/10.1007/978-3-032-23806-1

This Springer imprint is published by the registered company Springer Nature Switzerland AG
The registered company address is: Gewerbestrasse 11, 6330 Cham, Switzerland

For my family, friends, and students.

Preface

In regression analysis, the concepts to be learned are mathematical and require programming and intuition. The mathematical and intuition-based content can be taught in a traditional sense, but real-world scenarios are necessary to inspire students to remain interested. This book focuses on applied regression analysis for business students with an introduction to the Python programming language. It emphasizes illustrating and solving modern problems. Case studies from business are available within the text, along with clear and concise explanations of the theory, intuition, hands-on examples, and the coding required to use regression modeling. Each chapter contains the mathematical formulation and details of regression analysis and provides a practical in-depth breakdown using the Python programming language. Throughout the book, examples make use of a lightweight Python utility package developed for instructional clarity, built on standard Python libraries, to streamline repetitive tasks while preserving transparency in the modeling process.

My interest in applied regression analysis developed through project work for the US Department of Defense, where regression models were developed to estimate delivery timelines for requisitions originating in the United States and destined for combatant commands. The experience underscored the importance of building models that are not only statistically sound, but also interpretable and actionable in real operational settings.

This book stresses the practical usage of regression analysis, making it applicable to data sets encountered in everyday business. Specifically, while traditional books contain excessive theories and mathematical formulas, they can overwhelm business students without advanced mathematical knowledge. On the contrary, some books that stress intuitive lessons often omit the coding components, which prevents students from understanding the process. To address this gap, the book blends theory and application through fully worked real-world examples accompanied by detailed, reproducible code. Learning through relevant business cases is essential in today's business world. To that end, numerous practical applications and exercises are available in this book, allowing students to quickly apply complex math to the data sets they collect.

In this digital era, businesses have become highly dependent on data-driven analytics to guide their decision process. A survey by the GMAC even found that companies are increasingly looking to hire talent in data science and business analytics for advanced positions. Due to the increasing demand, the number of programs and courses in business analytics and data science has drastically grown in recent years. For example, Miami Herbert Business School at the University of Miami currently offers multiple annual sections dedicated to regression analysis. However, there is no single textbook that fully addresses applied regression analysis for business students using Python as the primary analytical tool.

Hence, the primary audience of this book will be (1) advanced undergraduate-level students studying business analytics and (2) graduate students studying business: including but not limited to students who study Master of Science in Business Analytics, Master of Science in Accounting, Master of Science in Tax, and Master of Business Administration. Nevertheless, this book suits advanced undergraduate-level students studying engineering, math, and statistics.

Coral Gables, FL, USA Daniel P. McGibney
January 15, 2026

Acknowledgments I would like to acknowledge the work of my reviewers, whose thoughtful suggestions and constructive feedback greatly contributed to the clarity, accuracy, and overall quality of this book. Special thanks are extended to the following individuals:

Jason Bugas
Aleksandra Gorzycka, Ph.D.
Isabella Gurdian
Emily Manchola
John Munitz
Kevin Ordet
Marissa Rodriguez
Heather Sasinowska, Ph.D.
Jesus Tuñon
Riley Yee
Noelle Colimon
Lina Graf
Katherine Hranowsky
Melanie Martinez
Nicole Niewiadomski
Mintra Putlek
Kaitlyn Salvi Cruz
Xavier Schore
Redmond Willis
Nefeli Zafeiri

I would also like to thank the following individuals for their support, insights, and helpful discussions, all of which contributed meaningfully to the development of this textbook:

Francisco Garcia
Masoud Kamalahmadi, Ph.D.
Anh Ninh, Ph.D.
Lan Wang, Ph.D.
Natalia Gavrilova
Doug Lehmann, Ph.D.
Jigar Patel, Ph.D.

A portion of the funding for this textbook was provided by the Deloitte Institute for Research and Practice in Analytics (DIRPA), which is housed within the Miami Herbert Business School at the University of Miami. I am also indebted to the staff at Springer Nature, particularly Jialin Yan (Associate Editor) and Amala Gobiraman (Production Editor).

Daniel P. McGibney, Ph.D.

Contents

1 Introduction 1
- 1.1 Introduction 1
- 1.2 History 2
- 1.3 Linear Regression, Machine Learning, and Data Science 2
- 1.4 Case Studies 3
- 1.5 R versus Python 4
- 1.6 Python Software 4
- 1.7 Python Packages for Regression Analysis 5
- 1.8 Ravix 7
  - 1.8.1 Advantages of Ravix 7
- 1.9 Datasets and Supplemental Material 8
- 1.10 Book Organization 8

2 Basic Statistics and Functions Using Python 9
- 2.1 Introduction 9
- 2.2 Basic Statistics 10
- 2.3 Sales Calls Application: Basic Statistics 11
- 2.4 Variable Assignment 13
- 2.5 Basic Operations in Python 14
  - 2.5.1 Arithmetic Operations 14
  - 2.5.2 Logical Operations 15
- 2.6 Installing and Importing Packages 16
  - 2.6.1 Importing Packages 17
  - 2.6.2 Importing Specific Functions 18
- 2.7 Lists, NumPy Arrays, and Dictionaries 18
  - 2.7.1 Lists 18
  - 2.7.2 NumPy Arrays 19
  - 2.7.3 Dictionaries 20
- 2.8 Indexing Lists and Arrays 20

2.9 Creating and Accessing Dataframes 21
2.9.1 The DataFrame Function 21
2.9.2 The read_csv Function 22
2.10 Dataframe Indexing 22
2.11 Accessing the Objects of a Dataframe in Python 23
2.11.1 The head Function 25
2.12 Basic Statistics in Python 25
2.12.1 The describe Function 25
2.12.2 The sum Function 26
2.12.3 The mean Function 27
2.12.4 The std Function 28
2.13 Sales Calls Application: Basic Statistics in Python 28
2.14 Ravix Fundamentals 29
2.14.1 The get_data Function 29
2.14.2 Formula Notation 30
2.14.3 Formula Examples 31
2.14.4 Other Ravix Arguments 32
2.15 Scatterplots 32
2.15.1 The plot Function 32
2.15.2 The plot Function for Multiple Scatterplots 33
2.16 Histograms 33
2.16.1 The hist Function 34
2.16.2 The hist Function for Multiple Histograms 34
2.17 Box Plots 34
2.17.1 The boxplot Function 35
2.18 Sales Calls Application: Plotting Using Python 35
2.19 Case Study: Top Companies 39
2.19.1 Problem Statement 39
2.19.2 Data Description 40
2.19.3 Scatterplots 42
2.19.4 Histograms 44
2.19.5 Case Conclusion 46

3 Regression Fundamentals 49
3.1 Introduction 49
3.2 Covariance 49
3.3 Correlation Coefficient 50
3.4 Coefficient of Determination 52
3.5 Sales Calls Application: Variable Correlations 52
3.6 Least Squares Criterion 55
3.7 Sales Calls Application: Simple Regression 58
3.8 Interpolation and Extrapolation 59
3.9 Sales Calls Application: Prediction 60
3.10 Explained Deviation 61

3.11 Case Study: Accounting Analytics 63
3.11.1 Problem Statement 63
3.11.2 Correlation and Scatterplot 64
3.11.3 Linear Regression Modeling 64
3.11.4 Audit Scenarios 66
3.11.5 Case Conclusion 68

4 Simple Linear Regression 73
4.1 Introduction 73
4.2 Simple Linear Regression Model 74
4.3 Model Assumptions 74
4.4 Model Variance 75
4.5 Application: Stock Revenues 76
4.6 Probability Functions 77
4.6.1 Probability Density Function 78
4.6.2 Cumulative Distribution Function 78
4.6.3 Percent Point Function 78
4.6.4 The Student's t-Distribution 79
4.7 Using the t-Distribution for Statistical Inference 80
4.7.1 The `ppf` Function for t 81
4.7.2 The `cdf` Function for t 82
4.8 Application: Using the cdf and ppf Functions 83
4.9 Hypothesis Testing: Student's t-Test 84
4.10 Employee Churn Application: Testing for Significance with t 85
4.11 Coefficient Confidence Interval 88
4.12 Employee Churn Application: Confidence Interval Hypothesis Testing 88
4.13 The F-Distribution 91
4.14 Hypothesis Testing: F-test 91
4.15 The `ppf` Function for F 92
4.16 The `cdf` Function for F 93
4.17 Employee Churn Application: Testing for Significance with F 94
4.18 Cautions About Statistical Significance 96
4.19 Case Study: Stock Betas 96
4.19.1 Problem Statement 96
4.19.2 Descriptive Statistics 97
4.19.3 Plots and Graphs 100
4.19.4 Finding Beta Values 103
4.19.5 Finding All the Betas 105
4.19.6 Recommendations and Findings 107
4.19.7 Case Conclusion 108

5 Multiple Regression 113
5.1 Introduction 113

5.2 Multiple Regression Model ... 114
5.3 Multiple Regression Equation ... 114
5.4 Website Marketing Application: Modeling ... 114
5.5 Significance Testing: t ... 117
5.6 Coefficient Interpretation ... 118
5.7 Website Marketing Application: Individual Significance Tests ... 118
5.8 Significance Testing: F ... 120
5.9 Multiple R^2 ... 121
5.10 Adjusted R^2 ... 122
5.11 Website Marketing Application: Multiple R^2 and Adjusted R^2 ... 122
5.12 Correlations in Multiple Regression ... 123
5.12.1 The `plot_cor` Function ... 123
5.13 Variance Inflation Factors ... 124
5.13.1 The `vif` Function ... 125
5.14 Case Study: Real Estate ... 125
5.14.1 Problem Statement ... 125
5.14.2 Data Description ... 126
5.14.3 Calculating VIF Values ... 128
5.14.4 Simple Linear Regression Models ... 129
5.14.5 Multiple Regression Model ... 131
5.14.6 Case Conclusion ... 135

6 Estimation Intervals and Analysis of Variance ... 139
6.1 Introduction ... 139
6.2 Expected Value ... 139
6.3 Confidence Interval of a Mean Response ... 140
6.3.1 The `intervals` Function for Confidence Intervals ... 141
6.4 House Prices Application: Confidence Interval ... 141
6.5 Prediction Intervals ... 146
6.5.1 The `intervals` Function for Prediction Intervals ... 146
6.6 House Price Application: Prediction Interval ... 147
6.7 Confidence Intervals Versus Prediction Intervals ... 148
6.8 Analysis of Variance ... 149
6.8.1 Mean of Squares Due to Regression ... 150
6.8.2 Mean Squared Error ... 151
6.8.3 The F Statistic ... 151
6.9 ANOVA Table ... 152
6.10 House Price Application: ANOVA Table ... 152
6.11 Generalized F Statistic ... 155
6.12 Case Study: Employee Retention Modeling ... 156
6.12.1 Problem Statement ... 156
6.12.2 Data Description ... 157
6.12.3 Multiple Regression Model ... 158
6.12.4 Predictions ... 161
6.12.5 Case Conclusion ... 164

7 Predictor Variable Transformations 169
7.1 Introduction 169
7.2 Categorical Variables 169
7.3 Employee Salary Application: Dummy Variables 170
7.4 Employee Salary Application: Dummy Variables 2 174
7.5 Multilevel Categorical Variables 176
7.6 Employee Salary Application: Dummy Variables with Multiple Levels 176
7.7 Coding Dummy Variables 178
7.8 Employee Salary Application: Dummy Variable Coding 181
7.9 Modeling Curvilinear Relationships 183
7.10 Sales Performance Application: Quadratic Modeling 183
7.11 Mean-Centering 187
7.12 Marketing Toys Application: Mean-Centering 188
7.13 General Linear Regression Model 193
7.14 Interactions 194
7.15 Marketing Toys Application: Interactions 195
7.16 Case Study: Video Engagement Modeling 198
7.16.1 Problem Statement 198
7.16.2 Data Description 199
7.16.3 Promoter A Model 202
7.16.4 Promoter B Model 205
7.16.5 Combined Model 206
7.16.6 Case Conclusion 208

8 Model Diagnostics 213
8.1 Introduction 213
8.2 Multiple Regression Model Revisited 213
8.3 Model Assumptions 214
8.4 Violations of the Model Assumptions 214
8.5 Residual Analysis 215
8.6 Sales Performance Application: Residual Analysis 216
8.7 Constant Variance 217
8.8 Tweet Engagement Application: Residual Variance 218
8.9 Response Variable Transformations 220
8.10 Logarithmic Transformations 220
8.11 Other Response Variable Transformations 223
8.12 Box-Cox Transformation 225
8.13 Tweet Engagement Application: Box-Cox 226
8.14 Assessing Normality 228
8.15 Assessing Independence 231
8.16 Outliers and Influential Observations 231
8.17 Residuals and Leverage 233
8.17.1 Leverage 233
8.17.2 Standardized Residuals 234

8.17.3 Studentized Residuals 234
8.17.4 Cook's Distance 234
8.18 Case Study: Lead Generation 235
8.18.1 Problem Statement 235
8.18.2 Data Description 236
8.18.3 Revenue by Lead Generation Method 237
8.18.4 Sales Versus Radio Ads 241
8.18.5 Sales Versus Robocalls 245
8.18.6 Sales versus Emails 248
8.18.7 Sales versus Cold-Calls 252
8.18.8 Recommendations and Findings 258
8.18.9 Case Conclusion 259

9 Variable Selection 261
9.1 Introduction 261
9.2 Parsimonious Models 261
9.3 Airbnb Pricing Application 262
9.4 Assessing Model Performance 265
9.4.1 Akaike Information Criterion 266
9.4.2 Bayesian Information Criterion 266
9.4.3 Mallows's C_p 267
9.5 Airbnb Pricing Application: Model Comparison 267
9.6 Backward Elimination 268
9.7 Airbnb Pricing Application: Backward Elimination 269
9.8 Forward Selection 271
9.9 Airbnb Pricing Application: Forward Selection 272
9.10 Stepwise Regression 273
9.11 Airbnb Pricing Application: Stepwise Regression 274
9.12 Best Subsets Regression 276
9.13 Airbnb Pricing Application: Best Subsets Regression 1 277
9.14 Airbnb Pricing Application: Best Subsets Regression 2 279
9.15 Airbnb Pricing Application: Best Subsets Regression 3 282
9.16 Stepwise and Best Subsets Regression 285
9.17 Case Study: Cancer Treatment Cost Analysis 285
9.17.1 Problem Statement 285
9.17.2 Data Description 286
9.17.3 Preliminary Analysis 286
9.17.4 Preprocessing 288
9.17.5 Best Subsets Model 290
9.17.6 Stepwise Methods 291
9.17.7 Exploratory Data Analysis 295
9.17.8 Modeling with Interactions 299
9.17.9 Coefficient Interpretation 303
9.17.10 Model Comparison 305
9.17.11 Charges Across BMI Categories 305
9.17.12 Charges Across BMI Categories for Nonsmokers 307

9.17.13 Income Across Plans ... 309
9.17.14 Recommendations and Findings ... 311
9.17.15 Case Conclusion ... 312

A Ravix Modeling Architecture ... 315
A.1 Purpose of the Modeling Module ... 315
A.2 Modeling Workflow in Ravix ... 316
A.3 Core Modeling Functions (Grouped by Role) ... 317
A.3.1 Estimation Functions ... 317
A.3.2 Inference and Reporting Functions ... 317
A.3.3 Model Selection Functions ... 318
A.4 Model Object Structure ... 318
A.5 Formula Processing Engine ... 319
A.6 Estimation Mechanics ... 319
A.7 Integration with Visualization and Diagnostics ... 320
A.8 Reproducibility and Analytical Consistency ... 320

References ... 323

Index ... 325

About the Author

Dr. Daniel P. McGibney is an Associate Professor of Professional Practice at the University of Miami Herbert Business School, where he teaches analytics and data science at both the graduate and undergraduate levels. His teaching spans applied statistics, regression analysis, machine learning, and data-driven decision making, with a particular emphasis on practical, business-oriented applications. He has taught applied linear regression to students in the MSBA, MBA, MST, and MAcc programs and actively mentors graduate capstone projects in analytics, collaborating with organizations such as Deloitte, Visa, Carnival, Citi, and Experian.

He currently serves as Co-Director of the MSBA program. Prior to joining Miami Herbert, he was a Visiting Assistant Professor of Mathematics at the College of William & Mary, where he taught and conducted research in statistics, analytics, and machine learning. He holds a Ph.D. in Systems Science and Mathematics from Washington University in St. Louis, as well as master's degrees in engineering and mathematics from Washington University in St. Louis and Southern Illinois University Edwardsville. His industry experience includes roles in operations research, engineering, and quantitative modeling at Northrop Grumman and NooterEriksen.

Chapter 1
Introduction

In God we trust. All others must bring data.

—W. Edwards Deming

1.1 Introduction

In today's data-driven world, businesses rely on regression analysis to make informed decisions. It supports a wide range of applications, from setting competitive prices to crafting effective marketing campaigns. Business analytics represents the broader framework within which regression operates, combining statistical modeling with computational methods to transform data into strategic insight. Regression analysis achieves these goals by predicting unknown values, assessing differences among groups, and checking the relationships among variables. When applied correctly to the right data, regression analysis can significantly enhance business profitability, whether the goal is predicting house sale prices, evaluating marketing strategies, or estimating the number of likes on a social media post.

To streamline and enhance the process of regression analysis, this book introduces the Ravix package. The package provides simplified and consistent output for common regression tasks, while relying on standard Python libraries for computation and inference.

This book includes numerous business examples where regression analysis, facilitated by Ravix, provides valuable insights. This chapter begins with a discussion of the history of regression analysis and its role in data science, machine learning, and artificial intelligence (AI). We also provide an overview of each of the eight case studies in this book, offering detailed analyses on how to use regression analysis and Ravix to obtain actionable business findings.

D. P. McGibney, *Applied Linear Regression for Business Analytics with Python*,
International Series in Operations Research & Management Science 378,
https://doi.org/10.1007/978-3-032-23806-1_1

1.2 History

Developed over two centuries ago, regression analysis boasts a rich and notable history. French mathematician Adrien-Marie Legendre published the first writings on regression in 1805, followed by German mathematician Johann Carl Friedrich Gauss in 1809. Legendre introduced the method of least squares, while Gauss contributed to its mathematical formulation and applied it to astronomical data. Later that century, Sir Francis Galton used the term "regression" to describe the tendency for extreme characteristics, such as very tall heights, to move closer to the average in subsequent generations—a concept known as "regression toward the mean." Galton's student, Karl Pearson, worked with Udny Yule to put regression into a general statistical framework. Another notable pioneer, Ronald Fisher, advanced statistical work by developing the analysis of variance (ANOVA) methodology and laying the foundations for modern statistics. Fisher's development of ANOVA provided a framework for comparing multiple groups, complementing regression analysis by allowing for more complex models and hypothesis testing. While regression analysis has roots in astronomy and biology, there are many modern applications in almost every academic field.

1.3 Linear Regression, Machine Learning, and Data Science

In modern statistics, the fundamental concepts of linear regression have not changed significantly over the past decades. More work has been done to develop and expand on these existing concepts, and regression continues to be an active area of research. Today, linear regression is often taught and applied within the broader context of machine learning or, even more broadly, artificial intelligence.

Traditional linear regression analysis can be used to make predictions and inferential understanding. In contrast, more recent machine learning techniques often use "black-box" models, which provide accurate predictions without revealing the underlying relationships between variables, making them less suitable for inferential analysis. Regression analysis, on the other hand, functions as an "open box," revealing how each component contributes. Despite this limitation, machine learning is gaining popularity due to the increasing need for predictions with the copious amounts of data being collected in the modern age.

Data scientists are required to have statistical, programming, and domain knowledge. To become a data scientist or develop expertise in business analytics, one must learn many powerful tools to understand and manipulate data. Business analytics, while closely related to data science, focuses more directly on applying these tools to solve business problems and guide strategic decisions. Among these tools are linear regression and the Python programming language.

1.4 Case Studies

The case studies in this book aim to show how analytics can solve numerous problems, particularly in business domains. Each case covers a different business application and demonstrates the relevance analytics can play when making important business decisions. The cases are listed below.

1. **Top 200 Companies**
   This case study focuses on visualizing and summarizing key financial metrics of the world's top 200 companies, providing insights into their performance. Specifically, we create visualizations and generate summary statistics from this data.
2. **Accounting Analytics**
   Using a dataset consisting of the adjusted gross income and the itemized deductions for a select list of subjects, we look at the relationship between the variables and investigate any irregularities. Regression analysis highlights observations where deductions deviate from expected patterns, which may warrant further audit review.
3. **Stock Betas**
   This dataset consists of monthly returns of seven stocks and the S&P 500. From this data, we calculate the stock "betas," which are valuable measures of the relationship between stocks and a market benchmark. Regression analysis is the foundation for estimating these betas.
4. **Housing Market Analysis**
   In this case study, we consider a dataset of 90 cities and investigate the housing market for each city. In addition, we test a claim about the impact tax rate has on house prices.
5. **Employee Retention**
   This case investigates two datasets: a historical dataset of employee retention records and another consisting of new applicants to the company. With these datasets, we predict the retention of new applicants and evaluate two potential employee staffing companies.
6. **Social Media Analysis**
   This case considers video posts advertised by one of two possible promoters. Here, we analyze which promoter we should use and how the sentiment and age of a video affect the number of likes.
7. **Lead Generation**
   Using a dataset consisting of the performance of 142 car dealerships across the country, we analyze the data to increase sales revenue. In particular, the dealerships could use radio ads, robocalls, emails, or cold calls as a means by which to increase sales. Here, we compare and contrast the different methods to determine the most effective lead-generation strategy.
8. **Cancer Treatment Cost Analysis**
   In this case study, we focus on the factors of healthcare costs that an insurance company can use to determine premiums. This case examines the data on charges

billed by health insurance for treating different types of cancer. By determining the variables that contribute to the costs, insurance companies can better account for these expenses.

Across these cases, regression analysis not only explains relationships among variables but also guides data-driven decision-making—linking theory, computation, and business application.

1.5 R versus Python

Business analytics and data science rely on several software tools, with R and Python being the most widely used today. In the past, languages such as C, C++, and Fortran were common in data analysis. However, R and Python are far more flexible, accessible, and easier to learn, making them the preferred choices for modern analytics.

R was designed specifically for statistical computing and data visualization. Its strong community support and extensive ecosystem of packages, such as dplyr and ggplot2, have helped it remain a favorite among statisticians and researchers. As data-driven decision-making has grown in importance, R's role in business analytics has continued to expand. Interestingly, many R users are now also adopting Python to complement their skill set.

While R excels in statistical modeling and visualization, Python has emerged as the most widely used in data science in recent years. Its rise is largely due to its ability to integrate seamlessly with other programming languages, databases, and business systems, making it especially suitable for applied analytics. Combined with powerful libraries such as Scikit-learn and Keras for machine learning, Python offers both flexibility and scalability for real-world applications.

Readers are encouraged to continue learning both Python and R beyond this material. Each language offers unique advantages—R in statistical processing and visualization, Python in machine learning and system integration—together forming a comprehensive toolkit for modern business analytics.[1]

1.6 Python Software

Setting up programming software can often be a cumbersome process, imposing unnecessary complexities on users. Fortunately, diving into Python doesn't require any local software installation hassle if you have internet access and a modern web browser like Chrome or Safari. By simply accessing Google Colab

[1] For a full treatment of regression analysis using R, see McGibney (2023).

(colab.research.google.com) and launching a new notebook, you can swiftly embark on your Python journey. While a Google account is a prerequisite for utilizing this option, the benefits are certainly worth the signup.

Within Google Colab, you have the flexibility to create either a code cell or a text cell. The content within these cells contributes to the formation of Jupyter notebooks, which are highly regarded and widely used by data scientists due to their versatility and practicality. Jupyter files have the file extension `.ipynb`, as they were originally called iPython Notebooks. The notebook you create connects to computing resources from Google, which can save valuable computing resources from the standpoint of the user.

While Google Colab is among the most popular environments for working with Jupyter notebooks, several alternative cloud-based platforms are also available, including:

- **jupyter.org**—the official Jupyter Project site, offering local and hosted notebook options.
- **cocalc.com**—a collaborative computation platform that supports Jupyter notebooks along with other programming and markup environments.
- **kaggle.com**—a Google-owned platform offering free GPU and TPU resources, commonly used for data science competitions and analysis.
- **databricks.com**—an enterprise-grade analytics platform integrating notebooks with large-scale data engineering and machine learning capabilities.

While quickly connecting to a Jupyter notebook should be a priority, you may choose to take on the additional task of installing Python to your personal computer. To install Python, you may want to consider one of the following distributions or integrated development environments (IDEs): Anaconda, PyCharm, VSCode, or RStudio.

Although platforms such as Google Colab already provide features like code completion and visualization, dedicated IDEs offer additional benefits for local development. These benefits include greater control over package management, integration with version control systems, and support for large or multi-file projects, which can enhance productivity and maintainability as analyses grow in complexity.

1.7 Python Packages for Regression Analysis

A package is a collection of related functions, data, and documentation that help you perform specific tasks without writing all the code yourself. There are a number of packages designed to assist in regression analysis. While some are specifically designed for this purpose, the following are essential for data handling and complementary processes.

1. **NumPy**
   NumPy is a fundamental package for numerical computations in Python. It provides support for large, multi-dimensional arrays and matrices, which are organized collections of numbers. NumPy also includes a collection of mathematical functions to operate on these arrays. NumPy arrays are a common structure for data manipulation and serve as the backbone for many other packages in the ecosystem.
2. **SciPy**
   Building on NumPy, SciPy extends its capabilities with additional functionality for scientific and technical computing. This includes modules for optimization, integration, interpolation, and more. SciPy is essential for performing complex mathematical computations that go beyond the basic operations provided by NumPy.
3. **Pandas**
   Pandas is the primary library for data manipulation and analysis. It provides data structures such as dataframes, which are akin to tables in a database or Excel sheets, and offers powerful tools for reading, writing, and processing data. Pandas dataframes allow data to be easily loaded, cleaned, and prepared.
4. **Scikit-learn**
   The Scikit-learn library (imported as 'sklearn') is a robust tool for machine learning and includes a variety of algorithms for predictive modeling and data analysis. It supports many regression techniques, such as linear regression, logistic regression, and more. While Scikit-learn is excellent for building and validating machine learning models, it lacks comprehensive tools for statistical inference and reporting.
5. **Statsmodels**
   For comprehensive statistical modeling and econometrics, statsmodels provides classes and functions for estimating and testing various types of statistical models, including linear and generalized linear models. This package excels in handling complex statistical tests and producing detailed reports of model parameters and diagnostics.
6. **Matplotlib**
   Inspired by the plotting capabilities of MATLAB, Matplotlib is the fundamental plotting library in Python. It offers a wide range of plotting functions to create various visualizations, such as line plots, scatter plots, bar charts, histograms, and more. This library is highly customizable, making it a preferred choice for creating detailed and publication-quality plots.
7. **Seaborn**
   Seaborn simplifies the creation of aesthetic and informative statistical graphics. It provides high-level functions to generate complex visualizations with less code and offers additional themes and color palettes that enhance the visual appeal of plots.

Refer to Sect. 2.6 for details on how to use and import these packages. All of these packages play an important role in regression analysis. They collectively form the core toolkit for regression analysis in Python, each contributing unique yet essential features that enhance the efficiency and effectiveness of data manipulation, statistical modeling, and visualization.

1.8 Ravix

The Ravix package aims to consolidate the strengths of various packages, providing enhanced capabilities for regression analysis. Ravix integrates capabilities from the broader Python analytics ecosystem while streamlining common modeling and visualization workflows.

It provides an R-style formula interface that supports intuitive model specification while extending regression output with enhanced interpretability and reporting structure. Although Ravix is primarily designed for regression analysis, it also utilizes Scikit-learn to integrate a variety of machine learning models. Graphs and visualizations are built on top of existing packages, aiming to provide standard graphs with concise syntax.

Ravix is intentionally designed to abstract lower-level customization in favor of speed, consistency, and clarity, substantially reducing the time required to perform regression analysis and data visualization for both beginners and experienced users.

1.8.1 Advantages of Ravix

- **Unified Interface:** Combines regression tools from multiple libraries into one consistent framework.
- **Simplified Syntax:** Uses concise, R-style formulas that are easier to read and write.
- **Enhanced Output:** Produces clear, detailed summaries and formatted results tables.
- **Integrated Modeling:** Leverages established statistical estimation frameworks within a cohesive analytical environment.
- **Visualization Tools for Regression:** Creates standard diagnostic and summary plots with minimal code.
- **Time Efficiency:** Reduces repetitive coding and accelerates the analysis process.
- **Beginner-Friendly:** Helps new users perform professional-quality analysis with minimal setup.

By reducing unnecessary code, Ravix allows students to focus on modeling decisions and interpretation rather than implementation details. Additional documentation and extended examples are provided through the book's supplemental resources.

1.9 Datasets and Supplemental Material

You can access the datasets showcased in this book through businessregression.com, conveniently arranged and grouped by individual chapters.[2] The associated website offers further insights into both the R and Python versions of this text and any updates on errata that may arise. Supplemental documentation, software updates, installation guidance, and extended examples for the Ravix package are available at ravixanalytics.com.[3]

1.10 Book Organization

Chapter 2 discusses some basic concepts in statistics and provides an introduction to Python programming. Chapters 3 and 4 provide the detailed calculation and theory of simple linear regression—the case in which the model includes a single predictor variable. Chapter 5 introduces multiple regression, which is regression with more than one predictor variable. Multiple regression is followed by Chap. 6, a chapter on additional theory and methods, particularly analysis of variance for regression and prediction/confidence intervals. Then, Chap. 7 provides a detailed discussion on predictor variable transformations, including dummy variables and squared predictor variables. Chapter 8 follows with a discussion of diagnosing regression models. Finally, Chap. 9 provides a discussion on variable selection methods for multiple regression models.

---

[2] For datasets and supplemental material associated with this textbook, see McGibney (2024).

[3] For supplemental documentation, software updates, installation guidance, and extended examples for the Ravix package, see McGibney (2026a).

Chapter 2
Basic Statistics and Functions Using Python

Numerical quantities focus on expected values, graphical summaries on unexpected values.

—John Tukey

2.1 Introduction

Data science represents a multifaceted discipline that combines statistical understanding, programming, and data manipulation, and the ability to explain the data, which is often done with one or more visualizations. Beyond statistics as the general subject matter within the branch of mathematics, the word "statistics" carries a second definition referring to the numeric values, which summarize a sample, such as the mean, median, standard deviation, and variance. Python is a preferred language among industry professionals and data scientists for its ease in manipulating data and calculating useful statistics. Python has many popular plots, but here we will focus on scatter plots, box plots, and histograms since they are necessary for regression analysis.

This chapter provides a basic overview of statistics, programming, and plotting. We review the aforementioned numeric quantities and a few other statistics. We introduce data manipulation in Python at a basic level so that, after the reader understands the basic statistics, he or she can input the data and calculate the values within Python. In addition, plots can provide graphical summaries of data samples and offer important tools to the field of statistics. We cover a few basic plots and how to create them using Python. The concepts are put into context with a simple application using data from a call center, which observes sales and the corresponding number of calls. In the final discussion of the chapter, we present and solve a case study utilizing a dataset consisting of the world's top 200 companies. This case study uses Python to generate descriptive graphs and basic statistics to understand the data.

D. P. McGibney, *Applied Linear Regression for Business Analytics with Python*,
International Series in Operations Research & Management Science 378,
https://doi.org/10.1007/978-3-032-23806-1_2

2.2 Basic Statistics

In basic statistics, the mean is used as a measure of center to summarize a random variable. If the random variable is X with n values in a sample $(x_1, x_2, ..., x_n)$, then the sample mean, denoted as $\bar{x}$, can be calculated from:

$$\bar{x} = \frac{\sum_{i=1}^{n} x_i}{n} \tag{2.1}$$

where x_i is the ith entry in X.

The variance of X is the expected squared deviation of X from its mean. Variance quantifies the spread of a random variable, indicating how dispersed the observations are within its distribution. The sample variance of a random variable X is denoted as s^2 and calculated:

$$s^2 = \frac{\sum_{i=1}^{n} (x_i - \bar{x})^2}{n - 1}. \tag{2.2}$$

While the variance is a useful measure of spread, the units are the square of the original X values, making it less intuitive. For computational purposes, calculating the sample variance from Eq. (2.2) is typically less convenient than calculating it from the mathematically equivalent equation:

$$s^2 = \frac{\sum_{i=1}^{n} x_i^2 - \left(\sum_{i=1}^{n} x_i\right)^2 / n}{n - 1}. \tag{2.3}$$

The denominator of Eqs. (2.2) and (2.3) is the degrees of freedom, which denotes the number of choices, or parameters, that are available in determining the sample variance (s^2). Since the value of $\bar{x}$ is known and used to calculate s^2, the number of x_i values needed are $n - 1$. More specifically, in Eq. (2.1), the value on the left side is known, so only $n - 1$ values of x_i are required. The remaining x_i can be found using simple algebra. It may seem odd that the denominator of the variance is $n - 1$ and not n; however, it should be noted that dividing by $n - 1$ denotes a better estimate of the unknown population variance. This is known as an "unbiased" estimator; however, a detailed discussion and proof are beyond the scope of this text.

The numerator of Eqs. (2.2) and (2.3) is commonly called the sum of squares of X, which represents the total variation in X. Pay special attention to this value—it is the foundation for many formulas in regression analysis. The formula for the sample sum of squares is:

$$SS_{xx} = \sum_{i=1}^{n} (x_i - \bar{x})^2, \tag{2.4}$$

where SS_{xx} denotes the sample sum of squares of X, x_i is the ith observation, and $\bar{x}$ is the sample mean. In practice, it is often more convenient to calculate the sum of squares using the equivalent expression:

$$SS_{xx} = \sum_{i=1}^{n} x_i^2 - \left(\sum_{i=1}^{n} x_i \right)^2 / n. \tag{2.5}$$

One popular measure of variation is the sample standard deviation, which can be found by simply taking the square root of the sample variance:

$$\begin{aligned} s &= \sqrt{s^2} \\ &= \sqrt{\frac{\sum_{i=1}^{n} (x_i - \bar{x})^2}{n-1}} \end{aligned} \tag{2.6}$$

Since the SS_{xx} is used to find the sample variance, it can also be used to find the sample standard deviation directly:

$$s = \sqrt{\frac{SS_{xx}}{n-1}}. \tag{2.7}$$

The following application captures the utility of these simple concepts in a business setting.

2.3 Sales Calls Application: Basic Statistics

Jordan manages retirement accounts for Enright Associates. She documented the number of sales calls that she made each day for the last 6 days.

Using Table 2.1, describe the distribution by performing the following calculations.

(a) Compute the mean.
(b) Compute the sum of squares.
(c) Compute the sample variance.
(d) Compute the sample standard deviation.

Table 2.1 Jordan's calls

Variable						
Calls (X)	12	18	5	25	15	8

Solution

(a) Plugging the values of X into Eq. (2.1), the sample mean is calculated to be

$$\begin{aligned}\bar{x} &= \frac{\sum_{i=1}^{n} x_i}{n} \\ &= \frac{12 + 18 + 5 + 25 + 15 + 8}{6} \\ &= 13.83.\end{aligned}$$

(b) The sum of all the values and the sum of all the x squared values will need to be calculated to find SS_{xx}:

$$\begin{aligned}\sum_{i=1}^{n} x_i &= 12 + 18 + 5 + 25 + 15 + 8 \\ &= 83\end{aligned}$$

$$\begin{aligned}\sum_{i=1}^{n} x_i^2 &= 12^2 + 18^2 + 5^2 + 25^2 + 15^2 + 8^2 \\ &= 1{,}407.\end{aligned}$$

Knowing these two values, Eq. (2.5) can now be utilized:

$$\begin{aligned}SS_{xx} &= \sum_{i=1}^{n} x_i^2 - \frac{\left(\sum_{i=1}^{n} x_i\right)^2}{n} \\ &= 1{,}407 - \frac{(83)^2}{6} \\ &= 258.833.\end{aligned}$$

(c) Since the numerator from Eq. (2.3) is SS_{xx}, the sample variance is easily solved using Eq. (2.3):

$$\begin{aligned}s^2 &= \frac{\sum_{i=1}^{n} x_i^2 - \frac{\left(\sum_{i=1}^{n} x_i\right)^2}{n}}{n - 1} \\ &= \frac{SS_{xx}}{n - 1} \\ &= \frac{258.833}{5} \\ &= 51.767.\end{aligned}$$

(d) The sample standard deviation can be found by taking the square root of the sample variance, as in Eq. (2.6):

$$\begin{aligned} s &= \sqrt{s^2} \\ &= \sqrt{51.767} \\ &= 7.195. \end{aligned}$$

The sample standard deviation of 7.19 suggests that the number of calls Jordan makes each day fluctuates by roughly seven around the average of 13.83. In other words, her daily calling activity shows a moderate amount of variability.

2.4 Variable Assignment

To begin our discussion of Python, we will start with variable assignments. In Python, variables are assigned using the equal sign. For example, to assign the value 24 to the variable **x**, use the equal sign as demonstrated below.

```
x = 24
```

Using this assignment method shown above creates a scalar, or singular value, named **x**.

Since the value assigned to **x** was an integer without mention of values after the decimal, Python automatically assigns the value to be an integer. An integer represents a whole value that can be positive, negative, or zero. The set of integers is $\{\ldots, -3, -2, -1, 0, 1, 2, 3, \ldots\}$. Because integers ignore digits past the decimal, storing a variable as an integer is often more efficient and precise than using a float.

A float is a number that includes a decimal point or represents a fractional value in decimal form. Examples include `3.0`, `-2.5`, `0.1`, and `7.25`. In many scenarios, restricting **x** to an integer should not cause problems, but if we want **x** to hold a fractional value, we can do so by running the following:

```
x = 24.0
```

In Python, arithmetic operations between integers and floats are allowed. When such operations are performed, Python automatically converts the integer to a float, and the resulting value is a float.

A string can be created with the use of single or double quotes. However, the opening and closing quotes must match. If we use a single quote to start and a double quote to end (or vice versa), then an error will be produced. Here, we assign y as a string of text characters.

```
y = "This is a string"
```

Note that `x` can be printed by simply entering `x` within the `print` function within a code cell of a Jupyter notebook.

```
print(x)
```

```
# 24.0
```

We need not use the `print` function if we simply state the variable in the last line of code in the cell.

2.5 Basic Operations in Python

Python allows users to perform arithmetic and logical operators directly inside a Jupyter notebook. By entering numbers and operators within the code cells, Python will perform the specified operations allowing for quick calculations.

2.5.1 Arithmetic Operations

Addition and subtraction are carried out with the + and - characters, respectively.

```
8 + 4 - 3
```

```
# 9
```

Here, we omitted the `print` function. The result is still displayed because it is the final expression in the code cell, and Jupyter automatically shows the output of the last line.

Multiplication, division, and exponentiation are carried out with *, /, and ** characters, respectively.

```
8 + 4       # Addition
8 - 4       # Subtraction
8 * 4       # Multiplication
8 / 4       # Division
8 ** 2      # Exponentiation
```

It is important to be mindful of the order of operations to avoid confusion. The orders of operation in Python follow the standard mathematical orders of operation (parentheses, exponents, multiplication/division, addition/subtraction). Consider the calculation below, where Python first performs exponentiation and then subtracts 1:

```
10 ** 2 - 1
```

```
# 99
```

When needed, parentheses can be used to specify the order of operations. In the following example, parentheses ensure that 1 is subtracted from the exponent before exponentiation:

```
10 ** (2 - 1)
```

```
# 10
```

From the previously created scalar x, we can subtract a value of 12.

```
x - 12
```

```
# 12.0
```

If we were to define a scalar variable y with a value of 12 and subtract y from x, it would yield the same result as above.

```
y = 12
x - y
```

```
# 12.0
```

2.5.2 *Logical Operations*

In addition to arithmetic operations, Python can perform logical operations that return either True or False (boolean values). These operations are summarized here.

```
x == y       # Test for equality
x != y       # Test for inequality
x < y        # Less than
x <= y       # Less than or equal to
x > y        # Greater than
x >= y       # Greater than or equal to
```

The first operation from the code above is the equivalency test, which can be performed using two equal signs. For instance, checking if two values, x and y, are equal can be done:

```
x == y
```

```
# False
```

Previously, we assigned **x** as a different value than y, so the result of this operation would be `False` since the values are not equal. Checking if two values are not equal can be done as follows:

```
x != y
```

```
# True
```

Checking to see if x is less than y can be done with the less than operator (<).

```
x < y
```

```
# False
```

Less than or equal requires the less than followed by the equal sign (<=).

```
x <= y
```

```
# False
```

Similarly, we can test if x is greater than y using the greater than operator (>).

```
x > y
```

```
# True
```

Greater than or equal yields the same result in this case. The greater than sign followed by the equal sign is how we achieve this operation (>=).

```
x >= y
```

```
# True
```

Logical operations are especially useful for selecting specific rows or columns within an array or other numerical collection.

2.6 Installing and Importing Packages

Common Python distributions come with preinstalled packages, which are ready to be imported and used. However, if a package is not preinstalled, it can be installed using `pip` from the command line. Within Jupyter notebooks, packages can be installed by accessing the system command line using the exclamation mark (`!`). Therefore, to install Ravix, run the following code inside a code cell.

```
!pip install ravix
```

When working on a local installation of Python, packages only need to be installed once. However, in Google Colab, each session runs in a temporary environment that resets when you log in, so packages must be reinstalled each time a new session starts. Once a package is installed in Python, it can be imported. Moving beyond the fundamental arithmetic and logical operations, packages can be imported in Python using the `import` command. The fundamental packages for regression analysis were mentioned in the previous chapter.

2.6.1 Importing Packages

Throughout this book, required packages are imported at the beginning of each section to allow examples to run independently.

In programming, syntax refers to the specific rules and structure that define how code must be written to be understood by the computer. The syntax for importing Ravix is the following:

```
import ravix as rav
```

Here we are not only importing the Ravix package but also creating an alias for it (`rav`). The alias allows for easy access to the package without having to specify the entire name of the package. However, aliases are not required and will not be used consistently throughout this text. After importing a package, it may be beneficial to verify the current version. If we import Ravix without an alias, then the version number can be returned as follows:

```
import ravix
ravix.__version__
```

```
# '1.0.0'
```

Some other common import commands are as follows:

```
import pandas as pd                    # Imports Pandas
import numpy as np                     # Imports NumPy
import matplotlib.pyplot as plt        # Imports Matplotlib PyPlot
import seaborn as sns                  # Imports Seaborn
import statsmodels.api as sm           # Imports statsmodels API
```

2.6.2 Importing Specific Functions

In some cases, only one function from a package is required. To import it directly, use the `from` and `import` commands. Importing specific functions helps to save memory and makes the code more readable by only including the necessary components. One such example is to import the `t` function from `scipy.stats`.

```
from scipy.stats import t
```

After running the above code, we can use the `t` function.

Beginning in Chap. 4, we will adopt this approach when working with Ravix by importing commonly used modeling functions explicitly rather than importing the entire package.

It is also possible to import all functions from a package in a single line. For example, with Ravix, we could write:

```
from ravix import *
```

This approach is discouraged because it obscures where functions originate and can lead to naming conflicts, reducing code clarity.

2.7 Lists, NumPy Arrays, and Dictionaries

2.7.1 Lists

A list of objects, whether they are numbers or text, can be created by listing the entries within square brackets. For instance, the numbers 1, 2, and 3 can be combined into a vector called `v1` by doing the following:

```
v1 = [1, 2, 3]
```

Lists are flexible collections of objects that can be modified after creation.[1] As mentioned, lists are not limited to numeric data. The letters a, b, and c can be combined into a character string vector `v2`. Double or single quotes around each letter identify each as a text string.

```
v2 = ["a", "b", "c"]
```

Numeric and character values can be brought together into a vector as well.

[1] Python also has *tuples*, which are similar to lists but cannot be changed after creation. They are defined using parentheses, e.g., `v = (1, 2, 3)`. While tuples are useful in some programming contexts, lists are generally preferred for data manipulation and statistical operations.

```
v3 = ["a", 2, "c"]
```

Similar to scalar values, lists can be printed in a Jupyter notebook by typing the name in the `print` function. To print `v1`, we write the following:

```
print(v1)
```

```
# [1, 2, 3]
```

2.7.2 *NumPy Arrays*

While lists are available in base Python, NumPy arrays have several advantages over standard lists. Particularly, NumPy arrays are optimized for numerical operations and, therefore, can be significantly faster than lists. This optimization is due to contiguous memory allocation and efficient implementation of numerical algorithms. First, we will need to import NumPy to create the arrays.

```
import numpy as np
```

Now that NumPy is imported, we can create a numeric array using the `array` function. Parentheses are used to call the function, and square brackets define the list of numbers passed to it:

```
v4 = np.array([1, 2, 3])
```

A NumPy array can also be created from an existing list such as `v1`.

```
v4 = np.array(v1)
```

An array of the text values a, b, and c can also be created.

```
v5 = np.array(["a", "b", "c"])
```

Also, the array can be directly created from the list as `v2`.

```
v5 = np.array(v2)
```

NumPy arrays are either stored as numeric or text; mixing and matching values in a list, as done with `v3`, will result in the elements being text vectors. First, observe

```
v3 = ["a", 2, "c"]
v3
```

```
# ['a', 2, 'c']
```

which contains the second element as a 2 without quotes. In contrast, when we convert the list to an array, the 2 is shown with quotes in the output.

```
v6 = np.array(["a", 2, "c"])
v6
```

```
# array(['a', '2', 'c'], dtype='<U21')
```

This behavior occurs because NumPy arrays must have a single data type.

2.7.3 Dictionaries

A dictionary is a grouping of labeled lists or arrays. Creating a dictionary is done using the `dict` function.

```
data_dict = dict(v1 = v1, v2 = v2, v3 = v3, v4 = v4, v5 = v5,
                 v6 = v6)
```

In this syntax, we are specifying `v1` as the name and `v1` as the first list or array by writing `v1=v1`. It is also common to use curly braces to define dictionaries. An identical dictionary can be defined as follows:

```
data_dict = {"v1": v1, "v2": v2, "v3": v3, "v4": v4, "v5": v5,
             "v6": v6}
```

In a dictionary, keys are the string identifiers (such as `"v1"`, `"v2"`, etc.), and values are the data associated with those keys. Each key–value pair maps a name to its corresponding list or array.

2.8 Indexing Lists and Arrays

Indexing is accomplished in Python by using square brackets immediately following the list or NumPy array. Specifically, the list `v2` has 3 values, the first of which is `a`. One important consideration is that Python begins indexing at 0, which means the second item in the list is given by the following:

```
v2[1]
```

```
# 'b'
```

The second and third values can be referenced as

```
v2[1:3]
```

```
# ['b', 'c']
```

When using a range like `1:3`, the last index (3) is not included, so this returns the items at positions 1 and 2 only. If the ending index is omitted, Python returns all elements starting from index 1 through the end of the vector:

```
v2[1:]
```

```
# ['b', 'c']
```

We can also return the last value in the list by specifying negative one as the index value.

```
v2[-1]
```

```
# 'c'
```

2.9 Creating and Accessing Dataframes

To use a dataset in Python, one must either input data or load data. Data can be input as a dictionary, and then a Pandas dataframe can be constructed using the `DataFrame` function. Importing Pandas can be done using the following:

```
import pandas as pd
```

2.9.1 The `DataFrame` Function

The `DataFrame` function from Pandas can be used to create a dataframe from a dictionary. While dictionaries can have lists or arrays of varying lengths, all vectors in a dataframe must have the same number of entries.

```
df1 = pd.DataFrame(data_dict)
```

This previous code assumes that we have a dictionary created called `data_dict`; however, it can also be done in one line of code.

```
df1 = pd.DataFrame({"v1": v1, "v2": v2, "v3": v3, "v4": v4,
                    "v5": v5, "v6": v6})
```

While small datasets can easily be used to create dataframes, it is more common that data exists within a file, such as a Comma Separated Value (CSV) file.

2.9.2 The `read_csv` Function

The `read_csv` function can be used to load in data from a CSV or other text files. The `read_csv` function is the most common way to import data with Pandas.

Although other file types can be read using functions such as `read_excel`, most standard data files can be saved as CSVs and easily imported with `read_csv`.

If a data file `test.csv` is in the current directory, then we can create a new dataframe in the following way.

```
df = pd.read_csv("test.csv")
```

For Python to easily locate the CSV file, it should be in the same directory as the Jupyter notebook. If it's not, provide the full file path. If the file is in the `location` folder in the C drive on a Windows computer, the file can be loaded in as follows:

```
df = pd.read_csv("C:/location/test.csv")
```

On a Mac, the drive name is not specified. If the file is in the `location` folder within the user's home directory, it can be loaded with the following code:

```
df = pd.read_csv("/Users/username/location/test.csv")
```

To locate the full file path, right-click the file in your operating system and select "Copy Path" (Windows) or "Get Info" (Mac). When pasting into Python, use either forward slashes (/) or double backslashes (\\) in the path to avoid errors.

2.10 Dataframe Indexing

Square brackets can be used to index dataframes. Dataframes have both rows and columns, which are referenced within the square brackets and separated by a comma. Returning the entry of a dataframe based on the indices can be done by using the `iloc` indexer.

```
df1.iloc[row, column]
```

To specify the first row and the second column of `df1`, we would use the `iloc` indexer:

```
df1.iloc[0, 1]
```

```
# 'a'
```

To specify all of the row values in the first column (`v1`) of `df1`, we could specify the "`:`" as the row.

```
df1.iloc[:, 0]
```

```
# 0    1
# 1    2
# 2    3
# Name: v1, dtype: int64
```

This output may seem like two columns; however, the left-most numbers are the dataframe index, and the second column is the vector `v1`.

We could also specify the second row from both columns by specifying the column value as `:`.

```
df1.iloc[1, :]
```

```
# v1    2
# v2    b
# v3    2
# v4    2
# v5    b
# v6    2
# Name: 1, dtype: object
```

Another common way of accessing rows within the dataframe is specifying the `loc` indexer. To return the third row (index 2) from `df1`, we could do the following:

```
df1.loc[2]
```

```
# v1    3
# v2    c
# v3    c
# v4    3
# v5    c
# v6    c
# Name: 2, dtype: object
```

Note that we need only specify the row for the `loc` function.

2.11 Accessing the Objects of a Dataframe in Python

The dataframes contain objects within them. The dataframe `df1` can be printed with the following:

```
df1
```

```
#     v1 v2 v3  v4 v5 v6
# 0    1  a  a   1  a  a
# 1    2  b  2   2  b  2
# 2    3  c  c   3  c  c
```

In df1, the object v1 can be accessed using the "." symbol.

```
df1.v1
```

Similarly, we can specify v1 within square brackets after the dataframe df1.

```
df1["v1"]
```

```
# 0    1
# 1    2
# 2    3
# Name: v1, dtype: int64
```

The output returns the indices in the first column and v1 in the second column. If we want multiple rows from a dataframe, such as all even numbered vectors, then we specify the rows in a list and use square bracket indexing.

```
even_vectors = ["v2", "v4", "v6"]
df1[even_vectors]
```

```
#    v2  v4 v6
# 0   a   1  a
# 1   b   2  2
# 2   c   3  c
```

The same operation is typically done in one line.

```
df1[["v2", "v4", "v6"]]
```

```
#    v2  v4 v6
# 0   a   1  a
# 1   b   2  2
# 2   c   3  c
```

Note the necessity of two sets of square brackets since a list is being input into the first set of square brackets.

The dataframe can also be appended easily. For instance, suppose we have the vector v7 which consists of the numbers 4, 5, and 6. To append the vector v7 to df1, we could input the code below.

```
v7 = [4, 5, 6]
df1['v7'] = v7
```

2.11.1 *The* head *Function*

While a dataframe can be printed to the console simply by typing the name of the dataframe, the head function can be a convenient way to observe the data. Specifically, the head function prints out the first five rows of the dataframe and all of the columns.

```
df1.head()
```

```
#    v1 v2 v3  v4 v5 v6  v7
# 0   1  a  a   1  a  a   4
# 1   2  b  2   2  b  2   5
# 2   3  c  c   3  c  c   6
```

In this particular example, the dataset only contains three rows and seven columns so only three rows are shown, but all columns are shown. Alternatively, the number of rows to be returned can be entered as the first argument. Here, we include only the first two rows of data.

```
df1.head(2)
```

```
#    v1 v2 v3  v4 v5 v6  v7
# 0   1  a  a   1  a  a   4
# 1   2  b  2   2  b  2   5
```

Observing the first five rows can reveal much about the dataframe without overwhelming the console with the entire dataset.

2.12 Basic Statistics in Python

In the previous sections, we discussed a few of the different types of data objects that can be used in Python, particularly the scalar, vector, and dataframe objects. In this section, we cover how to manipulate these objects to calculate some basic statistics. For convenience, we define the dataframe df2 as the numeric columns only.

```
df2 = df1[["v1", "v4", "v7"]]
```

2.12.1 *The* describe *Function*

The describe function can also help the user to understand the data of interest. The description of a dataframe will return the count, mean, standard deviation, minimum, 25th percentile, 50th percentile, 75th percentile, and the maximum for each numeric vector in the dataframe.

```
df1.describe()
```

```
#          v1   v4   v7
# count   3.0  3.0  3.0
# mean    2.0  2.0  5.0
# std     1.0  1.0  1.0
# min     1.0  1.0  4.0
# 25%     1.5  1.5  4.5
# 50%     2.0  2.0  5.0
# 75%     2.5  2.5  5.5
# max     3.0  3.0  6.0
```

The `describe` method will function in a similar manner when applied to a numeric vector. Note that `v2`, `v3`, `v5`, and `v6` are not numeric and therefore not shown in the output. Therefore, the output from describe for `df1` will be identical to that of `df2`.

```
df2.describe()
```

```
#          v1   v4   v7
# count   3.0  3.0  3.0
# mean    2.0  2.0  5.0
# std     1.0  1.0  1.0
# min     1.0  1.0  4.0
# 25%     1.5  1.5  4.5
# 50%     2.0  2.0  5.0
# 75%     2.5  2.5  5.5
# max     3.0  3.0  6.0
```

2.12.2 *The* `sum` *Function*

The `sum` function from NumPy will add up all of the values in a vector. This function can be applied to vector `v1`,

```
np.sum(v1)
```

```
# np.int64(6)
```

or within dataframes using the period (.) as follows:

```
np.sum(df1.v1)
```

```
# np.int64(6)
```

The vector `v1` can also be specified by using indices:

```
np.sum(df1["v1"])
```

```
# np.int64(6)
```

If all of the values in `df1` were numeric, the `sum` function could be used to sum up all of the values in the dataframe. We could also use the `sum` function on a dataframe, a method from Pandas.

```
df1["v1"].sum()
```

```
# np.int64(6)
```

2.12.3 *The* `mean` *Function*

The `mean` function calculates the mean or average of the vector.

```
np.mean(v1)
```

```
# np.float64(2.0)
```

Both NumPy and Pandas can compute the mean, but NumPy functions work on arrays and lists, while Pandas methods are designed for entire dataframes or individual columns. In practice, Pandas is often preferred for real datasets since it automatically handles column-wise operations. The mean function from Pandas can be used on the dataframe and will return the mean from each column.

```
df2.mean()
```

```
# v1    2.0
# v4    2.0
# v7    5.0
# dtype: float64
```

The above code, however, will return an error if we used `df1` in place of `df2`, since some of the columns contain text strings.

2.12.4 *The* `std` *Function*

The `std` function will calculate the standard deviation of a vector but can calculate the sample standard deviation by including the argument `ddof = 1`.

```
np.std(v1, ddof = 1)
```

```
# np.float64(1.0)
```

In Pandas, the sample standard deviation is the default for the `std` function. Similar to the mean function, it can be used to calculate the value for each column.

```
df2.std()
```

```
# v1    1.0
# v4    1.0
# v7    1.0
# dtype: float64
```

For consistency when working with datasets, Pandas is generally recommended.

2.13 Sales Calls Application: Basic Statistics in Python

Using Jordan's data from the previous application (Table 2.1), describe the distribution by doing the following with functions in Python.

(a) Compute the mean.
(b) Compute the sum of squares.
(c) Compute the sample variance.
(d) Compute the sample standard deviation.

Solution

(a) Before computing the mean, the data must be entered into a Pandas dataframe. Here we refer to the calls variable as `x`. Using a dictionary of `x` as an input of the `DataFrame` function, we create a dataframe called `dfx`.

```
import pandas as pd
dfx = pd.DataFrame({'x': [12, 18, 5, 25, 15, 8]})
```

The `mean` function from Pandas can be used to calculate the mean value.

```
dfx.mean()
```

```
# x    13.833333
# dtype: float64
```

(b) The sum of squares can be done by squaring each value of `x` individually, but Python allows for element-wise operations. Squaring each value within the vector can be accomplished using the ** operator.

```
(dfx**2).sum() - dfx.sum()**2/6
```

```
# x    258.833333
# dtype: float64
```

(c) Computing the sample variance can be done using the `var` function, or by squaring the result from the `std` function.

```
dfx.var()
```

```
# x    51.766667
# dtype: float64
```

(d) The sample standard deviation can be computed using the `std` function.

```
dfx.std()
```

```
# x    7.194906
# dtype: float64
```

2.14 Ravix Fundamentals

The Ravix package is primarily designed for regression analysis; however, with regression analysis, there are many important plots. The functions for plotting in Ravix have been streamlined to allow for easy syntax within Python. Most notably, the Ravix package implements an R-style formula interface for intuitive model specification within a unified analytical framework.

2.14.1 The `get_data` Function

There are multiple datasets available from Ravix and are easily attained using the `get_data` function. To view all datasets currently bundled with the package, import the Ravix package and run the function without specifying arguments.

```
import ravix
ravix.get_data()
```

```
# Available data files in Ravix (v1.0.0):
#    - airbnb.csv
#    - betas.csv
#    - charges.csv
#    - countries.csv
#    - employment.csv
#    - facebook.csv
#    - house_prices.csv
#    - hr_retention.csv
#    - insurance.csv
#    - job changes.csv
#    - loan_default.csv
#    - marketing_toys.csv
#    - price_size.csv
#    - salary_churn.csv
#    - sales.csv
#    - sales_ads.csv
#    - tire_sales.csv
#    - top200.csv
#    - tweet_engagement.csv
#    - vacation.csv
#    - video_engagement.csv
```

The function returns the list of data files included with the current Ravix version.

See businessregression.com for additional details regarding these datasets. After importing Ravix, the `get_data` function can be used to load a specific CSV file. To load in the `betas.csv` file as dataframe `df`, run:

```
df = ravix.get_data("betas.csv")
```

Recall that Ravix must be installed prior to importing, as mentioned in Sect. 2.6.

2.14.2 Formula Notation

The R formula is a symbolic method used to specify data in the R programming language which has been adopted in other programming languages, most notably in Python using the statsmodels formula API. It offers a concise way to define the relationship between a "response" variable and one or more "predictor" variables. The response variable is typically plotted on the y-axis while the predictor variables are typically plotted on the x-axis. In terms of graphing, this is the notable distinction, and often the y-axis variable is of particular interest. A detailed discussion and definition of both response and predictor variables reside in Chap. 3. The formula typically takes the form `"response ~ predictors"`, where the predictors may affect the response.

In the formula:

- The ~ operator separates the left-hand side (the response variable) from the right-hand side (the predictor variables).
- Plus (+) signs on the right-hand side denotes including all predictor variables separated by the plus sign.
- A period (.) on the right-hand side denotes that all of the columns of a dataframe other than the response should be used as predictor variables.
- The minus (-) sign on the right-hand side denotes the exclusion of a predictor variable.
- A colon (:) denotes interaction terms between variables.

This notation can also include functions of variables, such as logarithms or polynomials, to capture non-linear effects within a linear modeling framework. The formula interface is widely used across various functions and packages for fitting models, making it a fundamental concept for statistical analysis and data science.

2.14.3 Formula Examples

To specify that the `Y` list is on the y-axis and the `X` list is on the x-axis, one could simply use the formula:

```
"Y ~ X"
```

The formula needs to be contained within a string. Extra spacing is for formatting only and will be ignored within Ravix. Suppose there are two predictor variables in the form of lists or NumPy arrays, defined as `X1` and `X2`. The formula could be defined using both predictor variables with the following:

```
"Y ~ X1 + X2"
```

To specify all of the variables in the dataframe as predictor variables, other than the response variable (`Y`), use the period in the formula. This notation implies that a dataframe is specified.

```
"Y ~ ."
```

Alternatively, we can include all variables in the dataframe with the exception of just a few using the minus sign. To choose `Y` as the response, using all columns in the dataframe as predictor variables without the variable `X3`, we would use the following:

```
"Y ~ . - X3"
```

Variables specified in the formula, whether they are response or predictor variables, can either be in the dataframe or directly in the Python environment.

When variables are specified directly in the Python environment rather than within a dataframe, they must be of equal length. Otherwise, Ravix will return an error due to the length mismatch between variables.

2.14.4 Other Ravix Arguments

The formula details is an important part of the Ravix package, but there are many other arguments for the functions within the package. Most plotting functions in Ravix have the following arguments.

- formula—The aforementioned formula in a string format.
- data—The dataframe name, if any, to be plotted. If omitted, Ravix will search for the formula variables within the Python environment.
- xlab—The x-axis label.
- ylab—The y-axis label.
- title—The title of the plot.
- color—Color of the variable plots.

Additional arguments may be allowed depending on the function.

2.15 Scatterplots

Scatterplots are an important type of plot to visually demonstrate the relationship between two numeric variables. Since the goal of regression analysis is to determine the relationship between numeric variables, the scatterplot is of particular interest to analysts performing a regression. In particular, scatterplots can be used to assess the quality of a regression as discussed in Chap. 8.

Drawing a scatterplot consists of a simple process, whereby the x-axis and y-axis are drawn and then points are plotted. To generate a scatterplot using Ravix, the `plot` function can be used.

2.15.1 The `plot` Function

The `plot` function is used for generating scatter plots within Python, which makes use of Seaborn and Matplotlib. The syntax for the `plot` function given two numeric vectors x and y in a dataframe is as follows:

```
import ravix
ravix.plot("y ~ x", data = df1)
```

Note that the vectors `x` and `y` do not need to be inside a dataframe. If that were the case, the code is simplified further.

```
ravix.plot("y ~ x")
```

Since `df1` was defined earlier, the `v4` vector can be placed on the y-axis while the `v1` vector can be placed on the x-axis. Furthermore, it is not necessary to specify the `data` argument since the vectors are defined outside of the dataframe and also within the dataframe. The labels for `x` and `y` are specified using the `xlab` and `ylab` arguments, respectively. The `title` argument allows for a title of the graph.

```
ravix.plot("v4 ~ v1", data = df1, xlab = "Vector 1",
           ylab = "Vector 4", title = "Title")
```

In Fig. 2.1 a sample scatterplot is shown which uses the `plot` function.

2.15.2 *The* `plot` *Function for Multiple Scatterplots*

Ravix also provides a convenient way to view multiple scatterplots simultaneously. This is an invaluable method when observing datasets containing more than two variables. The `plot` function within Ravix is built on top of Seaborn functions and designed specifically with regression analysis in mind. The syntax for the case when we have a dataframe `df` with columns `Y`, `X1`, `X2`, `X3`, and `X4` is intuitive and can be specified as follows:

```
ravix.plot("Y ~ X1 + X2 + X3 + X4", data = df, color = "blue",
          model = "line", lcolor = "black")
```

To accept all of the default functions and only specify the variables within the Python environment, the syntax is simplified greatly.

```
ravix.plot("Y ~ X1 + X2 + X3 + X4")
```

2.16 Histograms

A histogram refers to an important plot that analysts use to identify the shape of a distribution. Most notably, a frequency histogram separates numeric values into bins of equal length, and a bar is drawn for each bin where the height of the bar is determined by the frequency of values within the corresponding bin. The common bell-shaped normal curve represents one popular shape identified by histograms. Like the scatterplot, histograms can be used in regression analysis to verify the validity of a regression model, as discussed in Chap. 8. Describing variables is an essential step in understanding data, whether in a business context or another field.

2.16.1 The `hist` Function

To create a histogram using Ravix, the `hist` function can be used. To get a histogram of the vector `v1` contained in the global environment, we code the following:

```
import ravix
ravix.hist(v1)
```

To get a histogram of the vector `v1` contained within `df1`, we code the following:

```
ravix.hist(df1.v1)
```

Similar to the `plot` function, the `xlab` and `title` arguments can be used here. While the `ylab` argument can also be used here, the y-axis label defaults to "Frequency" which is typically a satisfactory label.

```
ravix.hist(v1, title = "Histogram Title", xlab = "Vector 1")
```

2.16.2 The `hist` Function for Multiple Histograms

The `hist` function can create a matrix of histograms from the numeric variables in the dataset.

```
ravix.hist(formula, data, color = "red", norm = False,
           layout = "column")
```

The `norm` argument, as in the `hist` function, places the normal curve over the distribution for comparison. The `layout` argument defaults to `"column"`, which stacks the histograms. To accept all of the default functions and only specify the variables within the Python environment, the syntax is simplified greatly.

```
ravix.hist("Y ~ X1 + X2 + X3 + X4")
```

2.17 Box Plots

A box plot, sometimes called a box-and-whiskers plot, is great for visualizing the variability and the center of variables simultaneously. While histograms are very useful in describing a single variable, box plots are helpful in comparing variables such as different daily stock performance measures (see Sect. 4.19.3). The overall spread of each variable is shown by the total height of the box-and-whiskers, while the box itself represents the middle 50% of the data. The height of each box is referred to as the Interquartile Range (IQR) and can be calculated by finding the 75th percentile (Q_3) and subtracting the 25th percentile (Q_1):

$$IQR = Q_3 - Q_1. \tag{2.8}$$

If observations fall outside of $Q_1 - 1.5(IQR)$ and $Q_3 + 1.5(IQR)$, then they are deemed outliers, displayed in the box plot as unfilled circles above or below the whiskers in the standard `boxplot` function. Also, the median, a measure of center, can be observed by a black line within each box.

2.17.1 *The* `boxplot` *Function*

The `boxplot` function produces a box plot. To get a box plot of a vector, simply specify the vector within the `boxplot` function.

```
ravix.boxplot(df1.v1)
```

To get a box plot of a dataframe with all numeric values, specify the dataframe as the first argument. This will produce a box plot for each of the vectors within the dataframe together on the same plot. For instance, assume we have a dataframe with all numeric values called `df`. A box plot can be generated using the following code.

```
ravix.boxplot(df, title = "Boxplot", xlab = "Vector 1",
              ylab = "Vector 4")
```

2.18 Sales Calls Application: Plotting Using Python

Jordan, from the previous application, also collected the number of sales that she made each day for the last 6 days, denoted as Y.

Using Table 2.2, investigate the relationship between X and Y by doing the following:

Table 2.2 Jordan's calls and sales

Variable						
Calls (X)	12	18	5	25	15	8
Sales (Y)	10	15	4	21	14	6

(a) Create vectors for X and Y.
(b) Create a dataframe containing calls and sales.
(c) Use the `plot` function to create a scatter plot of calls and sales. Interpret the plot.
(d) Use the `hist` function to create histograms of both calls and sales. Interpret the plots.
(e) Use the `boxplot` function to create a box plot of calls and sales on separate plots. Interpret the plots.
(f) Use the `boxplot` function to create a box plot of calls and sales on the same plot. Interpret the plot.

Solution

(a) Vectors are created as lists using square brackets.

```
x = [12, 18, 5, 25, 15, 8]
y = [10, 15, 4, 21, 14, 6]
```

(b) Using the `x` and `y` lists as inputs of the `DataFrame` function, we create a dataframe called `df`. Here we label `x` as `Calls` and `y` as `Sales`.

```
import pandas as pd
df = pd.DataFrame({'Calls': x, 'Sales': y})
```

(c) Using the `plot` function, we input the `x` and `y` lists. Here we label the axes and give the plot a title.

```
import ravix
ravix.plot("Sales ~ Calls", df, xlab = "Calls", ylab = "Sales",
           title = "Scatterplot")
```

From the scatterplot in Fig. 2.1, we note a clearly defined relationship between calls and sales. In particular, as the number of calls increases, the number of sales also increases. A positive linear trend line could be used to accurately describe the relationship between the two variables.

(d) Here we use the `hist` function to create a histogram of Calls and a histogram of Sales while specifying that we want five bars using the `bins` argument.

```
ravix.hist("Sales ~ Calls", df, norm = False, bins = 5)
```

From both histograms of Fig. 2.2, we see that the highest frequency bars correspond to the left-most bar in each graph each with a frequency of two. The remaining bins have frequencies of one each.

(e) Plugging in the `x` and `y` lists into the `boxplot` function individually, we have the following:

```
ravix.boxplot(df.Calls)
ravix.boxplot(df.Sales)
```

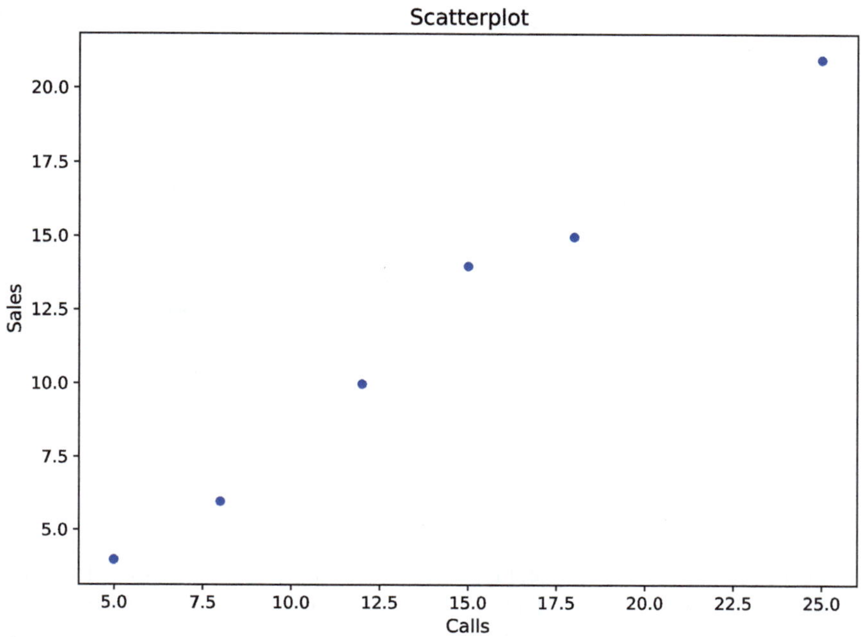

Fig. 2.1 Jordan's Scatterplot

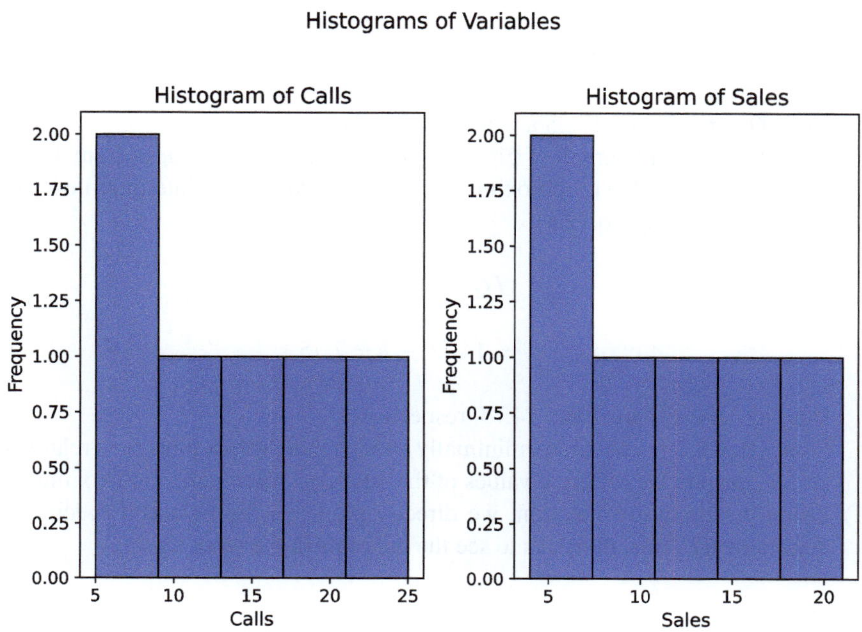

Fig. 2.2 Jordan's histograms

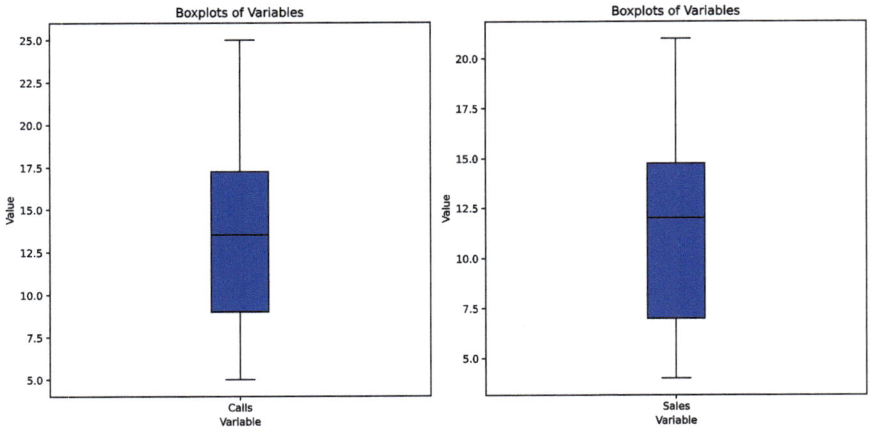

Fig. 2.3 Jordan's box plots

From the left-side graph of Fig. 2.3, we see that the number of calls ranges from 5 to 25 and the median value is approximately 13.5. Furthermore, the Interquartile Range (IQR) is given by Eq. (2.8):

$$\begin{aligned} IQR &= Q_3 - Q_1 \\ &= 17.25 - 9 \\ &= 8.25 \end{aligned}$$

since Q_1 and Q_3 are 17.25 and 9, respectively.
From the right-side graph of Fig. 2.3, we see that the sales ranges from 4 to 21 and the median value is approximately 12. Furthermore, the Interquartile Range (IQR) is given by Eq. (2.8):

$$IQR = Q_3 - Q_1$$

$$= 14.75 - 7 = 7.75$$

since Q_1 and Q_3 are 7 and 14.75, respectively.
Notice that the y-axis differs minimally enough that the box plots look relatively similar despite the different values of the quartiles, minimum, and maximum.

(f) Using the **boxplot** function, we directly plug in a formula and specify the dataframe **df**. This allows us to see the boxplots in the same figure.

```
ravix.boxplot("Sales ~ Calls", data = df)
```

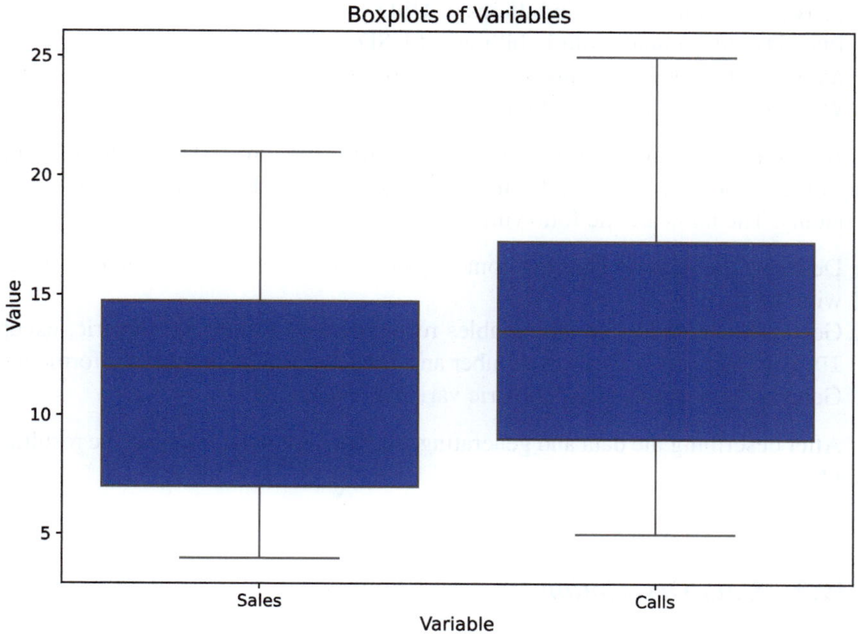

Fig. 2.4 Jordan's box plots

While the interpretation is similar to part (e), in Fig. 2.4, we can visualize the difference between sales and calls. Clearly the number of calls is higher for each quartile, the minimum, and the maximum.

2.19 Case Study: Top Companies

2.19.1 Problem Statement

Creating informative visualizations is a crucial aspect of business analytics and is essential for understanding data across various business domains. Understanding data can lead to actionable results in business. This case study presents a dataset of 200 of the world's top companies and offers a quick analysis to facilitate data comprehension.

To understand the overall global economy, we can observe the world's largest public companies using four metrics: assets, market value, sales, and profits. The dataset for the year 2022 consists of 6 columns and 200 rows. The columns in the data are as follows:

- `ID`—numeric integer corresponding to the row number
- `Country`—the country in which the company resides

- `Sales`—the annual sales in billions of USD
- `Profit`—the annual profit in billions of USD
- `Assets`—the total company assets in billions of USD
- `Value`—the company's total market value in billions of USD

As an analyst working for an investment firm, you are tasked with analyzing the top 200 global companies from 2022 to get some understanding of the overall economy. The tasks are the following:

1. Describe the data by running some Python commands to familiarize yourself with the data.
2. Generate scatterplots of all variables recognized by Python as numeric, except `ID`, as it only serves as a row number and does not provide valuable information.
3. Generate histograms of all numeric variables except `ID`.

After describing the data and generating the plots, we must interpret the resulting graphs.

2.19.2 Data Description

Loading in the `Top200.csv` dataframe can be achieved using Ravix's `get_data` function. Simply import Ravix and then use the function specifying the name of the file.

```
import ravix
df = ravix.get_data("Top200.csv")
```

The first step in analyzing a dataset is to get familiar with the dataset of interest. After loading the data, examine the first five observations using the `head` function. Note that three of the companies listed in the `head` output are from the United States, with the other two coming from China and Japan.

```
df.head()
```

```
#     ID Country   Sales  Profit   Assets    Value
# 0    1      US  15.264   6.261  234.234   55.157
# 1    2      US  31.454  11.662   12.398    6.204
# 2    3   China  36.713  16.065   46.544  184.417
# 3    4   Japan   0.743  -6.856    0.765   29.102
# 4    5      US   2.511   7.644   26.513   48.270
```

Although Python recognizes `ID` as numeric, it is not a quantitative variable but rather a qualitative one, and its summary statistics do not provide meaningful information. The quantitative variables are `Sales`, `Profit`, `Assets`, and `Value` within the dataframe. Here, we reference these variables in the dataframe by passing a list inside the dataframe.

```
numeric_df = df[['Sales', 'Profit', 'Assets', 'Value']]
```

The resulting `numeric_df` is a dataframe consisting of the numeric columns excluding `ID`. The `describe` function provides insight into the data by displaying some summary statistics for the data perceived as numeric. Instead of describing `df`, we choose to describe `numeric_df` to avoid viewing the statistics for `ID`.

```
numeric_df.describe()
```

```
#              Sales      Profit      Assets       Value
# count  200.00000  200.000000  200.000000  200.000000
# mean    19.81952    6.295585   87.662165   44.728445
# std     18.34595    6.651592   84.816835   46.858284
# min      0.03400   -9.619000    0.351000    0.131000
# 25%      6.57225    1.260000   26.327500   12.813500
# 50%     14.93200    6.058000   64.045000   31.363500
# 75%     27.53775   10.380250  123.857250   58.948500
# max    126.62600   28.638000  507.224000  263.381000
```

The output shows the numeric columns' count, mean, standard deviation (std), minimum (min), first quartile (25%), median (50%), third quartile (75%), and maximum (max) values. From observing this data, note the minimum values are at or near 0 for `Sales`, `Assets`, and `Value`. On the other hand, `Profit` has a negative minimum value. Further note that the third quartiles are much different than the maximum values for these numeric variables, indicating a trailing tail—that is, a few unusually large values extending the distribution to the right. Since the third quartile is the 75th percentile, we employ the `quantile` function to verify the third quartile from the output statistics given above for `Sales`. The first argument in the `quantile` is the percentile.

```
Q3Sales = df['Sales'].quantile(0.75)
```

The value of a variable can be displayed by passing its name to the `print` function. For demonstration, we print `Q3Sales` by typing the Python code.

```
Q3Sales
```

```
# np.float64(27.53775)
```

Notice that 27.53775 is shown as the third quartile for `Sales`, which is consistent with the output from the `describe` function.

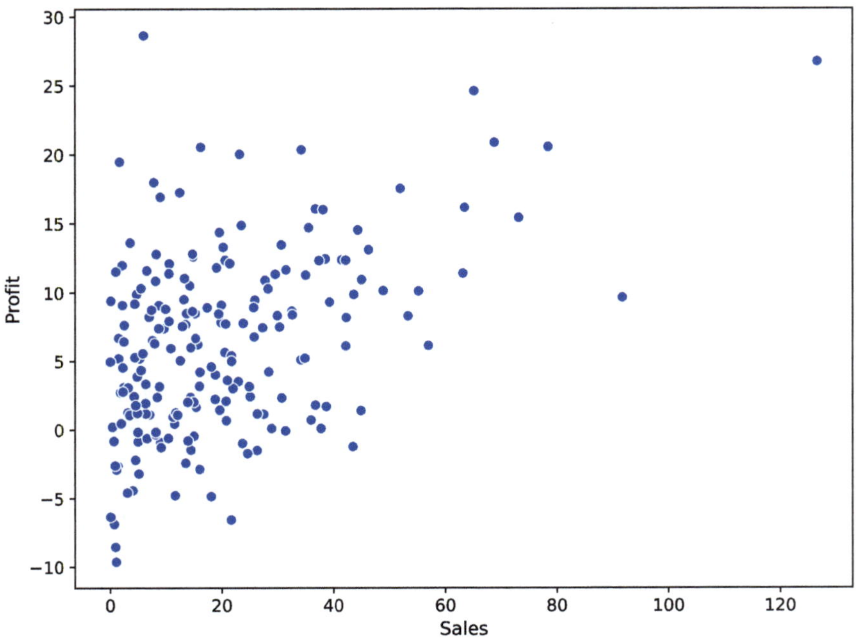

Fig. 2.5 Scatterplot of the top 200 companies (unlabeled)

2.19.3 Scatterplots

To generate a plot of `Sales` and `Profit`, the `plot` function is executed from the Ravix package. Here we choose `Profit` to be on the *y*-axis and `Sales` on the *x*-axis by writing the formula `"Profit ~ Sales"`. These values are contained within the dataframe `df` previously loaded.

```
ravix.plot("Profit ~ Sales", data = df)
```

Although Fig. 2.5 is correct, it labels the *x*-axis as `Sales` and the *y*-axis as `Profit`. To improve these variable names, we can use the following notation to plot the data. The `xlab`, `ylab`, and `title` arguments are specified here in order to properly label the axes and provide a title. Since only `Profit` and `Sales` are plotted, the `data` argument can be specified as `df` or `numeric_df`.

```
ravix.plot("Profit ~ Sales", data = df,
           xlab = "Sales (Billions $)",
           ylab = "Profits (Billions $)",
           title = "Scatterplot of the Top 200 Companies")
```

From the plot in Fig. 2.6, each circle represents an observation in the data.

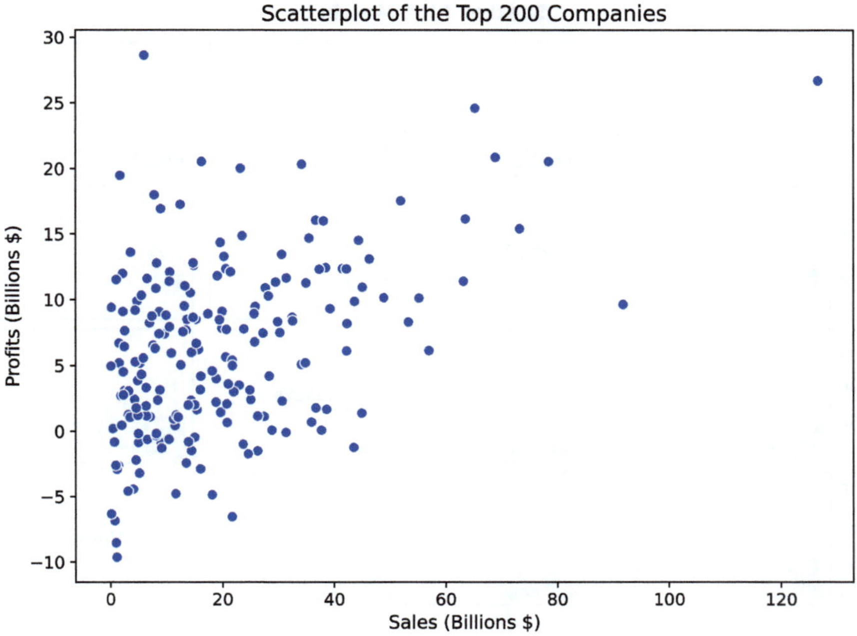

Fig. 2.6 Scatterplot of the top 200 companies

Next, we generate a scatterplot matrix of all numeric variables (excluding `ID`). Passing `"Sales ~ ."` as the formula and first argument indicates we would like Sales highlighted at the top of the matrix. Specifying the data as `numeric_df` into the `plots` function is another argument required to produce the matrix of scatterplots.

```
ravix.plot("Sales ~ .", numeric_df)
```

Along the diagonal of Fig. 2.7, we see the distribution or density function of the variables. Several structural patterns are also evident:

- Because `Sales` appears on the left side of the tilde in the formula, it is treated as the response variable and therefore appears on the y-axis in all top-row scatterplots.
- All scatterplots in the first column have `Sales` on the x-axis.
- The first row and second column shows a scatterplot of `Profit` versus `Sales`.

This matrix can help assess the relationship between variables quickly. However, the usefulness declines when too many variables are included, since the graphs will become smaller and less visible. Even though the scatterplots provide limited insight, it can be seen that `Profit` does tend to increase with `Sales` and `Value`.

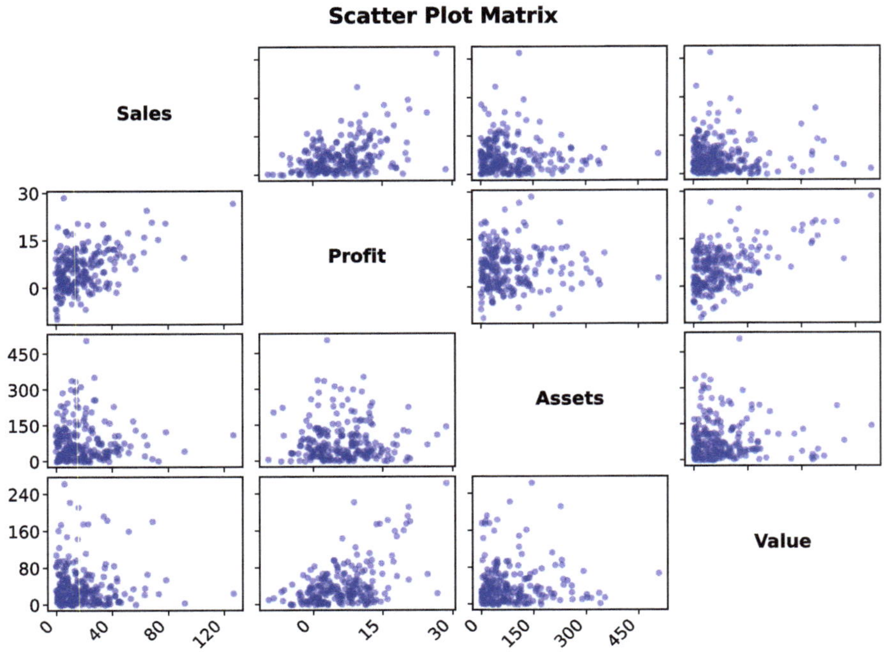

Fig. 2.7 Scatterplot matrix of the top 200 companies

2.19.4 *Histograms*

Using the `hist` function from Ravix, we generate a histogram of `Sales` by passing the `df.Sales` as seen in Fig. 2.8.

```
ravix.hist(df.Sales)
```

Next, we employ the `hist` function from Ravix to plot a histogram of each numeric vector in the dataframe `numeric_df`. We specify the formula using the `"."` shortcut to include all other columns in the dataframe.

```
ravix.hist("Sales ~ .", numeric_df)
```

As mentioned in the interpretation of the summary statistics, all quantitative variables in this dataset have a tail leading to the right, referred to as a right skew. We see from Fig. 2.9 that a bulk of sales is below 60 billion USD, a majority of assets are below 150 billion USD, and a majority of market value is below 100 billion USD. Profit has most values between 0 and 20 billion, but there are also many companies with negative profits.

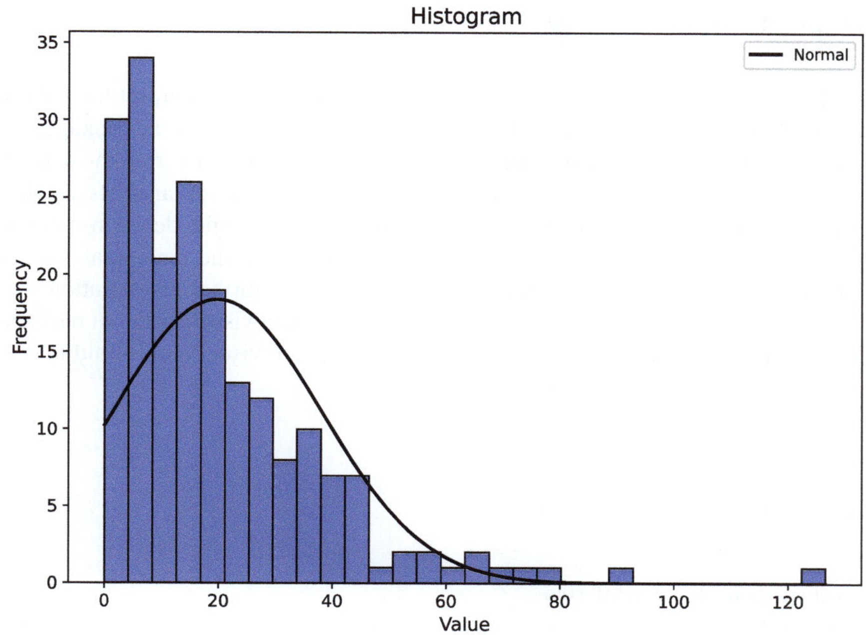

Fig. 2.8 Histogram of sales of the top 200 companies

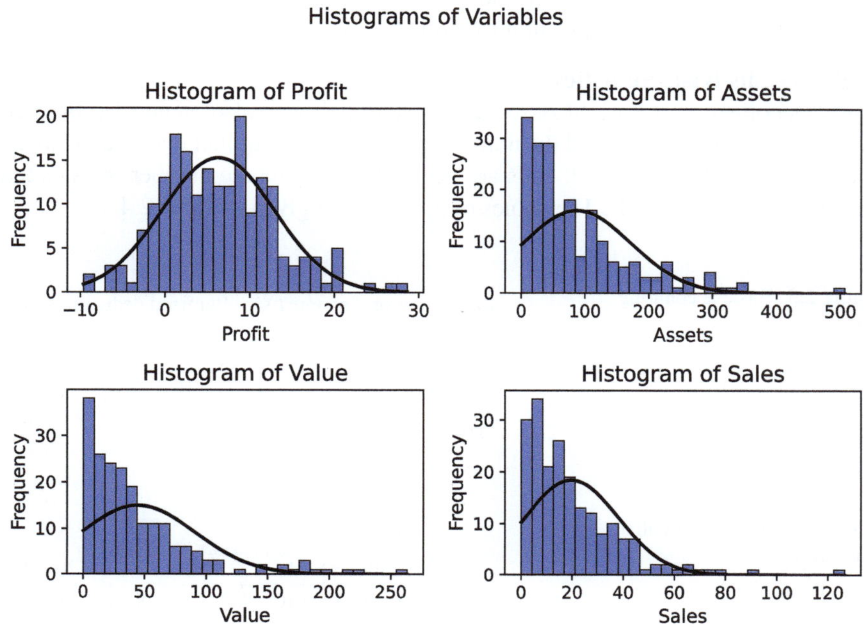

Fig. 2.9 Histograms of the top 200 companies

2.19.5 *Case Conclusion*

In this analysis, we utilized visualizations to characterize and interpret the data of the top 200 global companies from 2022. By leveraging the Ravix package, we produced scatterplots and histograms that highlighted key insights from the dataset.

These visualizations and summary statistics are crucial for analysts as they provide a foundation for deeper understanding and strategic decision-making. By understanding the distribution and relationships within the data, analysts can identify trends, outliers, and potential areas of interest for further investigation.

Overall, this case study demonstrates the power of data visualization in business analytics, emphasizing the need for clear and informative visual tools to aid in data interpretation and decision-making processes.

Problems

1. Accounting Basic Statistics
Using Table 2.3, answer the questions below. Both variables are in thousands of US dollars.

Without the use of a computer:

a. Calculate the mean for both X and Y.
b. Calculate the standard deviation for both X and Y.

2. Sales Calls Basic Statistics
Jordan's colleague at Enright Associates is Abdullah who also manages retirement accounts. Jordan documented the number of sales calls Abdullah made each day for the last week. The number of calls are denoted by X, and the number of sales that resulted are denoted as Y. The values of X and Y are given in Table 2.4.

Table 2.3 Basic accounting data

Adjusted gross income (X)	Itemized deductions (Y)
51	11
142	31
66	11
67	14
35	12

Table 2.4 Abdullah's calls and sales

Variable							
Calls (X)	18	23	25	15	8	20	12
Sales (Y)	10	11	15	8	3	11	7

Without the use of a computer:

a. Calculate the mean for both X and Y.
b. Calculate the standard deviation for both X and Y.
c. Compute the variance for both X and Y.

3. Sales Calls Basic Statistics in Python
Investigate the relationship between X and Y from the previous problem by doing the following. While using a computer (with Python):

a. Create vectors for X and Y.
b. Calculate the mean for both X and Y.
c. Calculate the standard deviation for both X and Y.
d. Compute the variance for both X and Y.

4. Sales Calls Plots in Python
Investigate the relationship between X and Y from the previous problem by doing the following. While using a computer (with Python):

a. Create a dataframe containing calls and sales.
b. Use the `plot` function to create a scatter plot of calls and sales. Interpret the plot.
c. Use the `hist` function to create histograms of both calls and sales. Interpret the plots.
d. Use the `boxplot` function to create a boxplot of calls and sales. Interpret the plots.

5. Top 200 Companies Basic Statistics
In the chapter case study, the summary function was used to provide summary statistics of the variables. Use the Top 200 dataset to answer the following questions:

a. Calculate the mean of each numeric variable.
b. Calculate the standard deviation of each numeric variable.
c. Find the sum of each numeric variable.

Chapter 3
Regression Fundamentals

There is bound to be a regression toward the mean.

—Charlie Munger

3.1 Introduction

Although data collection has expanded dramatically in recent years, transforming raw information into actionable insight remains a central challenge. Linear regression, developed more than 200 years ago, continues to serve as a cornerstone of statistics and business analytics and lies at the heart of many predictive methods, including modern machine learning models.

In this chapter, we examine the core concepts of linear regression. The discussion of linear regression will be limited to two variables in order to get a clear understanding of the calculations and theory. The concepts are put into context with a simple application using data from a call center, predicting successful sales by the number of calls.

In the final sections of this chapter, we present and solve an accounting case study involving the prediction of the adjusted gross income of an individual based on the amount of their itemized deductions. This case study makes use of the Python programming language with descriptive graphs and relevant source code.

3.2 Covariance

When analyzing a single random variable, the mean and standard deviation are often the only statistics needed since they represent the center and spread of the variable's distribution. When two random variables are analyzed for a relationship, the covariance can be calculated. For random variables X and Y with data pairs (x_i, y_i), the joint sum of squares of X and Y can be calculated as follows:

D. P. McGibney, *Applied Linear Regression for Business Analytics with Python*,
International Series in Operations Research & Management Science 378,
https://doi.org/10.1007/978-3-032-23806-1_3

$$SS_{xy} = \sum_{i=1}^{n} (x_i - \bar{x})(y_i - \bar{y}) \tag{3.1}$$

or equivalently as:

$$SS_{xy} = \sum_{i=1}^{n} x_i y_i - \frac{\left(\sum_{i=1}^{n} x_i\right)\left(\sum_{i=1}^{n} y_i\right)}{n}. \tag{3.2}$$

The covariance between X and Y is thus:

$$Cov(X, Y) = \frac{SS_{xy}}{n-1}. \tag{3.3}$$

While the covariance shows how two variables vary together, the correlation coefficient provides a more interpretable measure of association.

3.3 Correlation Coefficient

Knowing which variables are related to the unknown variable of interest is helpful for prediction. For example, knowing that height and weight are correlated is useful if you would like to approximate the weight of an individual based on their height. It is also useful if the degree of correlation between two random variables can be measured. The Pearson correlation coefficient, referred to as the correlation coefficient herein, provides such a measurement. For height and weight values, the correlation coefficient would be relatively high since there is a relationship between the two variables.

Similar to the covariance, the correlation coefficient describes the relationship between two or more random variables. The correlation coefficient differs from the covariance because the correlation is scaled and therefore unitless.

The sample correlation coefficient can be calculated using the equation:

$$r = \frac{Cov(X, Y)}{s_x s_y}, \tag{3.4}$$

where s_x and s_y are the sample standard deviations of X and Y, respectively. The sample correlation coefficient can also be solved in terms of the sum of squares:

$$r = \frac{SS_{xy}}{\sqrt{SS_{xx} SS_{yy}}}. \tag{3.5}$$

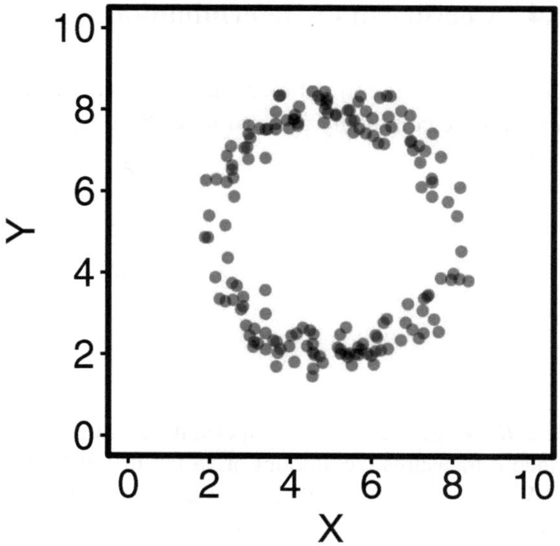

Fig. 3.1 Non-linear correlation

The correlation coefficient has several properties:

1. It is a unitless measurement between -1 and 1 inclusive ($-1 \leq r \leq 1$).
2. If $r = 1$, there is perfect positive linear correlation. If $r = -1$, there is perfect negative linear correlation. If $r = 0$, there is no linear correlation.
3. Positive values of r imply that as x increases, y tends to increase. Negative values of r imply that as x increases, y tends to decrease.
4. The value of r is the same for the pairs (x, y) and the corresponding pairs (y, x).
5. The value of r does not change when either variable is converted to different units.
6. The correlation coefficient only measures the degree of linear correlation.

While the correlation coefficient is very useful in finding a measure of correlation between variables, it is limited since it only measures the degree of linear correlation between variables. Some may wrongfully assume that if there is a high correlation coefficient, then the relationship must be linear. In fact, there are many relationships that seem to be linear, but are actually non-linear.

Another misinterpretation of the correlation coefficient occurs in the case of a low value of r. If the value of r is 0, it does not mean there is no relation between X and Y, but rather that there is no *linear* relationship between X and Y. While this may seem counterintuitive, consider the example in Fig. 3.1 where a clear relationship between X and Y exists. Because the relationship is circular, the correlation coefficient is 0.

3.4 Coefficient of Determination

The square of the correlation coefficient (r) is called the coefficient of determination and is more commonly referred to as R^2. The value of R^2 is obtained by squaring equation

$$R^2 = \left(\frac{SS_{xy}}{\sqrt{SS_{xx} SS_{yy}}} \right)^2 = \frac{SS_{xy}^2}{SS_{xx} SS_{yy}} \tag{3.6}$$

The R^2 value signifies an important measure of how well the model fits the data since it measures the amount of variability in Y that is explained by the variability in X. For convenience and interpretability, R^2 is often converted to a percentage and represents the percentage of the explained variability in Y.

If R^2 represents the fraction of explained variability, it follows that the value $1 - R^2$ is the ratio of *unexplained* variability in Y. This unexplained variation can be the result of random chance and/or lurking variables. Obviously, one desires to maximize R^2 and thus minimize $1 - R^2$.

Many students inquire: what constitutes a high R^2 or a high r value? Although no precise threshold differentiates high from low correlation, careful observation helps to generally distinguish between them. As shown in Fig. 3.2, the first scatterplot at the top-left of the page illustrates a perfect positive linear correlation with an r value of 1. The next three scatterplots on the left represent r values of 0.8, 0.5, and 0.3. The last scatterplot at the bottom of the figure corresponds to a r value of 0. Note that the data trend is still discernible at $r = 0.8$, but the trend slowly degrades as r decreases. In the same figure, in the upper right, the values of r are $-1, -0.8, -0.5,$ and -0.3. From these plots, it should be evident that the magnitude of r denotes the degree of linear relationship, whereas the sign denotes whether the linear trend is increasing or decreasing. The sign of the slope for a linear regression is always in agreement with the slope of the r value. The value of R^2 does not take this upward or downward trend into account as seen in the figure (Fig. 3.2); particularly, the spread around the line is relatively similar for $r = 0.8$ and $r = -0.8$.

3.5 Sales Calls Application: Variable Correlations

Here, we investigate Jordan's data further. Using Table 2.2 from the previous chapter, investigate the relationship between X and Y by doing the following:

(a) Calculate the sum of squares of X and Y.
(b) Calculate the sample covariance.

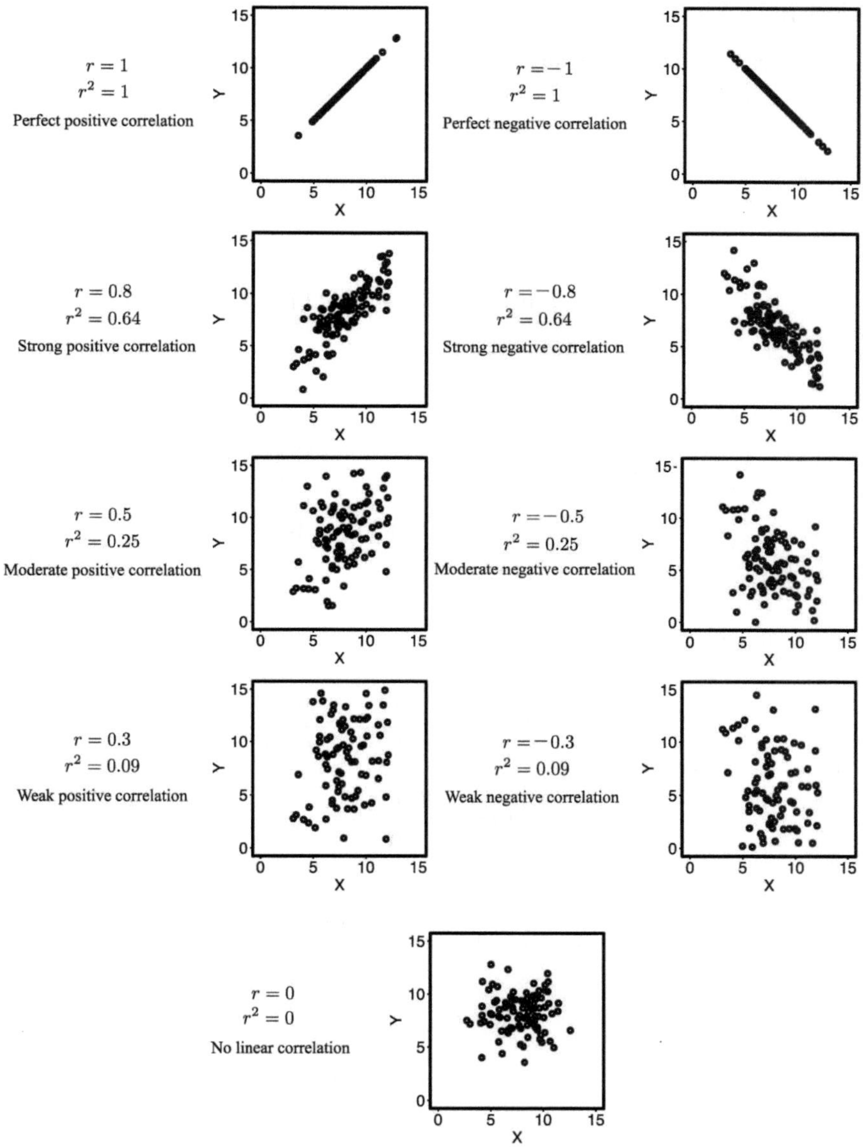

Fig. 3.2 Correlation values

(c) Compute the sample correlation coefficient.
(d) Compute the coefficient of determination.
(e) Plot a scatter plot.

Solution:

(a) First, calculate the following values required to obtain SS_{xy}:

$$\sum_{i=1}^{n} x_i = 12 + 18 + 5 + 25 + 15 + 8$$
$$= 83$$
$$\sum_{i=1}^{n} y_i = 10 + 15 + 4 + 21 + 14 + 6$$
$$= 70$$
$$\sum_{i=1}^{n} x_i y_i = (12 \times 10) + (18 \times 15) + (5 \times 4)$$
$$+ (25 \times 21) + (15 \times 14) + (8 \times 6)$$
$$= 1{,}193.$$

Knowing these three values, Eq. (3.2) can now be utilized:

$$SS_{xy} = \sum_{i=1}^{n} x_i y_i - \frac{\left(\sum_{i=1}^{n} x_i\right)\left(\sum_{i=1}^{n} y_i\right)}{n}$$
$$= 1{,}193 - \frac{(83)(70)}{6}$$
$$= 224.67.$$

(b) The sample covariance can be calculated from Eq. (3.3):

$$\text{Cov}(X, Y) = \frac{SS_{xy}}{n-1}$$
$$= \frac{224.67}{5}$$
$$= 44.93.$$

(c) To calculate the sum of squares for Y, we must first calculate:

$$\sum_{i=1}^{n} y_i^2 = 10^2 + 15^2 + 4^2 + 21^2 + 14^2 + 6^2 = 1014.$$

From this calculation and (a), we have:

$$SS_{yy} = \sum_{i=1}^{n} y_i^2 - \frac{\left(\sum_{i=1}^{n} y_i\right)^2}{n}$$

$$= 1{,}014 - \frac{(70)^2}{6}$$

$$= 197.33.$$

From Sect. 2.3, recall $SS_{xx} = 258.83$ and $SS_{xy} = 224.67$. The correlation coefficient is calculated from Eq. (3.5),

$$r = \frac{SS_{xy}}{\sqrt{SS_{xx} SS_{yy}}}$$

$$= \frac{224.67}{\sqrt{258.83 \times 197.33}}$$

$$= 0.994.$$

(d) The coefficient of determination can be found by squaring the value of r:

$$R^2 = 0.994^2 = 0.988$$

indicating that 98.8% of the variation in Y (sales) can be explained by the variation in X (calls).

(e) The scatter plot is shown in Fig. 3.3. From this plot, we see the relationship between calls and sales. Particularly, if calls increase, then the sales increase.

3.6 Least Squares Criterion

To better understand the relationship between two variables, we examine how well a line can describe their association. The least squares criterion is a method that determines the best line between the variables. The best line is achieved by making the distance between the line and the points as small as possible (Fig. 3.4). These vertical distances are referred to as residuals, and the ith residual is expressed as:

$$e_i = y_i - \hat{y}_i \tag{3.7}$$

where y_i represents the ith value of the response and $\hat{y}_i$ denotes the value of the response on the least squares line.

Summing the residuals directly is not meaningful since some are positive and others are negative. In order to capture the magnitude, we square each residual and then sum them. We minimize the squared distances by minimizing the sum of the squares of the vertical distances from the data points (x_i, y_i) to the line, which is referred to as the Sum of the Squared Error (SSE), expressed as:

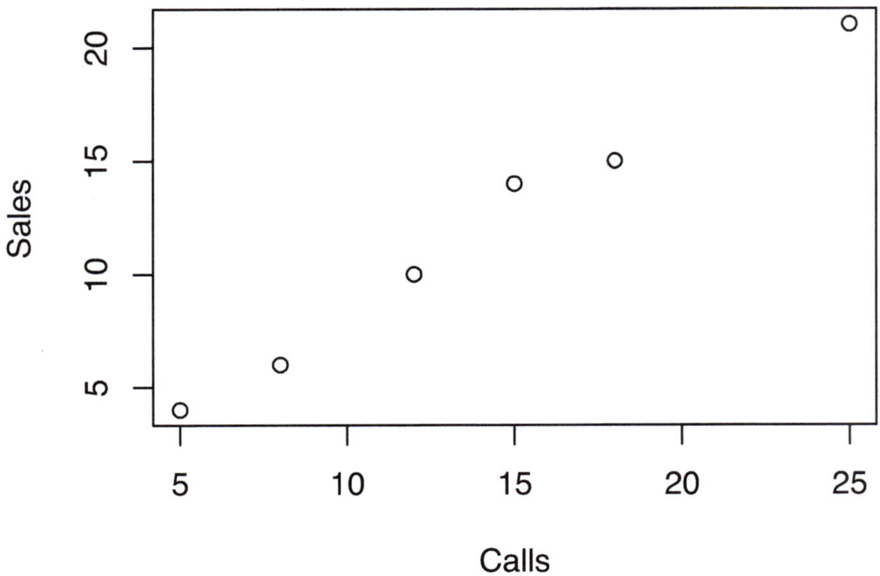

Fig. 3.3 Scatterplot of Sales vs. Calls

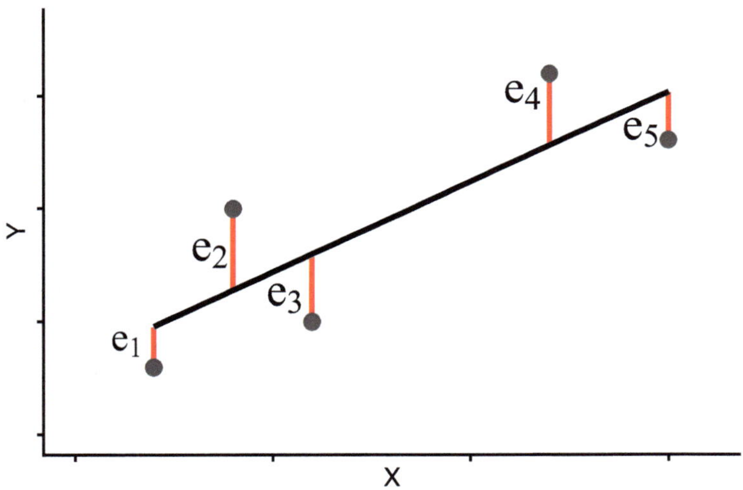

Fig. 3.4 Least squares depiction

$$
\begin{aligned}
SSE &= \sum_{i=1}^{n} e_i^2 \\
&= \sum_{i=1}^{n} (y_i - \hat{y}_i)^2
\end{aligned}
\tag{3.8}
$$

Using the least squares criterion, one can find the equation of a line, which can be referred to as the regression line. The equation of the regression line is simply called the regression equation and is given by:

$$\hat{Y} = \hat{\beta}_0 + \hat{\beta}_1 X \tag{3.9}$$

where

- $\hat{Y}$ denotes an estimate of Y,
- $\hat{\beta}_0$ and $\hat{\beta}_1$ estimate the Y-intercept and the slope, respectively.

The values of $\hat{\beta}_0$ and $\hat{\beta}_1$ are solved for by finding the minimum value of Eq. (3.8) while substituting in Eq. (3.9) for $\hat{y}$:

$$
\begin{aligned}
SSE &= \sum_{i=1}^{n} \left(y_i - (\hat{\beta}_0 + \hat{\beta}_1 x_i)\right)^2 \\
&= \sum_{i=1}^{n} \left(y_i - \hat{\beta}_0 - \hat{\beta}_1 x_i\right)^2
\end{aligned}
\tag{3.10}
$$

To minimize Eq. (3.10) and thereby find equations for $\hat{\beta}_0$ and $\hat{\beta}_1$, the partial derivative of the SSE with respect to $\hat{\beta}_0$ and $\hat{\beta}_1$ should be found and set equal to 0 as is common practice in calculus. The resulting equations are referred to as the "normal" equations.

The result of solving the normal equations is:

$$\hat{\beta}_1 = \frac{SS_{xy}}{SS_{xx}} \tag{3.11}$$

$$\hat{\beta}_0 = \bar{y} - \hat{\beta}_1 \bar{x} \tag{3.12}$$

where $\bar{x}$ and $\bar{y}$ denote the sample means of X and Y, respectively.

The regression line has the following properties:

1. The point $(\bar{x}, \bar{y})$ always lies on the linear regression line.
2. The value $\hat{Y}$ is read "Y hat," where the "hat" denotes an estimate.
3. The variable X is referred to as the independent variable, predictor, or explanatory variable.

4. The variable Y is referred to as the dependent variable, response, or target variable.
5. The values of X can be plugged into the regression equation to get estimates of Y denoted as $\hat{Y}$.

With the knowledge to properly solve the regression equation, we are equipped with the necessary tools to continue the sales calls application. Previously, we identified a positive linear correlation between calls and sales, which suggested further analysis.

3.7 Sales Calls Application: Simple Regression

From the data used in the previous application (Table 2.2), the least squares regression line can be computed. Further investigate the relationship between X and Y by doing the following:

(a) Find the regression equation that best fits the data.
(b) Interpret the slope of the regression equation.

Solution:

(a) Recall from Sects. 2.3 and 3.5 that $SS_{xx} = 258.83$ and $SS_{xy} = 224.67$. Using these values and Eq. (3.11), the slope is

$$\begin{aligned}\hat{\beta}_1 &= \frac{SS_{xy}}{SS_{xx}} \\ &= \frac{224.667}{258.833} \\ &= 0.868.\end{aligned}$$

Using the slope and the sample means ($\bar{x} = 13.8333$ and $\bar{y} = 11.6667$), the intercept is given from Eq. (3.12):

$$\begin{aligned}\hat{\beta}_0 &= \bar{y} - \hat{\beta}_1\bar{x} \\ &= 11.6667 - 0.868 \times 13.8333 \\ &= -0.3406.\end{aligned}$$

As a final step, the intercept and slope are plugged into Eq. (3.9)

$$\begin{aligned}\hat{Y} &= \hat{\beta}_0 + \hat{\beta}_1 X \\ &= -0.3406 + 0.868X.\end{aligned}$$

(b) Since the slope is 0.868, if the number of calls (X) is increased by 1, then the number of sales (Y) is expected to increase by 0.868. This interpretation can be expanded to 10 calls; that is, if the number of calls increases by 10, then the number of sales is expected to increase by 8.68.

3.8 Interpolation and Extrapolation

When we predict the response Y for a value of X that falls within the range of observed X values, the prediction is referred to as interpolation. If the value of X is outside the range of observed X values, then the resulting prediction is called an extrapolation.

Because interpolation occurs within the range of known X values, predictions are generally more reliable compared to extrapolation, where x_i falls outside that range. Extrapolation should hence be met with skepticism since the available data is not similar enough to the X value of interest. Furthermore, a pattern may appear linear, but considering extreme values can reveal a nonlinear trend. This is why caution is needed when extrapolating. For instance, if Jordan makes 100 calls in a day, she may not be able to get as many sales as she would if she made 10 calls in a day. While it is possible that Jordan could maintain her sales performance, the lack of data for the 100 calls prevents a well-informed guess.

In Fig. 3.5a, the minimum value of X can be deciphered as 5. To the left of $X = 5$, there are no data values, which means that the interval of $X < 5$ is outside the range of X. The area to the left of $X = 5$ is therefore denoted in pink (in Fig. 3.5a) since it is outside the range of X. Similarly, the maximum value of X is 25, and therefore, the area to the right of $X = 25$ is in pink.

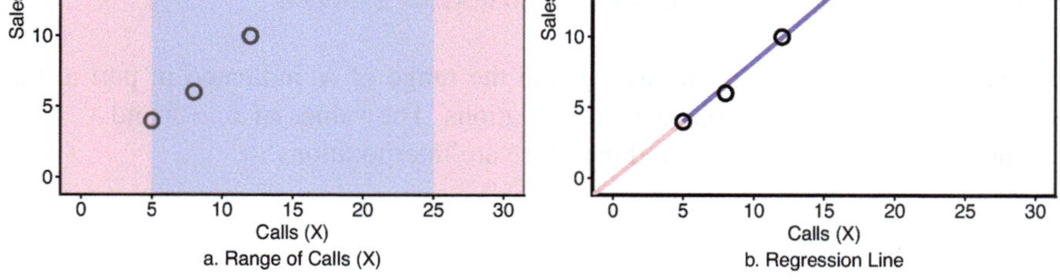

Fig. 3.5 Interpolation and Extrapolation

Using the range information from Fig. 3.5a along with the regression line, Fig. 3.5b shows which portions of the regression line are extrapolation and which are interpolation. As mentioned previously, the extrapolated values are values predicted with X values outside of the range of X.

3.9 Sales Calls Application: Prediction

Jordan would like to use her regression model to evaluate several potential work scenarios. On days when she is sick, she expects to make only three calls. If she works a half day on Saturday, she could make five calls. She would like to complete 15 calls on Tuesday, and she also wants to see what would happen if she increases her calls to 30. Using Table 2.2, investigate the relationship between X and Y by doing the following.

(a) Find the range of X.
(b) Predict the corresponding Sales for $x = 3, 5, 15, 30$.
(c) Indicate whether each prediction from (b) is an interpolation or an extrapolation.

Solution:

(a) The smallest value of X is 5, and the largest is 25. Therefore, the range is between 5 and 25 inclusive:

$$5 \leq X \leq 25.$$

Figure 3.5a depicts the range of X for this application.

(b) The least squares regression line was computed in the previous application:

$$\hat{Y} = -0.3406 + 0.868X.$$

Using the regression line, the predictions are made as follows:

- When $x = 3$, then $\hat{y} = -0.3406 + 0.868(3) = 2.2634$.
- When $x = 5$, then $\hat{y} = -0.3406 + 0.868(5) = 3.9994$.
- When $x = 15$, then $\hat{y} = -0.3406 + 0.868(15) = 12.6794$.
- When $x = 30$, then $\hat{y} = -0.3406 + 0.868(30) = 25.6994$.

(c) Since $x = 3$ and $x = 30$ are outside the range of X indicated in part a, the corresponding predictions are extrapolations. The values of $x = 5$ and $x = 15$ are within the range of X, and therefore are interpolations.

3.10 Explained Deviation

To understand how R^2 measures the percentage of explained variation, it's important to examine the components of total variation, specifically the deviations that contribute to it.

Figure 3.6 depicts the explained, the unexplained, and the total deviations for one particular observed value (x, y) which is above the regression line. From the depiction, the total deviation from the mean of Y $(\bar{y})$ is $y - \bar{y}$. This deviation, or difference, consists of both explained and unexplained deviation.

For the observed value of x, the linear regression prediction of y would be $\hat{y}$. Therefore, the amount of deviation explained by linear regression would be $\hat{y} - \bar{y}$. The amount of unexplained deviation is $y - \hat{y}$.

Since the total deviation can be broken down into unexplained and explained deviations, the deviations can be expressed for the particular value of Y as:

$$(y - \bar{y}) = (\hat{y} - \bar{y}) + (y - \hat{y}), \tag{3.13}$$

or for the entire vector Y as:

$$(Y - \bar{y}) = (\hat{Y} - \bar{y}) + (Y - \hat{Y}). \tag{3.14}$$

Here, "deviation" refers to the difference between observed and predicted values, while "variation" represents the squared sum of these deviations across the dataset. While the deviation for a single observation provides insight into its individual contribution, the total variation is calculated by considering all observations in the

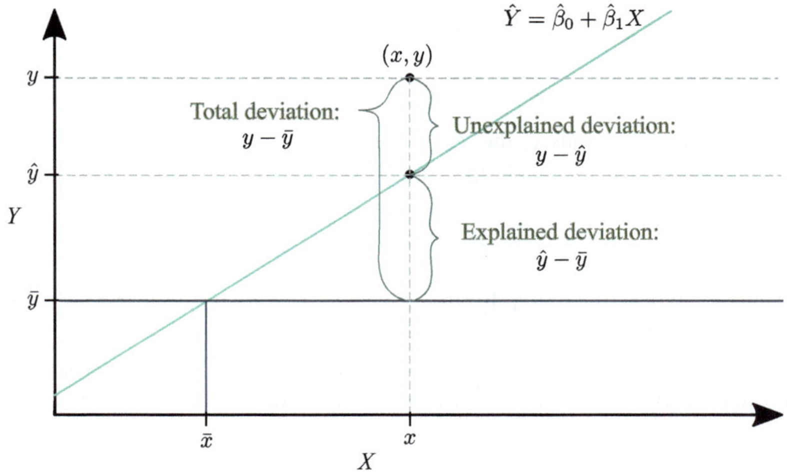

Fig. 3.6 Explained vs. Unexplained Deviation

dataset. By squaring and summing the deviations that comprise the total deviation of each observation, a measure of the total variation is:

$$SST = \sum_{i=1}^{n} (y_i - \bar{y})^2, \tag{3.15}$$

which is referred to as the Sum of Squares Total (SST).

The variation of the Sum of Squares due to Regression (SSR) is:

$$SSR = \sum_{i=1}^{n} (\hat{y}_i - \bar{y})^2, \tag{3.16}$$

which is often referred to as the explained variation because it is accounted for by the regression line.

The variation captured by the Sum of Squares due to Error (SSE) is:

$$SSE = \sum_{i=1}^{n} (y_i - \hat{y}_i)^2, \tag{3.17}$$

which can be called the unexplained variation.

Similar to Eq. (3.14), the SST can be expressed in terms of the SSR and SSE:

$$SST = SSR + SSE. \tag{3.18}$$

This equation is equivalent to:

$$\sum_{i=1}^{n} (y_i - \bar{y})^2 = \sum_{i=1}^{n} (\hat{y}_i - \bar{y})^2 + \sum_{i=1}^{n} (y_i - \hat{y}_i)^2. \tag{3.19}$$

As mentioned previously, the R^2 value is the amount of explained variation. Therefore, using SSR and SST, the value of R^2 is:

$$R^2 = SSR/SST. \tag{3.20}$$

Equation (3.20) is mathematically equivalent to Eq. (3.6), which is left for the reader to prove as one of the problems at the end of the chapter.

The SSE, when divided by its associated degrees of freedom ($n - 2$), yields the mean squared error (MSE), which quantifies the average variance in the errors:

$$MSE = \frac{SSE}{n - 2}. \tag{3.21}$$

The mean squares due to regression (MSR) is the explained variation and can be calculated as:

$$MSR = \frac{SSR}{p}, \tag{3.22}$$

where p is the number of predictors. In the next sections, concepts from this chapter are brought together to solve a case study in accounting analytics.

3.11 Case Study: Accounting Analytics

3.11.1 Problem Statement

Making predictions and analyzing data are valuable skills to have across many different business domains. For this reason, the skills required for entering the job market have evolved, with analytics now being one of the most widely sought-after skill sets. Analytics has especially changed the field of accounting, in which irregularities in the data can be quickly found using simple analytical methods.

As an accounting analyst, you are investigating the amount of itemized deductions that are reasonable considering someone's adjusted gross income. You are handed the tax return documents of eight individuals, and you are tasked with analyzing this data. From this data, you are to generate a scatter plot, calculate the correlation coefficient, fit a linear regression model, and use your fitted model to predict the itemized deductions for an individual who is at risk of an audit. The data are listed in Table 3.1 where both variables are in thousands of US dollars.

Table 3.1 Accounting data

Adjusted Gross Income (in $1,000s)	Itemized Deductions (in $1,000s)
77	16
96	15
50	8
54	10
130	21
67	13
135	24
114	24

3.11.2 Correlation and Scatterplot

From the data in Table 3.1, the variable that will be predicted is the amount of itemized deductions. Therefore, the itemized deductions variable is the response variable, which will be assigned as Y by convention. The variable used to predict Y is known as the predictor variable and is conventionally known as X, which in this example is the adjusted gross income. Lists are created for both x and y using square brackets. In this example, lists are used for simplicity, but numpy arrays provide optimized performance.

```
x = [77, 96, 50, 54, 130, 67, 135, 114]
y = [16, 15, 8, 10, 21, 13, 24, 24]
```

While the relationship between X and Y is often observed in a plot, it is beneficial to quantify the relationship using the correlation coefficient (r), and also, the coefficient of determination (R^2). Using the `corrcoef` function from NumPy allows for the calculation of the correlation coefficient:

```
import numpy as np
np.corrcoef(x,y)[0,1]
```

```
# np.float64(0.9426904764082495)
```

The value of 0.9427 denotes a strong positive linear correlation between the response and the predictor variable. The coefficient of determination (R^2) can be found by squaring the correlation coefficient:

```
np.corrcoef(x,y)[0,1]**2
```

```
# np.float64(0.8886653343108125)
```

In interpreting the model, one should note that the R^2 value is relatively high at 0.8887. From this value, one can conclude that 88.87% of the variation in Y can be explained by the variation in X. More specifically, the variation in the adjusted gross income can explain 88.87% of the variation in the itemized deductions.

3.11.3 Linear Regression Modeling

One can further analyze the relationship between adjusted gross income and itemized deductions using linear regression. If a linear relationship exists between X and Y, or even if the relationship is approximately linear, the simple linear regression model can be used. The coefficients of the simple linear regression model can be calculated using the `ols` function from the Ravix package.

First, we import Pandas and construct the dataframe using the `DataFrame` function. The `pd` alias is a common convention, and we pass a dictionary, created with curly braces, to the `DataFrame` function.

```
import pandas as pd
df = pd.DataFrame({"y": y, "x": x})
```

Next, import the Ravix package with the `import` command. Then, create a regression object using the `ols` function, which takes a formula string as input, specifying the response and predictor variables. Assign the result to the name `reg` using the following code:

```
import ravix
reg = ravix.ols("y ~ x", data = df)
```

A summary of the linear regression can be obtained using the `summary` method:

```
reg.summary()
```

```
# Summary of OLS Regression Analysis:
# ======================================================
#
# Coefficients:
# ------------------------------------------------------
#                 Estimate  Std. Error  t-value    p-value
# Intercept        0.79430      2.3825    0.333     0.7502
# x                0.17240    0.024912    6.920     0.0005 ***
#
# Model Statistics:
# ------------------------------------------------------
# Residual Std. Error: 2.2044
# R-squared:       0.8887            AIC: 37.0487
# Adj. R-squared: 0.8701             BIC: 37.2076
# F-statistic: 47.8916 on 1 and 6 DF, p-value: 0.000451
# ======================================================
```

From the summary output of the regression line, the coefficient estimates are given in the "Estimate" column. From this column, the estimate of the intercept is displayed as 0.7943, and the estimate of the slope is displayed as 0.1724. The resulting regression equation from these coefficients is:

$$\hat{Y} = 0.7943 + 0.1724X$$

As a further note from the summary, the "R-squared" value matches that of the R^2 value from squaring the r value found using the `corrcoef` function from

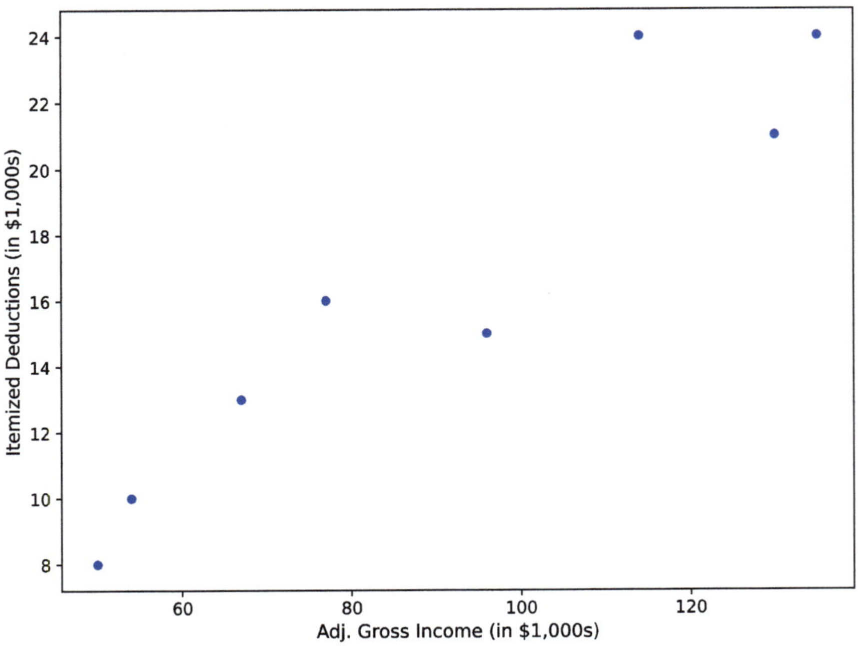

Fig. 3.7 Accounting Data Scatterplot

NumPy. The value of R^2 is referred to as Multiple R^2 when there are multiple predictor variables.

The scatter plot from Fig. 3.7 can be modified to include the linear regression line within the plot using the `regplot` function within the Seaborn package. Within the Ravix package, we can streamline the plotting by automatically choosing default graphing options. As a result, we can generate the plot with a single line of code using `plot`:

```
ravix.plot("y ~ x", data = df,
           xlab = "Adj. Gross Income (in $1,000s)",
           ylab = "Itemized Deductions (in $1,000s)")
```

3.11.4 *Audit Scenarios*

After analyzing the data, your manager identifies four individuals who will most likely be audited due to irregularities in their income data and the amount of itemized deductions that they have reported. These individuals were not part of the original dataset. The adjusted gross incomes for these new individuals are shown as follows:

1. $30,000
2. $78,000
3. $214,000
4. $120,000

The model can predict the appropriate number of itemized deductions using the `predict` function. The values will need to be combined into a dataframe in order to pass them through the `predict` function. A dataframe can be created by importing Pandas and using the `DataFrame` function. In this function, we pass a dictionary where column names are paired with lists of data. Here, we create a dataframe with a single column `x` using the following code:

```
import pandas as pd
df_new = pd.DataFrame({"x": [30, 78, 214, 120]})
```

After creating the new dataframe (`df_new`), the `predict` function can be used to approximate the amount of itemized deductions using the fitted model (`reg`). The `predict` function requires the fitted model object as the first argument and the new dataframe containing predictor values as the second argument. This will output the predicted values ($\hat{y}$) for the specified data points.

```
ravix.predict(reg, df_new)
```

```
# 0     5.966315
# 1    14.241543
# 2    37.688022
# 3    21.482367
# dtype: float64
```

Note the third value in `df_new` is $214,000, which is outside the range of X, and therefore means the third prediction (37.688022 or $37,688) represents an extrapolation.

After reporting your predictions to your manager, she notes that the reported itemized deductions for each individual are:

1. $6,849
2. $12,862
3. $24,690
4. $42,409

Interpreting the output, the first and second individuals' actual itemized deductions are close to the predicted values, indicating a reasonable fit. However, the third and fourth individuals show significant deviations, suggesting potential anomalies or the need for further investigation. The third individual corresponds to the extrapolation, which has a large difference between the observed value and the predicted value. However, more data should be gathered before assuming the third individual's itemized deductions are irregular. The fourth individual has approximately twice the predicted itemized deductions and should be scrutinized. While there are many more intricacies to tax and audit analysis, this case study shows the value that predictive analytics can play in the area of accounting.

3.11.5 *Case Conclusion*

In this analysis, we demonstrated the value of analytics by fitting a model to predict the amount of itemized deductions using the adjusted gross income. The tax code has many other important variables that one can consider for a detailed analysis. For this reason, accounting professionals have sought analytical skills as a means to have an advantage over their competition. Analytics will continue to play an important role in accounting and accounting-related applications.

Simple linear regression—regression with a single predictor—is fundamental to understanding regression analysis. This case study demonstrates its application using core Python tools, with Ravix serving as the primary interface alongside NumPy and Pandas. These tools are used throughout the text for more advanced analyses.

Problems

1. Accounting Analytics Basic Statistics
In the chapter case study, the problem solved involved the use of Python programming with Ravix. The same analysis can be done without the use of a computer as well. Using Table 3.1, answer the questions below. Both variables are in thousands of US dollars.

Without the use of a computer:

a. Find the correlation coefficient and the coefficient of determination for X and Y.
b. Find the equation of the least squares line between X and Y.

2. Accounting Analytics Prediction
From the Accounting Analytics Case Study, it was shown that the regression equation was:

$$\hat{Y} = 0.7942958 + 0.1724006X$$

Without the use of a computer:

a. Find the range of X using the values of X from Table 3.1.
b. Predict the corresponding itemized deductions for $x = 30, 55, 115, 300$.
c. Indicate whether each prediction from part b is an interpolation or an extrapolation.

3. Sales Calls Problem

Jordan's colleague at Enright Associates is Abdullah who also manages retirement accounts. Jordan documented the number of sales calls Abdullah made each day for the last week. The number of calls are denoted by X and the amount of sales that resulted are denoted as Y. The values of X and Y are given in Table 2.4.

Without the use of a computer:

a. Find the correlation coefficient and the coefficient of determination for X and Y.
b. Find the equation of the least squares line between X and Y.
c. From Jordan's data, a slope of 0.868 was noted. How does Abdullah's slope compare with Jordan's?

While using a computer (with Python):

d. Verify your results from parts a and b.
e. Generate a summary of the regression line between X and Y.
f. Plot a scatterplot of X and Y.

4. Social Media Influencers

The CEO of Zen Sports Apparel contracts social media influencers to generate clicks to their products. The data shown in Table 3.2 reflect the amount paid (in dollars) and the resulting number of clicks.

Without the use of a computer:

a. Find the correlation coefficient and the coefficient of determination for the pay (X) and clicks (Y).
b. Find the equation of the least squares line between pay (X) and clicks (Y).
c. What does the slope coefficient indicate?

While using a computer (with Python):

d. Verify your results from part a.
e. Generate a summary of the regression line between pay (X) and clicks (Y).

Table 3.2 Social Media Influencers Data

Pay	Clicks
1000	26855
250	2230
5000	99323
800	23321
2500	47916
780	3274
100	4686
980	15591
400	7174
4200	92447
270	10219
3600	59221

Table 3.3 Automotive Earnings and Price Data

Rank	Name	Ticker	EPS	Price
1	Douglas Dynamics	PLOW	0.0021	55.00
2	Volkswagen AG	VWAGY	2.6600	19.29
3	Tesla	TSLA	−4.9200	418.33
4	Ford	F	0.0100	9.30
5	Penske Automotive Group	PAG	5.2800	50.22
6	General Motors	GM	4.6200	36.60

f. Plot a scatterplot of pay (X) and clicks (Y).
g. Would it be more beneficial to hire several low-cost social media influencers or hire only one social media influencer with the same amount of money?

5. Automotive Stocks

Table 3.3 displays the data for six automotive stocks. The data indicate the Earnings Per Share (EPS) reported on 12/31/2019 and the corresponding stock closing price on the same day.

Without the use of a computer:

a. Find the correlation coefficient and the coefficient of determination for the EPS (X) and price (Y).
b. Find the equation of the least squares line between EPS (X) and price (Y).
c. Based on the value calculated for r, describe the linear relationship between earnings and price.

While using a computer (with Python):

d. Verify your results from part a.
e. Generate a summary of the regression line between EPS (X) and price (Y).
f. Plot a scatterplot of EPS (X) and price (Y).
g. Based on the regression equation, calculate the predicted price given an EPS of 2.

6. Age and Net Worth of Billionaires

The age and net worth of 10 billionaires are shown in Table 3.4, where "Age" is in years, and "Net Worth" is in billions of US dollars.

Without the use of a computer:

a. Find the correlation coefficient and the coefficient of determination for the age (X) and net worth (Y).
b. Find the equation of the least squares line between age (X) and net worth (Y).
c. Based on the value calculated for R^2, is there a significant relationship between age and net worth?

While using a computer (with Python):

d. Verify your results from part a.

Table 3.4 Billionaires Data

Rank	Age	Net Worth	Source
12	82	1.3	Technology
27	58	2.6	Casinos
61	83	2.7	Software
85	82	4.0	Logistics
100	69	8.6	Hedge Funds
123	39	10.3	Metals & Energy
192	49	10.4	Business Software
383	58	18.7	Video Games
763	63	21.4	Consulting
944	99	26.4	Telecom

Table 3.5 Bitcoin Data

Month	BTC	SPY
1	0.142	−0.006
2	0.363	0.028
3	0.305	0.042
4	−0.020	0.056
5	−0.354	0.007
6	−0.061	0.019
7	0.188	0.028
8	0.133	0.030
9	−0.072	−0.050
10	0.400	0.074
11	−0.070	−0.008
12	−0.077	0.043

e. Generate a summary of the regression line between age (X) and net worth (Y).
f. Plot a scatterplot of age (X) and net worth (Y).

7. Bitcoin and S&P 500
The returns of Bitcoin in US Dollars (BTC-USD) is said to relate closely with the overall stock market. Using the returns of the S&P 500 market benchmark (SPY), analyze the returns of Bitcoin with the S&P 500. The data given in Table 3.5 are the monthly returns from 2021.

Without the use of a computer:

a. Find the correlation coefficient for the returns of Bitcoin (X) and the S&P 500 (Y).
b. Find the equation of the least squares line between the returns of Bitcoin (X) and the S&P 500 (Y).
c. Based on the value calculated for r, describe the linear relationship between the returns of Bitcoin (X) and the S&P 500 (Y).
d. Calculate and interpret the coefficient of determination.

While using a computer (with Python):

e. Verify your results from part a.
f. Generate a summary of the regression line between the returns of Bitcoin (X) and the S&P 500 (Y).
g. Plot a scatterplot of the returns of Bitcoin (X) and the S&P 500 (Y).

8. Residual Summation Proof
Show that both sides of Eq. (3.14) are equivalent:

$$(Y - \bar{y}) = (\hat{Y} - \bar{y}) + (Y - \hat{Y}).$$

9. Sum of Squares of X Proof
Show that the equations for SS_{xx} (Eqs. (2.4) and (2.5)) are equivalent. In particular, show that:

$$\sum_{i=1}^{n}(x_i - \bar{x})^2 = \sum_{i=1}^{n} x_i^2 - \left(\sum_{i=1}^{n} x_i\right)^2 /n.$$

10. Sum of Squares of X and Y Proof
Show that the equations for SS_{xy} (Eqs. (3.1) and (3.2)) are equivalent. In particular, show that:

$$\sum_{i=1}^{n}(x_i - \bar{x})(y_i - \bar{y}) = \sum_{i=1}^{n} x_i y_i - \frac{\left(\sum_{i=1}^{n} x_i\right)\left(\sum_{i=1}^{n} y_i\right)}{n}.$$

11. Normal Equation Proof Part 1
Show that the estimated intercept value is given by

$$\hat{\beta}_0 = \bar{y} - \hat{\beta}_1 \bar{x}$$

by taking the partial derivative of the SSE with respect to $\hat{\beta}_0$ and setting it equal to 0.

12. Normal Equation Proof Part 2
Prove that the partial derivative of the SSE with respect to $\hat{\beta}_1$ results in

$$\hat{\beta}_1 = \frac{SS_{xy}}{SS_{xx}}.$$

Hint: You will need to use the result from the previous problem.

Chapter 4
Simple Linear Regression

If you can't explain it simply, you don't understand it well enough.

—Albert Einstein

4.1 Introduction

Albert Einstein's quote above stresses the paramount importance of simplicity. In regression analysis, focusing on only two variables illustrates the concepts in a clear and simple way. Thus, in Chap. 3, we calculated the least squares line by using two variables. We also plotted scatter plots and calculated correlation coefficients to further assess the linear relationship. From this analysis, we obtained a detailed understanding of the relationship between two variables. Upon understanding a linear relationship, other more complicated processes become easier to grasp.

Simple linear regression refers to linear regression with one predictor variable and one response variable. In the previous chapter, our primary focus entailed the calculation and fundamental programming concepts of linear regression. The goal of this chapter is to guide the reader through some of the understanding and analysis details of the simple linear regression.

We begin with a discussion of the mathematical model of simple linear regression, the assumptions required to make inferences from the fitted model, and how to measure statistical significance of a fitted model. We then cover confidence intervals for the model coefficients, followed by an application. In the final discussion, we present a stock market case study and solve it in detail. The case study involves utilizing simple linear regression to calculate stock *betas*, which measure the performance of a stock as a function of the overall stock market. This case study uses Python and Ravix with descriptive graphs and the relevant source code.

D. P. McGibney, *Applied Linear Regression for Business Analytics with Python*, International Series in Operations Research & Management Science 378,
https://doi.org/10.1007/978-3-032-23806-1_4

4.2 Simple Linear Regression Model

A mathematical model is an equation that describes the behavior of a system or concept mathematically. It can be used to draw conclusions, such as the time it takes an object to fall to the ground or the speed of an orbiting satellite. In finance, stock betas are calculated from a mathematical model as seen in the chapter case study (Sect. 4.19.1). Mathematical models are applicable to a wide array of subject areas. In regression analysis, one example is the simple linear regression model which is defined as:

$$Y = \beta_0 + \beta_1 X + \varepsilon, \tag{4.1}$$

where β_0 and β_1 are the model parameters. These parameters represent the theoretical coefficients. The ε term is a random variable with a mean of 0, and is called the irreducible error or the error term. The error term, ε, in the equation represents the deviation between the true response value and the model's predicted value at the population level. In practice, the deviation between the actual and predicted response value is estimated by the residual, $\hat{\varepsilon}$, which accounts for this discrepancy within the sample data.

In using the least-squares approach as outlined in Chap. 3, the simple linear regression model was automatically assumed although not explicitly stated.

4.3 Model Assumptions

Since we use the least-squares criterion, the values $\hat{\beta}_0$ and $\hat{\beta}_1$ result in the best line that fits the data. For this model to be appropriate, the following four assumptions must hold true.

1. The relationship between the response and the predictor variable is linear.
2. Each value of ε_i is independent.
3. The random variable, ε, is normally distributed.
4. The variance of ε, denoted as σ^2, is the same for all observations.

Assumption 1, often referred to as the linearity assumption, is equivalent to assuming that the mean of each ε_i is 0. Mathematically, these assumptions can be summarized by the following:

$$\varepsilon_i \overset{\text{i.i.d.}}{\sim} \mathrm{N}(0, \sigma^2).$$

This notation is read as: the ith error (ε_i) is independent and identically distributed (i.i.d.) and is distributed according to a normal (or Gaussian) distribution with a mean of 0 and a variance of sigma squared (σ^2). Chapter 8 provides a more in-depth discussion of the model assumptions for linear regression.

4.4 Model Variance

As mentioned in the model assumptions, the irreducible error of the model, ε, has a variance of σ^2, implying the standard deviation is assumed to be σ. A large value of σ indicates the data are widely dispersed about the regression line, and a small value of σ indicates the data lie relatively close to the regression line. In Fig. 4.1, the variance in the error is relatively constant as x increases. This means that the vertical spread of the data points around the regression line stays approximately the same for all values of x. Having a constant variance about the least-squares line upholds assumption 4 of the model as listed above. If this assumption were not met, some of the analysis associated with the variance may not be correct.

While the parameter σ^2 is the unknown theoretical value of the variance, an estimate can be obtained by dividing the SSE by the degrees of freedom:

$$s^2 = \frac{SSE}{n-2}. \tag{4.2}$$

The degrees of freedom are given by $n - 2$ when there are n observations in the dataset and two coefficients in the model. The simple linear regression model has two coefficients (β_0 and β_1) as shown in Eq. (4.1). In simple linear regression, it is always the case that the degrees of freedom for the SSE is $n - 2$. From the approximation of the variance, an approximation of the standard deviation can be found by taking the square root:

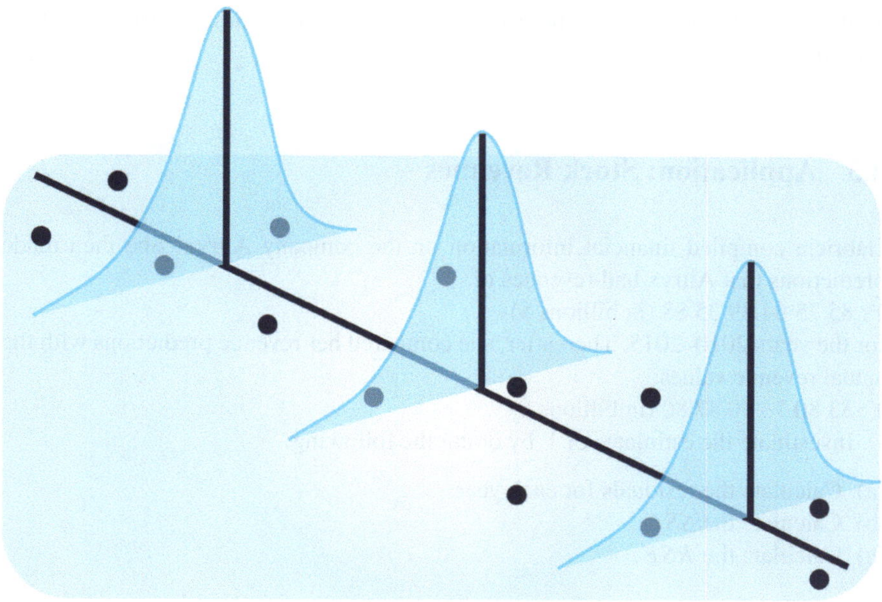

Fig. 4.1 Variance in the error

$$s = \sqrt{s^2}. \tag{4.3}$$

Plugging in Eq. (4.2) for s^2, the value becomes:

$$s = \sqrt{\frac{SSE}{n-2}}. \tag{4.4}$$

This standard deviation is often referred to as the residual standard error (RSE), the root mean squared error ($RMSE$), or the standard error of the regression. The RSE provides an overall measure of the spread of residuals around the regression line. A lower RSE suggests that the model fits the data closely, while a higher RSE indicates a poorer fit. By assessing the variability of the residuals, the RSE helps evaluate the accuracy of the regression model and can guide further refinement and improvement to achieve better predictive performance.

Recall from Eq. (3.17) in Chap. 3 that the SSE is

$$SSE = \sum_{i=1}^{n} (y_i - \hat{y}_i)^2. \tag{4.5}$$

If the simple regression equation is plugged in, the SSE becomes:

$$SSE = \sum_{i=1}^{n} (y_i - \hat{\beta}_0 - \hat{\beta}_1 x_i)^2. \tag{4.6}$$

In the next section, we use these equations to calculate these metrics for stock revenues.

4.5 Application: Stock Revenues

Gabriele compiled financial information on the company Altryx. She then made predictions that Altryx had revenues of
$\hat{Y}$: 85 75 64 89 35 88 (in billions \$)
for the years 2010–2015. Thereafter, she compared her revenue predictions with the actual revenue values:
Y: 83 80 73 90 47 80 (in billions \$)

Investigate the estimates of Y by doing the following:

(a) Calculate the residuals for each year.
(b) Calculate the SSE.
(c) Calculate the RSE.

Table 4.1 Stock residuals

y	$\hat{y}$	$y-\hat{y}$	$(y-\hat{y})^2$
83	85	−2	4
80	75	5	25
73	64	9	81
90	89	1	1
47	35	12	144
80	88	−8	64

Solution

(a) The calculated residuals are displayed in Table 4.1 within column 3. Recall from Chap. 4 the residuals are given by $y - \hat{y}$.

(b) The squared residuals are in column 4 of Table 4.1. From these squared residuals and Eq. (3.10), the SSE is

$$\begin{aligned} SSE &= \sum_{i=1}^{n}(y_i - \hat{y}_i)^2 \\ &= 4 + 25 + 81 + 1 + 144 + 64 \\ &= 319. \end{aligned}$$

(b) From Eq. (4.2), the variance estimate is

$$\begin{aligned} s^2 &= \frac{SSE}{n-2} \\ &= \frac{319}{6-2} \\ &= 79.75. \end{aligned}$$

From the variance estimate, the RSE is given by:

$$\begin{aligned} s &= \sqrt{s^2} \\ &= \sqrt{79.75} \\ &= 8.93. \end{aligned}$$

4.6 Probability Functions

At the core of probability and statistical theory are functions that describe a probability distribution. Describing the behavior of a random variable can provide an understanding of the variable and its relationships. Three of the primary random variable functions are described in the following subsections.

4.6.1 Probability Density Function

A probability density function (PDF), sometimes called a probability distribution function, is typically denoted as $f(x)$ where x is a value of the random variable X. A PDF denotes the behavior of a continuous probability distribution as opposed to probability mass functions (PMF), which describe discrete probability distributions. The value (x) of the random variable can be plotted on the x-axis and $f(x)$ on the y-axis such that the probability is shown as the area under the plot. Given this definition, the PDF takes on the following properties:

1. Any value of $f(x)$ must be non-negative.
2. The total area of the distribution under the curve is one.
3. The probability that X is between a and b can be expressed as an integral using the following notation:

$$P(a \leq X \leq b) = \int_a^b f(x)\,\mathrm{d}x. \tag{4.7}$$

4.6.2 Cumulative Distribution Function

The cumulative distribution function (CDF) provides additional utility of the PDF by making additional assumptions. Particularly, the CDF is the probability that X is less than or equal to a particular value:

$$P(X \leq b) = \int_{-\infty}^b f(x)\,\mathrm{d}x. \tag{4.8}$$

Figure 4.5 provides a graphical summary showing the area $(P(T \leq q))$, which is the CDF evaluated at q for the random variable T. This function provides a practical way to use random variables, particularly for computing probability estimates.

4.6.3 Percent Point Function

The percent point function (PPF), also known as the quantile function, is the inverse of the CDF. It is denoted as $Q(p)$, where p is a probability, and it returns the value b such that:

$$Q(p) = b \quad \text{if} \quad P(X \leq b) = p. \tag{4.9}$$

The PPF is particularly useful in the following contexts:

1. In hypothesis testing, the PPF is used to find critical values corresponding to specific significance levels. For example, the critical value for a 95% confidence level can be found using the PPF of the standard normal distribution.
2. The PPF is used to determine the boundaries of confidence intervals for coefficient parameters.

The PPF is a fundamental tool in statistical analysis and provides a direct way to translate probabilities into data values, which is essential for various inferential statistical methods.

4.6.4 The Student's t-Distribution

The Student's t-distribution arises when the population standard deviation is unknown and must be estimated from sample data. This situation occurs frequently in regression analysis, making the t-distribution fundamental to statistical analysis.

The shape of the t-distribution depends on the degrees of freedom (v) as depicted in Fig. 4.2. With fewer degrees of freedom, the distribution has heavier tails, reflecting greater uncertainty in the estimate of the standard deviation. As the degrees of freedom increase, the t-distribution converges to the standard normal distribution.

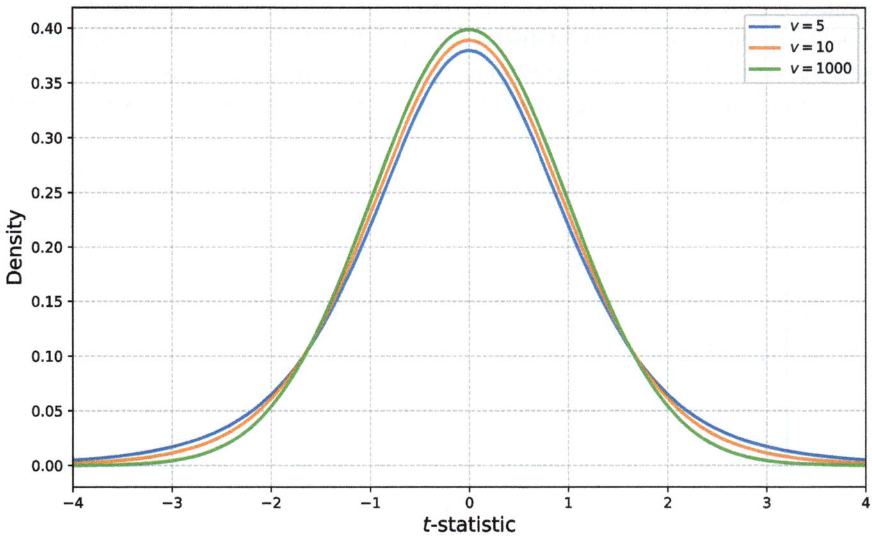

Fig. 4.2 Student's t-distributions for selected degrees of freedom

4.7 Using the *t*-Distribution for Statistical Inference

Inferential statistics refers to the practice of drawing conclusions from data. The practice signifies an important topic not only in simple linear regression but also in regression analysis in general. In fact, regression analysis addresses two major goals: prediction and inference. The purpose of prediction lies in estimating a variable.

The importance of prediction is often clear. For instance, decision-makers would select the option with the highest predicted profit.

The value of inference, on the other hand, may not be as obvious. One example of inference is interpreting how variables influence a predicted value, which can help explain a dataset. For example, knowing that a product is sold for every five sales calls illustrates the practical value of inference.

Hypothesis testing is an important tool in inferential statistics. When performing a hypothesis test, we assume a null hypothesis (H_0) and calculate a test statistic assuming the null hypothesis is true. If the statistic is within an acceptable range, then the null hypothesis is reasonable. However, if the test statistic is extreme, the null hypothesis is unlikely to be correct. In this case, we would reject the null hypothesis. Hypothesis testing of a mean is usually explained in introductory statistics, and the logic remains the same for more advanced hypothesis testing.

To efficiently conduct a hypothesis test, we introduce the concept of a p-value. Assuming H_0 is true, the probability that the test statistic will take on values as extreme as or more extreme than the observed test statistic is called a p-value.

Determining the results of a hypothesis test by using the p-value is called the p-value method, whereas comparing a test statistic with a critical value is referred to as the critical region method. In regression analysis, the p-value method is preferred because it provides a probabilistic measure of how strongly we can reject the null hypothesis. Furthermore, test statistics are typically less intuitive than probabilities. A depiction of the p-value corresponding to a test statistic, t, is shown in Fig. 4.3. In this figure, the density area shaded in blue represents the p-value, which results

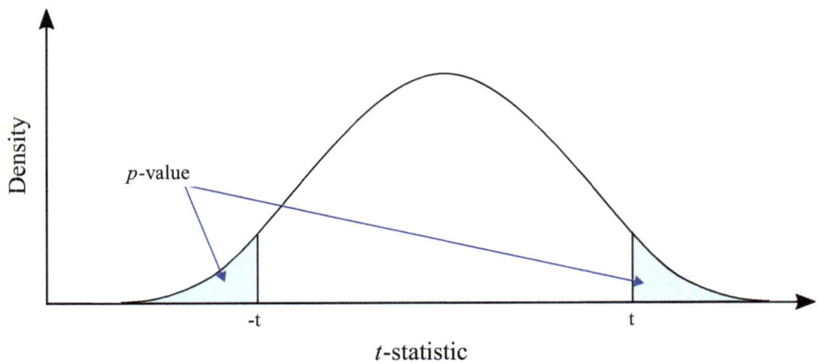

Fig. 4.3 A two-tailed p-value

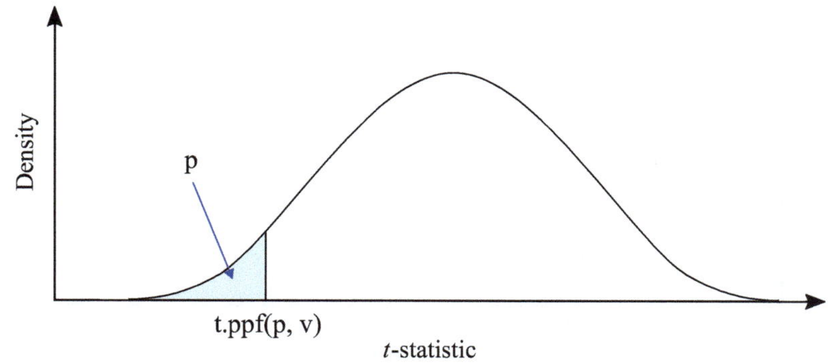

Fig. 4.4 The t.ppf function

from a two-tailed hypothesis test using a Student's t distribution. This distribution is commonly referred to as a t-distribution. The t-distribution is particularly suited for cases when the standard deviation of the population is unknown and must be estimated from the sample, making it more appropriate for the inferential statistics commonly used in regression analysis.

4.7.1 The `ppf` Function for t

Traditionally, the critical value or p-value necessary to conduct a t-test was found using statistical tables that are widely used and easily obtained online. However, we may also easily and more accurately find this information using Python. The `ppf` function for t in Python is used to find quantile values of the t-distribution and, therefore, allows us to find the t critical value.

The `ppf` function for t can be found within `scipy.stats` and is easily imported using the `import` command. This function has two required arguments and the syntax as follows:

```
from scipy.stats import t
t.ppf(p, v)
```

The first argument (`p`) is a probability or a vector of probabilities and the second argument (`v`) refers to the degrees of freedom. Figure 4.4 depicts the resulting t value that can be found using the `ppf` function for t.

By entering a probability and the degrees of freedom, the `ppf` function for t returns the t-value that corresponds to that probability (`p`). By default, the probability for `p` will correspond to the area on the left side of the quantile value. To find the critical value for a 2-tailed t-test, $\alpha/2$ will be the first argument and $n - 2$ will be the second argument in the `ppf` function for t. In this case the `ppf` function

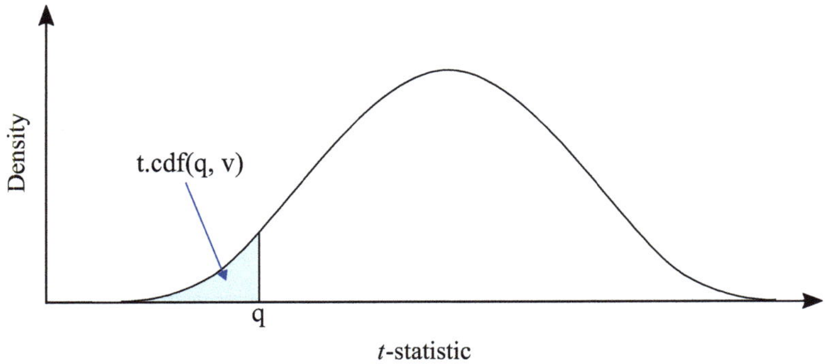

Fig. 4.5 The t.cdf function

for t will yield the negative critical value ($-t_{\alpha/2}$). If we have the variables `alpha` and `n` defined for us, then the left-side critical value is:

```
t.ppf(alpha/2, n-2)
```

Since the t-distribution is symmetrical, $1 - \alpha/2$ could be entered for `p`, which would result in the positive critical value ($t_{\alpha/2}$). Using `alpha` and `n` we find the right-side critical value as:

```
t.ppf(1 - alpha/2, n-2)
```

4.7.2 *The* `cdf` *Function for t*

In Python, the `cdf` function returns the probability associated with a specific quantile value. After importing `t` from `scipy.stats`, the `cdf` function can be accessed using the syntax:

```
from scipy.stats import t
t.cdf(q, v)
```

where `q` refers to the quantile (t-value) and `v` denotes the degrees of freedom. Figure 4.5 illustrates the resulting cumulative probability density value (`p`) that can be found using the `cdf` function for t.

We can use the `cdf` function for t to find the p-value corresponding to a test statistic. To find the p-value for a hypothesis test, the test statistic should be entered as the first argument (`q`) and the degrees of freedom should be entered as the second argument (`v`). It is important to note the `cdf` function for t will return the area to the left of `q`; thus, the area will be less than 0.5 if `q` is negative but greater than 0.5 for a positive value of q. If there is a positive test statistic, the area more extreme than q is:

```
1 - t.cdf(q, v)
```

When using t-tests in regression analysis, it will almost always be the case that the hypothesis test is two-tailed. For example, the area in the t-distribution that is more extreme than 2 is not only the tail above 2 but also the tail below -2. When finding the p-value in a two-tailed test, the results of the cdf function for t will need to be multiplied by 2 when q is negative:

```
2*t.cdf(q, v)
```

In the case when q is non-negative, we have:

```
2*(1 - t.cdf(q, v))
```

Replacing `t.cdf(q, v)` with `1-t.cdf(q, v)` accounts for the t-value being on the right-side, as the `t.cdf` method returns the area to the left.

4.8 Application: Using the cdf and ppf Functions

Use your knowledge of the cdf and ppf functions to answer the following questions without the use of Python. While the answers to these questions can be found using Python, it is best to test your understanding of these concepts here.

Using ppf *for t*: What will the following code return?

(a) `t.ppf(0.5, 10)`
(b) `t.ppf(0.5, 100)`
(c) `t.ppf(0.025, 10000)`
(d) `t.ppf(-0.025, 10000)`
(e) `t.ppf(100, 10000)`

Using cdf *for t*: What will the following code return?

(a) `t.cdf(0, 5)`
(b) `t.cdf(0, 25)`
(c) `t.cdf(-1.96, 10000)`
(d) `t.cdf(-100, 10000)`
(e) `t.cdf(100, 10000)`

Solution

Using ppf *for t*:

(a) Since having an area of 0.5 on the left of the graph represents the center of the t-distribution and the center of the distribution is at 0, this ppf function for t

will return 0. The degrees of freedom dictate the size of the distribution's tails, which is irrelevant in this case since the distribution is symmetrical.

(b) The degrees of freedom dictate the size of the distribution's tails which is irrelevant in this case since the distribution is symmetrical. Having an area of 0.5 on the left side of the graph corresponds to a t value of 0.

(c) When the degrees of freedom are large, then the t-distribution emulates that of a standard normal distribution. Recalling the properties of the standard normal distribution from the empirical rule, having an area of 0.025 on the left side of the graph corresponds to a t value of -1.96.

(d) The first argument within the `ppf` function for t represents a probability or an area, neither of which can be negative. Therefore, this function will return an error.

(e) Having a probability of 100 is impossible and will return an error.

Using `cdf` *for t*

(a) Since 0 represents the center of the distribution, the probability to the left of 0 is 50% or half. The degrees of freedom dictate the size of the distribution's tails, which is irrelevant in this case since the distribution is symmetrical.

(b) The probability to the left of 0 is 50% or half. Even though the degrees of freedom change from the values in part (a), the distribution with 25 degrees of freedom is also symmetrical and centered about 0.

(c) When the degrees of freedom are large, then the t-distribution emulates that of a standard normal distribution. With this in mind, the area to the left of -1.96 in a standard normal distribution is 0.025 or 2.5%. For verification of this result, review the empirical rule from an introductory statistics source.

(d) As in (c), we note that the large degrees of freedom result in the distribution being approximately equal to a standard normal distribution. The area to the left of -100 in a standard normal distribution is approximately 0 since -100 is an extreme value in the left tail.

(e) As in (c) and (d), the large degrees of freedom result in the distribution being approximately equal to a standard normal distribution. The area to the left of 100 in a standard normal distribution is approximately 1 since 100 is an extreme value in the right tail.

4.9 Hypothesis Testing: Student's t-Test

When testing the significance of a slope parameter β_1, it may prove valuable to check if the parameter is zero, since a zero β_1 means that there is no relationship between the response and the predictor variable. Therefore, a test can be created to check if β_1 is zero or nonzero. Setting up the hypotheses, the null hypothesis would stipulate that the parameter is 0, and the alternative hypothesis would therefore be that the parameter would be nonzero. Using statistical symbols, this can be written as:

$$H_0 : \beta_1 = 0$$

$$H_1 : \beta_1 \neq 0$$

where H_0 and H_1 denote the null and alternative hypotheses, respectively.

To find a statistic for this test, $\hat{\beta}_1$ is used as the estimate of β_1. The value of $\hat{\beta}_1$ is divided by the standard error of $\hat{\beta}_1$:

$$t_1 = \frac{\hat{\beta}_1}{s_{\hat{\beta}_1}}. \quad (4.10)$$

The standard error of $\hat{\beta}_1$ is given by:

$$s_{\hat{\beta}_1} = \frac{s}{\sqrt{SS_{xx}}}. \quad (4.11)$$

Note that the standard deviation of the error is used to determine the standard error of $\hat{\beta}_1$.

The value of t_1 represents a student's t-distribution with $n-2$ degrees of freedom. Therefore, the null hypothesis can be rejected if the p-value corresponding to t_1 lies below the level of significance (α). Although somewhat arbitrary, it is common to set α to 0.05. This value can usually be assumed if there is no mention of α.

4.10 Employee Churn Application: Testing for Significance with t

Karen, from human resources at Global IT Enterprises, is trying to find the number of employees that will leave the company by department. The number of employees in a department (X) can be used to predict the number of employees that will churn (Y). The data are shown here in Table 4.2.

Using the formulas from Chap. 3, we calculate the regression equation as

$$\hat{y} = 2.1818 + 1.6452x.$$

Investigate the relationship between X and Y by doing the following:

(a) Calculate the SSE.
(b) Calculate the standard deviation of the error.

Table 4.2 Churn data

Variable							
Employees (in 100s) (X)	2	2	5	1	3	0.5	7
Churns (Y)	7	5	12	4	8	1	12

(c) Calculate the standard error of the slope.
(d) Calculate the test statistic for the slope.
(e) Compute the p-value.
(f) Make the decision.
(g) Interpret the slope.

Solution

(a) The predicted values are easily obtained by plugging in each value of X into the regression equation:

$$\hat{y}_1 = 2.1818 + 1.6452(2) = 5.4722$$
$$\hat{y}_2 = 2.1818 + 1.6452(2) = 5.4722$$
$$\hat{y}_3 = 2.1818 + 1.6452(5) = 10.4078$$
$$\hat{y}_4 = 2.1818 + 1.6452(1) = 3.827$$
$$\hat{y}_5 = 2.1818 + 1.6452(3) = 7.1174$$
$$\hat{y}_6 = 2.1818 + 1.6452(.5) = 3.0044$$
$$\hat{y}_7 = 2.1818 + 1.6452(7) = 13.6982$$

Plugging in the predicted values ($\hat{y}$) and the actual values of y into Eq. (3.10), the SSE is:

$$SSE = \sum_{i=1}^{n}(y_i - \hat{y}_i)^2.$$

$$\begin{aligned} SSE &= (7 - 5.4722)^2 + (5 - 5.4722)^2 + (12 - 10.4078)^2 + (4 - 3.8270)^2 \\ &\quad + (8 - 7.1174)^2 + (1 - 3.0044)^2 + (12 - 13.6982)^2 \\ &= 12.80266 \end{aligned}$$

(b) From Eq. (4.2), the variance of the error is:

$$\begin{aligned} s^2 &= \frac{SSE}{n-2} \\ &= \frac{12.80266}{7-2} \\ &= 2.560532. \end{aligned}$$

The standard deviation is therefore:

$$s = \sqrt{2.560532}$$
$$= 1.600166.$$

(c) To calculate the standard error of the slope, the standard deviation of the error and the sum of squares of X (SS_{xx}) are required. The sum of squares of X is:

$$SS_{xx} = \sum_{i=1}^{n} x_i^2 - \frac{\left(\sum_{i=1}^{n} x_i\right)^2}{n}$$
$$= 92.25 - \frac{(20.5)^2}{7}$$
$$= 32.21429,$$

and the standard deviation of the error was calculated in the previous part. Using Eq. (4.11), the standard deviation of the slope estimate is therefore:

$$s_{\hat{\beta}_1} = \frac{s}{\sqrt{SS_{xx}}}$$
$$= \frac{1.600166}{\sqrt{32.21429}}$$
$$= 0.2819297.$$

(d) The test statistic for the slope, as stated in Eq. (4.10), is calculated to be

$$t_1 = \frac{\hat{\beta}_1}{s_{\hat{\beta}_1}}$$
$$= \frac{1.6452}{0.2819297}$$
$$= 5.835497.$$

(e) The p-value corresponding to $t_1 = 5.835497$ and 5 degrees of freedom can be found using the Python function:

```
from scipy.stats import t
2*t.cdf(-5.835497, 5)
```

```
# np.float64(0.0020899559661924634)
```

(f) Since the p-value of 0.002089956 is less than $\alpha = 0.05$, the null hypothesis should be rejected at the $\alpha = 0.05$ level of significance. The slope is therefore statistically significant.

(g) A slope of 1.6452 means that for every 100 employees in a department, approximately 1.6452 are expected to churn.

4.11 Coefficient Confidence Interval

Confidence intervals give us an interval estimate for our approximation assuming a confidence level of $1-\alpha$. The confidence interval for β_1 in a simple linear regression model is:

$$\hat{\beta}_1 \pm t_{\alpha/2} s_{\hat{\beta}_1}, \tag{4.12}$$

where $\hat{\beta}_1$ is the point estimate of β_1, $t_{\alpha/2}$ is the t-value such that $\alpha/2$ is in the upper tail with degrees of freedom of $n-2$, and $s_{\hat{\beta}_1}$ is the standard error of the slope.

The confidence interval of a coefficient can be displayed along with other important values using the `summary` method from a Ravix model. If there is a regression object, `reg`, available in the Python environment, the confidence interval can be calculated as:

```
reg.summary(out = "statsmodels")
```

The statsmodels summary shows a full model summary with coefficient intervals. We can modify the `summary` arguments to get a smaller table with the coefficients using the `out` argument.

```
reg.summary(out = "confint", level = 0.95)
```

The `summary` function requires a regression object as input and defaults to a 95% confidence interval. The optional `level` argument can be specified to return a different level of confidence.

In the previous sections, a test statistic was used to conclude the hypothesis test for t. Additionally, confidence intervals of the coefficients can be used to test the hypotheses used in the previous t-test. For instance, when $\alpha = 0.05$, if the value $\beta_1 = 0$ lies outside the 95% confidence interval, then H_0 is rejected because the null hypothesized value of β_1 is not included in the interval. In the application below, we conduct an equivalent hypothesis test using a confidence interval for β_1.

4.12 Employee Churn Application: Confidence Interval Hypothesis Testing

Using Karen's churn data (Table 4.2), find out if the number of employees in each department (X) is a significant variable in determining the number of employees that will churn (Y).

Recall that the regression equation is

$$\hat{y} = 2.1818 + 1.6452x.$$

Investigate the relationship between X and Y by doing the following.

(a) Set up a statistical test for β_1.
(b) Calculate the $t_{\alpha/2}$ critical value.
(c) Find a 95% confidence interval for β_1.
(d) Conclude the test using the confidence interval.

Solution

(a) The null hypothesis states that β_1 is 0 and the alternative hypothesis is nonzero:

$$H_0 : \beta_1 = 0$$

$$H_1 : \beta_1 \neq 0.$$

(b) Since α is split into two equal parts at the beginning and end of the distribution, the probability value of 0.025 is the probability argument for the first critical value. Keeping in mind that the `ppf` function for t takes as arguments areas to the left, we plug in 0.975 as the probability argument to get the right-side critical value. The values corresponding to $\pm t_{\alpha/2}$ with 5 degrees of freedom and $\alpha = 0.05$ can be found using the Python code:

```
from scipy.stats import t
t.ppf(0.025, 5)
t.ppf(0.975, 5)
```

```
# -2.570581835636315

# 2.570581835636314
```

From this result, the positive value is returned:

$$t_{\alpha/2} = 2.570582.$$

(c) Using Eq. (4.12) and the values calculated in the previous application, the confidence interval becomes:

$$\hat{\beta}_1 \pm t_{\alpha/2} s_{\hat{\beta}_1}$$

$$1.6452 \pm 2.570582 \times 0.2819297,$$

or written as:

$$0.9204766 \le \beta_1 \le 2.369923.$$

As confirmation, the 95% confidence intervals for both β_0 and β_1 can be calculated in Python using Ravix by creating the dataframe, fitting a regression model and outputing the summary. Constructing the dataframe using Pandas consists of importing Pandas, creating the lists for x and y, creating a dictionary, and creating the dataframe with the `DataFrame` function.

```
import pandas as pd
x = [2, 2, 5, 1, 3, .5, 7]
y = [7, 5, 12, 4, 8, 1, 12]
df = pd.DataFrame({"x": x,"y": y})
```

Fitting a regression model and outputting the summary using Ravix requires that we import Ravix and run the `summary` function in Ravix on a regression object created using the `ols` function.

```
from ravix import ols, summary
reg = ols("y ~ x", data = df)
reg.summary(out = "confint")
```

```
# Confidence Intervals:
# ======================================================
#             Estimate 95% CI Lower 95% CI Upper
# Intercept    2.18182    -0.449095      4.81273
# x            1.64523     0.920509      2.36996
# ======================================================
```

By specifying `out = "confint"` within the summary function, we can limit the summary output to the coefficients table. On the right side of this table, the 95% confidence interval is displayed. A confidence level of 95% is the default, but other levels can be specified within the summary function using the `level` argument. Note that the confidence interval limits for β_1 calculated by Python are consistent with the earlier calculation when ignoring minor rounding errors.

(d) Since 0 is not included in the confidence interval for β_1, the null hypothesis is rejected. The estimate $\hat{\beta}_1$ is therefore said to be statistically significant.

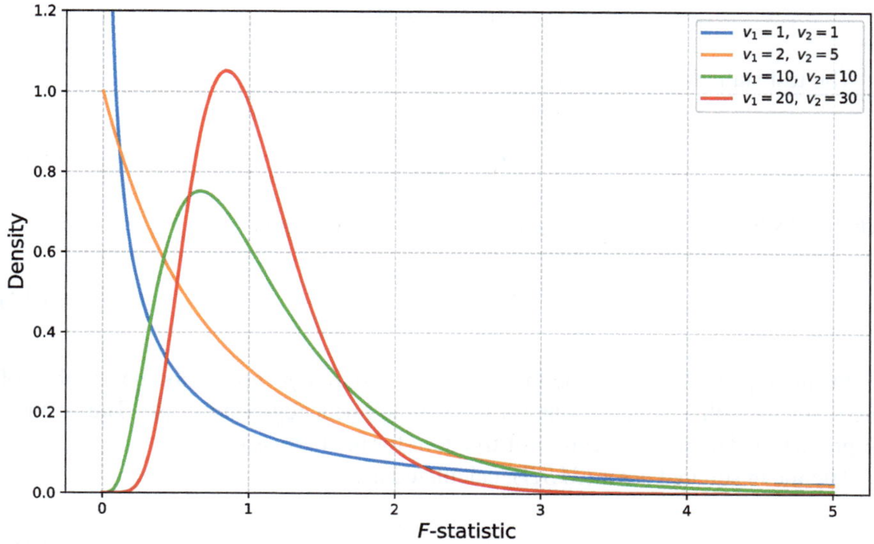

Fig. 4.6 F-distributions for selected numerator and denominator degrees of freedom

4.13 The F-Distribution

Beyond the t-test, the F-distribution arises as the ratio of two scaled Chi-square distributions and plays a crucial role in regression analysis as the basis of the F-test.

Unlike the t-distribution, the F-distribution is not symmetric and is defined only for nonnegative values. The extreme values in this distribution are represented by the right-side tail. However, the specific shape of the F-distribution varies greatly based on the degrees of freedom in the numerator (v_1) and in the denominator (v_2). Figure 4.6 shows the F-distribution with varying degrees of freedom.

4.14 Hypothesis Testing: F-test

Generally, the F-test assesses whether at least one of the predictor variables in a model significantly explains the variation in the response variable. In simple linear regression, where there is only one predictor variable, the F-test and t-test yield equivalent results, as both assess the significance of the same predictor. For models with more than one predictor variable, once the F-test deems the model significant, the t-test may be used to determine whether individual predictor variables are significant.

In simple linear regression, the hypotheses for the F-test are:

$$H_0 : \beta_1 = 0$$

$$H_1 : \beta_1 \neq 0$$

The test statistic for F is given as:

$$F = \frac{MSR}{MSE}, \tag{4.13}$$

where MSR is the mean squares due to regression which is

$$MSR = \frac{SSR}{p}. \tag{4.14}$$

In this equation, SSR is defined in Eq. (3.16) and p represents the number of predictor variables, which is one for the simple linear regression case. The number of predictors denotes the degrees of freedom for the regression.

The mean squared error (MSE), calculated as:

$$MSE = \frac{SSE}{n - p - 1}, \tag{4.15}$$

represents the average of the squared differences between observed and predicted values, as described in Eq. (3.21). The denominator in Eq. (4.14) reflects the $n-p-1$ degrees of freedom associated with SSE.

The rejection rule holds that one should reject H_0 if F is greater than or equal to the critical value for a given α. An equivalent method constitutes verifying that the p-value of the test statistic occurs at or below α. In particular, reject H_0 if p-value $\leq \alpha$ or if $F \geq F_\alpha$ where the critical value F_α is based on an F-distribution with p degrees of freedom in the numerator and $n - p - 1$ degrees of freedom in the denominator.

The null hypothesis of the F-test is that none of the predictor variables are linearly related to the response variable. The alternative hypothesis posits a linear relation between at least one predictor variable and the response variable.

4.15 The `ppf` Function for F

We use the `ppf` function for F in Python to find quantile values of the F distribution. Similar to the `ppf` function for t, the `ppf` function for F can return the critical value of an F-distribution if the significance level is specified. In contrast to the t-distribution, the F-distribution requires two degrees of freedom values since this distribution is the ratio of two χ^2 distributions.

The `ppf` function for F has three required arguments. The syntax to import and run this function is as follows:

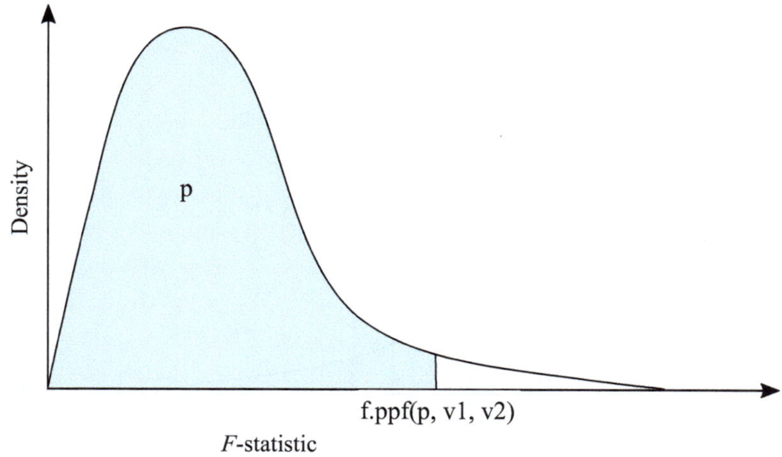

Fig. 4.7 The f.ppf function

```
from scipy.stats import f
f.ppf(p, v1, v2)
```

The first argument (p) refers to a probability or a vector of probabilities, whereas the second and third arguments (v1 and v2) are the degrees of freedom in the numerator and denominator, respectively. By entering a probability and the degrees of freedom, the ppf function for F returns the F-value that corresponds to that probability, as depicted in Fig. 4.7. To find the critical value, $1 - \alpha$ will be the first argument since the function requires the area to the left. In simple linear regression, the degrees of freedom in the numerator are 1, which is attributed to the number of predictor variables, and the degrees of freedom in the denominator are $n - 2$. Note that the F-distribution shape varies with different degrees of freedom combinations.

4.16 The cdf Function for *F*

The cdf function for F in Python returns the probability corresponding to a quantile value for the F-distribution. The syntax is:

```
from scipy.stats import f
f.cdf(f, v1, v2)
```

where f denotes the quantile (F-value) and the second and third arguments (v1 and v2) are the degrees of freedom in the numerator and denominator, respectively. Note that cdf function for F will return the area to the left of f by default. Figure 4.8 shows a depiction of the cdf function for F.

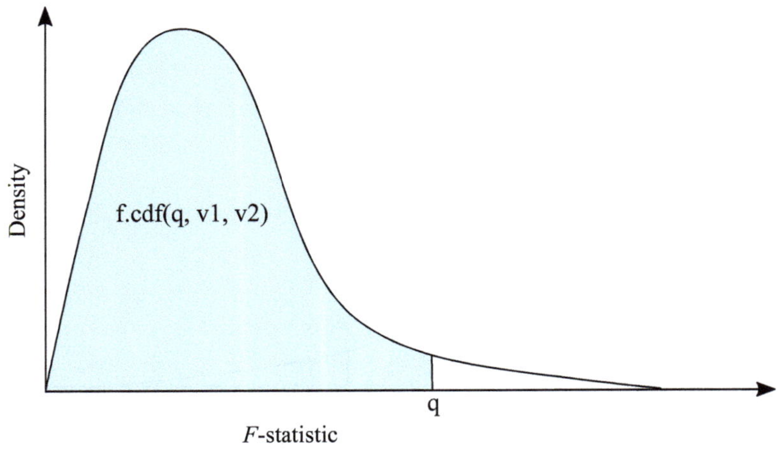

Fig. 4.8 The f.cdf function

4.17 Employee Churn Application: Testing for Significance with F

Again using Karen's churn data, find out if the number of employees in each department (X) is a significant variable in determining the number of employees that will churn (Y).

Recall that the regression equation is

$$\hat{y} = 2.1818 + 1.6452x.$$

Investigate the relationship between X and Y by doing the following:

(a) Calculate the F critical value using $\alpha = 0.05$.
(b) Calculate the F-test statistic.
(c) Make the F-test decision.
(d) Find the p-value from the F-test statistic.
(e) Comment on the p-value.
(f) Interpret the results.

Solution

(a) The F critical value is 1 degree of freedom in the numerator (v_1) and $n - 2$ degrees of freedom in the denominator (v_2). The `ppf` function for F gives:

```
from scipy.stats import f
f.ppf(0.95, 1, 5)
```

```
# np.float64(6.607890973703367)
```

(b) To find the F statistic, one must first calculate the SSR. Using Eq. (3.16), the SSR is:

$$\begin{aligned}
SSR &= \sum_{i=1}^{n} (\hat{y}_i - \bar{y})^2 \\
&= (5.4722 - 7)^2 + (5.4722 - 7)^2 + (10.4078 - 7)^2 + (3.8270 - 7)^2 \\
&\quad + (7.1174 - 7)^2 + (3.0044 - 7)^2 + (13.6982 - 7)^2 \\
&= 87.19386.
\end{aligned}$$

From this value of SSR, the MSR is:

$$\begin{aligned}
MSR &= \frac{SSR}{v_1} \\
&= \frac{87.19386}{1} \\
&= 87.19386.
\end{aligned}$$

Using the value of the MSE from the previous application, the F statistic is:

$$\begin{aligned}
F &= \frac{MSR}{MSE} \\
&= \frac{87.19386}{2.560532} \\
&= 34.05302.
\end{aligned}$$

(c) Since the F statistic is greater than the critical value, the null hypothesis is rejected. More specifically,

$$F > F_\alpha$$

$$34.05302 > 6.607891.$$

(d) The p-value from F with 1 degree of freedom in the numerator (v_1) and $n - 2$ degrees of freedom in the denominator (v_2) can be found with the `cdf` function for F:

```
1 - f.cdf(34.05302, 1, 5)
```

```
# np.float64(0.0020899566811585224)
```

Note that when we use the F-distribution, we only consider the right tail.

(e) There are two important findings from the p-value:

1. The p-value corresponding to the F statistic is equivalent to the one for the t statistic. In simple linear regression, the F-test is equivalent to the t-test which results in the same p-values.
2. The hypothesis test can be concluded by comparing the p-value with α. This test gives the same result as if the critical value were compared to the F-test statistic (as in part c).

(f) The F value is more extreme than the critical value, and therefore, the model is significant.

4.18 Cautions About Statistical Significance

In practice, if a t-test or F-test shows that rejecting the null hypothesis is the appropriate course of action, it does not enable us to definitively conclude that a cause-and-effect relationship exists between x and y. Concluding a test as shown in this chapter may suggest correlation, but correlation between variables does not imply causation. Also, finding statistical significance in the aforementioned tests does not enable us to conclude that there is a linear relationship between x and y. A nonlinear model may be a better fit regardless of the significance.

4.19 Case Study: Stock Betas

4.19.1 Problem Statement

In studying the stock market, data analysis can provide insights about the selection of stocks, sector performance, and overall market behavior. Understanding how an individual stock performs relative to the market is particularly important. One such measure of this performance is referred to as *beta*. As shown in the next section, beta can be derived by fitting a regression model. This case demonstrates how regression analysis can be used in finance, particularly stock market analysis.

The beta value for a particular stock can be found by fitting a simple linear regression model, which can be written as

$$R_a = \alpha + \beta R_b + \varepsilon, \tag{4.16}$$

where

- R_a is the return of an asset (or stock),
- R_b is the return of a market benchmark, and
- ϵ is the error of the model.

The remaining model parameters α and β denote the intercept and slope of the linear regression model. The estimate of β can be interpreted as "correlated relative volatility," but in simpler terms, this means that for every percentage increase in the market benchmark (R_b), the R_a increases by β percent.

For the market benchmark in this case study, we use the S&P 500, which is commonly used as a measure of the market's total performance. Using several individual stocks, we will fit a regression model and calculate the resulting coefficients.

As an analyst working for a hedge fund, you are tasked with analyzing data on seven stocks to obtain the beta values for each. The seven stocks, which are of particular interest to management, are listed here, along with their tickers.

- Apple Incorporate (AAPL)
- Caterpillar Incorporated (CAT)
- Johnson & Johnson (JNJ)
- McDonald's Corporation (MCD)
- Procter & Gamble Company (PG)
- Microsoft Corporation (MSFT)
- Exxon Mobil Corporation (XOM)

Specifically, you will need to use data from the 36 months prior to September 1, 2021. Using the data, you will then find each stock's beta value and create some visualizations. From the calculated beta values, you will then need to make some investment recommendations to management.

4.19.2 Descriptive Statistics

The case study data consist of 36 monthly returns of a selection of stocks. The first code you will want to do prior to any analysis is importing the packages you will be using with appropriate aliases. The packages used in this case study are Pandas, Ravix, and NumPy, which are imported with their respective aliases.

```
import pandas as pd
from ravix import get_data
import numpy as np
```

After importing the packages, the next step in analyzing a clean dataset is to fully understand the dataset of interest. After loading in the data with Ravix's `get_data` function, examine the first five observations using the `head` function.

```
df = get_data("Betas.csv")
df.head()
```

```
#     Unnamed: 0      AAPL       CAT       JNJ       MCD
# 0   2018-09-01 -0.004825  0.098236  0.032673  0.037628
# 1   2018-10-01 -0.030477 -0.204407  0.013172  0.057445
# 2   2018-11-01 -0.184045  0.125409  0.049361  0.065630
# 3   2018-12-01 -0.113616 -0.063389 -0.115917 -0.052225
# 4   2019-01-01  0.055154  0.047927  0.031228  0.006814

#           PG      MSFT       XOM       SPY
# 0   0.003376  0.022079  0.071435  0.001412
# 1   0.065481 -0.066101 -0.062809 -0.064891
# 2   0.075159  0.038199 -0.002259  0.018549
# 3  -0.027404 -0.080090 -0.133569 -0.093343
# 4   0.049499  0.028158  0.074644  0.086373
```

The dataset from `Betas.csv` was assigned to the `df` dataframe using the `get_data` function from Ravix. The first column corresponds to the date of each return, while the next seven columns consist of the monthly returns for seven stocks, and the last column represents the S&P 500 (denoted as SPY) index. The first date value is September 1, 2018, and the second date value is October 1, 2018, indicating each date corresponds to the first of the month. On closer inspection, the dates range from September 1, 2018, to August 1, 2021. The dataframe can be modified using simple indexing to exclude the first column. The column names can be found using the `columns` method:

```
df.columns
```

```
# Index(['Unnamed: 0', 'AAPL', 'CAT', 'JNJ', 'MCD', 'PG',
#        'MSFT', 'XOM', 'SPY'],
#       dtype='object')
```

The `drop` function can be used by specifying the column we want dropped in the columns argument. Rather than typing the name, we use square brackets to denote column 0 should be dropped. The `inplace = True` argument allows the dataframe to be modified directly.

```
df.drop(columns = df.columns[0], inplace = True)
```

Datasets can also be understood by finding descriptive statistics and graphs that summarize the information. The `describe` command provides a quick and powerful way to compute descriptive statistics in Python.

```
df.describe()
```

Summary statistics are shown for a subset of variables for readability using the iloc method and specifying all the rows and the first five columns.

```
df.iloc[:, :5].describe()
```

```
#              AAPL        CAT        JNJ        MCD         PG
# count  36.000000  36.000000  36.000000  36.000000  36.000000
# mean    0.032999   0.017261   0.010656   0.014158   0.018292
# std     0.095151   0.082262   0.053878   0.054946   0.043768
# min    -0.184045  -0.204407  -0.115917  -0.143098  -0.086000
# 25%    -0.035885  -0.037298  -0.019163  -0.009394  -0.010842
# 50%     0.055795   0.039862   0.020563   0.017790   0.017859
# 75%     0.101165   0.061480   0.041866   0.051348   0.054529
# max     0.214380   0.186986   0.144208   0.134321   0.096596
```

From the summary, the data becomes easier to understand. One particularly helpful measure of spread is the standard deviation, which will help assess the variability of the stock returns. To get the standard deviation of the Microsoft returns, the std function from NumPy can be used. Recall from Sect. 2.12 that the ddof argument must be specified to return the sample standard deviation.

```
MSFT_stdev = np.std(df.MSFT, ddof = 1)
print(MSFT_stdev)
```

```
# 0.05361294136868017
```

Rather than find the standard deviation of each stock return separately, the standard deviations can be calculated using the std function by passing the dataframe and the axis argument of 0. Specifying the axis argument to 0 simply specifies to calculate the standard deviations of the columns.

```
stdev = np.std(df, ddof = 1, axis = 0)
print(stdev)
```

```
# AAPL    0.095151
# CAT     0.082262
# JNJ     0.053878
# MCD     0.054946
# PG      0.043768
# MSFT    0.053613
# XOM     0.102811
# SPY     0.054134
# dtype: float64
```

The stdev created in the previous code is a Pandas Series. To create a Pandas transposed dataframe, use the DataFrame function and input a list of stdev.values. The stdev.index consists of the column names from the dataframe. This restructuring of the standard deviations will allow for easy integration into Ravix's barplot function as demonstrated in the following section.

```
sd_df = pd.DataFrame([stdev.values], columns = stdev.index)
print(sd_df)
```

```
#          AAPL       CAT       JNJ       MCD        PG  \
# 0  0.095151  0.082262  0.053878  0.054946  0.043768
#
#          MSFT       XOM       SPY
# 0  0.053613  0.102811  0.054134
```

The means can be calculated in a similar manner:

```
means = np.mean(df, axis = 0)
means_df = pd.DataFrame([means.values], columns = means.index)
print(means_df)
```

```
#          AAPL       CAT       JNJ       MCD        PG  \
# 0  0.032999  0.017261  0.010656  0.014158  0.018292
#
#          MSFT       XOM       SPY
# 0  0.030299 -0.000624  0.015307
```

The result is a dataframe consisting of standard deviations (sd_df) and a dataframe consisting of the means (means_df).

4.19.3 Plots and Graphs

A bar graph can depict the mean monthly returns for each stock which is simple yet informative. The barplot function in Ravix can generate this visualization with relative ease. The first argument is a formula string "SPY ~ .", which indicates that the SPY column will treated differently, but we will include all of the other columns in the means_df dataframe. The xlab and ylab arguments are optional but allow for x and y-axis labels. The title argument allows for a title.

```
from ravix import barplot
barplot("SPY ~ .", means_df, xlab = "Stock Ticker",
        ylab = "Mean Monthly Returns",
        title = "Mean Monthly Stock Returns")
```

Defining the means_df helped define the barplot, however, the aggregation can be done inside of the barplot function. Specifying agg = "mean" and using df as the dataframe input we would produce a barplot identical to that of Fig. 4.9:

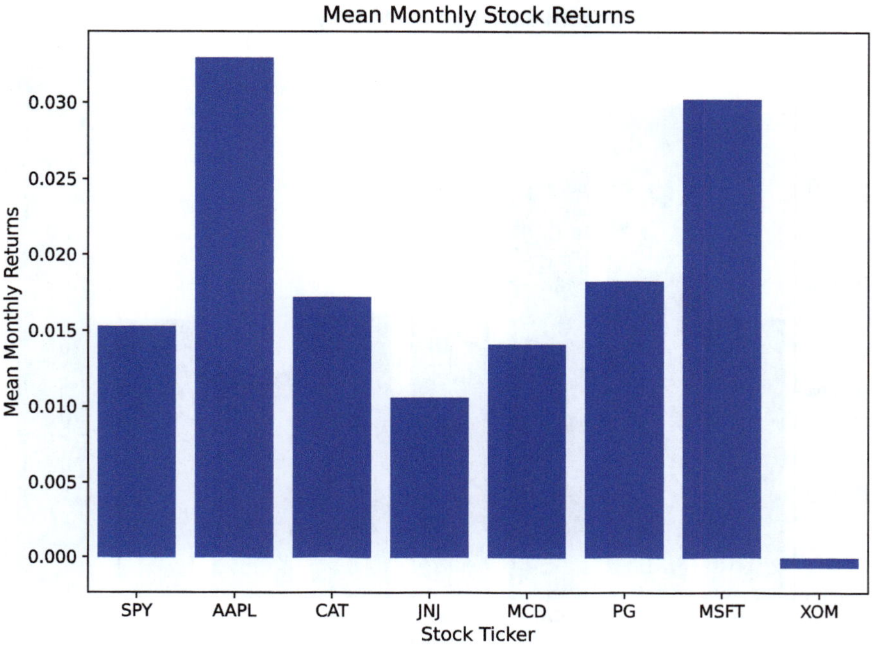

Fig. 4.9 Barplot of means

```
barplot("SPY ~ .", df, agg = "mean", xlab = "Stock Ticker",
        ylab = "Mean Monthly Returns",
        title = "Mean Monthly Stock Returns")
```

The default color is blue, but the `color` argument can be modified by specifying a string for all the bars or a list for each bar.

From the bar plot in Fig. 4.9, we see that Apple (AAPL) has the highest mean return among the stocks, closely followed by Microsoft (MSFT) and then Procter & Gamble (PG). Depending on the purpose of our analysis of these stocks, we may find importance in noting the lowest two stocks are Exxon Mobil (XOM) and Johnson & Johnson (JNJ), respectively.

Previously, a vector of standard deviations was calculated and printed. A bar plot of the standard deviations provides a useful visualization.

```
barplot("SPY ~ .", sd_df, xlab = "Stock Ticker",
        ylab = "Standard deviation of Monthly Returns",
        title = "Standard deviation of Monthly Stock Returns")
```

Recall the standard deviation is a measure of variability and is often associated with risk. As shown in the bar plots above, Exxon Mobil has the largest standard deviation (Fig. 4.10) and the lowest mean of the stocks listed (Fig. 4.9). While Exxon

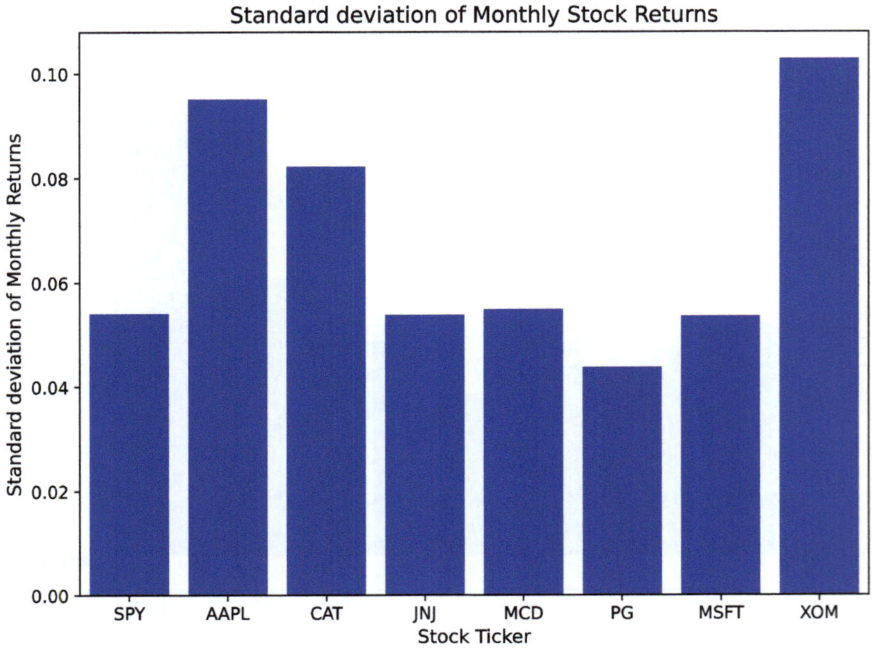

Fig. 4.10 Barplot of standard deviations

Mobil's low mean indicates it did not perform well, the variability shown by the standard deviation would indicate it is a risky investment.

In contrast to Exxon Mobil's performance, it is often the case that a large standard deviation is consistent with a large mean, which is shown from Apple's relatively large mean and standard deviation. This relation between high mean performance and high standard deviation reinforces the notion that with greater risk comes greater reward.

The S&P 500 often has low variability since it is an index consisting of 500 stocks, as indicated by a low standard deviation. However, from Fig. 4.10, we observe that Procter & Gamble has an even lower standard deviation. To investigate this unlikely result, we use another plot to observe the variability further.

As discussed in Chap. 2, the `boxplot` function within Ravix is an important Python tool to analyze data. Box plots are great for visualizing the variability and the center of each variable simultaneously.

```
from ravix import boxplot
boxplot("SPY ~ .", data = df, xlab = "Stock Ticker",
        ylab = "Monthly Returns",
        title = "Boxplot of Monthly Returns of Stocks")
```

From Fig. 4.11, several conclusions can be made. For example, we can see that Exxon Mobil has the highest variability, which is in agreement with the standard

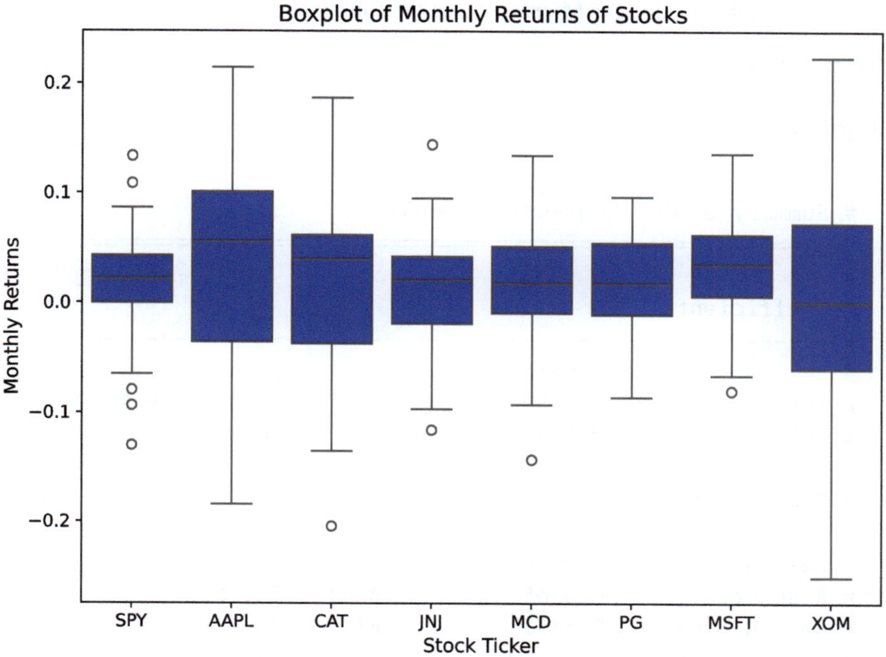

Fig. 4.11 Boxplot of returns

deviation. From the standard deviations, we noted that Procter & Gamble had the lowest standard deviation. However, the S&P 500 appears to have a lower IQR variability indicating that the middle 50% of the observations have lower variability than the middle 50% of observations from Procter & Gamble. The S&P 500 generally has lower variability since it represents an index of 500 stocks, but it appears a few outliers were able to drive the standard deviation higher than Procter & Gamble as seen in Fig. 4.10.

4.19.4 Finding Beta Values

The beta value of Microsoft can be found easily by using the `ols` function which can be imported from Ravix. The syntax for the `ols` function is as follows:

```
from ravix import ols
ols("y ~ x", data = df)
```

where `y` is the response, `x` is the predictor variable, and `df` is the dataframe where the `y` and `x` variables are located.

To find the beta value for Microsoft, use the following:

```
from ravix import ols
regression = ols("MSFT ~ SPY", data = df)
regression.summary()
```

```
# Summary of OLS Regression Analysis:
# =========================================================
#
# Coefficients:
# ---------------------------------------------------------
#                 Estimate  Std. Error  t-value    p-value
# Intercept      0.019050   0.0063231    3.013     0.0049 **
# SPY             0.73482     0.11387    6.453   2.24e-07 ***
#
# Model Statistics:
# ---------------------------------------------------------
# Residual Std. Error: 0.036469
# R-squared:        0.5505           AIC: -134.308
# Adj. R-squared: 0.5373             BIC: -131.141
# F-statistic: 41.6431 on 1 and 34 DF, p-value: 2.24e-07
# =========================================================
```

From the coefficient summary, the value of the slope coefficient (beta) for Microsoft is 0.734825. The regression equation is therefore:

$$\text{MSFT} = 0.019050 + 0.734825 \times \text{SPY}, \tag{4.17}$$

where MSFT is the predicted value of the Microsoft returns, as predicted by the S&P 500 returns (SPY). This regression equation makes use of the model in Eq. 4.16.

Since the slope coefficient represents the desired result, we may attain it by typing:

```
regression.params['SPY']
```

```
# np.float64(0.7348248411767134)
```

The R^2 value can also be found from the regression by typing:

```
regression.rsquared
```

```
# np.float64(0.5505208433351765)
```

4.19.5 Finding All the Betas

We can find all of the beta values by getting the slope from each model which can be automated with a loop. For example, we may want to find the betas of multiple stocks for the purpose of generating a barplot. The basic steps to get the slopes of the models are initializing a vector, looping through to calculate beta for every stock, and then if necessary, formatting the result for input into the barplot.

The stock names can be easily ascertained by indexing `df.columns` so that we exclude the SPY which is the item in the list with index 7 which corresponds to column 8.

```
stocks = df.columns[:7]
print(stocks)
```

```
# ['AAPL', 'CAT', 'JNJ', 'MCD', 'PG', 'MSFT', 'XOM']
```

Note that we could initialize a vector of beta values as a list or a dictionary. In this case, we will initialize a dataframe not only because it is a popular data structure but also the end result doesn't require additional formatting. Use the `DataFrame` function from Pandas to accomplish this setting the columns as the stocks with an index of 0.

```
betas_df = pd.DataFrame(columns = stocks, index = [0])
print(betas_df)
```

```
#   AAPL  CAT  JNJ  MCD   PG MSFT  XOM
# 0  NaN  NaN  NaN  NaN  NaN  NaN  NaN
```

As shown, the dataframe has columns consisting of the stock tickers with one row which we index as 0. The values in the dataframe are currently not a number (`NaN`) values.

A `for` loop can provide a quick and easy way to look through all the stock values. The `for` loop creates a temporary variable `stock`, which takes on an item within the `stocks` vector. Within the `for` loop, a formula is created to predict the stock as a function of the SPY. Using this formula, Ravix's `ols` function constructs the desired model. As a last step, the slope is set as a value within the `betas_df` dataframe.

```
for stock in stocks:
    formula = f'{stock} ~ SPY'
    model = ols(formula, data = df)
    betas_df[stock] = model.params['SPY']
```

Observing the dataframe can be done using the print command or typing the variable name.

```
betas_df
```

```
#        AAPL       CAT       JNJ       MCD        PG  \
# 0  1.226589  0.804454  0.672668  0.605301  0.405789
#
#        MSFT       XOM
# 0  0.734825  1.396555
```

Similarly, a dataframe containing the R^2 values can be created with the code:

```
r_squared_df = pd.DataFrame(columns = stocks, index = [0])

for stock in stocks:
    formula = f'{stock} ~ SPY'
    model = ols(formula, data = df)
    r_squared_df[stock] = model.rsquared
```

A bar plot of the betas provides a visual depiction:

```
from ravix import barplot
barplot("AAPL ~ .", betas_df, color = "darkblue",
        xlab = "Stock Ticker", ylab = "Beta",
        title = "Beta Values by Stock")
```

If a stock beta is 1, then the stock increases at the same rate as the market benchmark. If a stock beta is above 1, then the stock is considered to be more volatile than the market. From the stock beta plot, Fig. 4.12, we notice Apple and Exxon Mobil are more volatile than the market. Stocks with betas below 1 are considered to be less volatile than the market.

A visual representation of the R^2 values is shown in Fig. 4.13.

```
barplot("AAPL ~ .", r_squared_df, color = "darkred",
        xlab = "Stock Ticker", ylab = "Percent Return",
        title = "Percent of Return Explained by Market")
```

By looking at the stock R^2 values in Fig. 4.13, we note that Exxon Mobil is the stock most correlated with the market. To investigate this matter further, we plot a scatter plot to see the relationship.

```
from ravix import plot
plot("XOM ~ SPY", df, xlab = "S&P 500 Benchmark Return",
     ylab = "Exxon Mobil Return")
```

In Fig. 4.14, we note that there is a clear relationship with the S&P 500. However, the vertical axis has a larger range than the horizontal axis. This indicates further evidence that Exxon Mobil, the vertical axis in the figure, is more variable than the market.

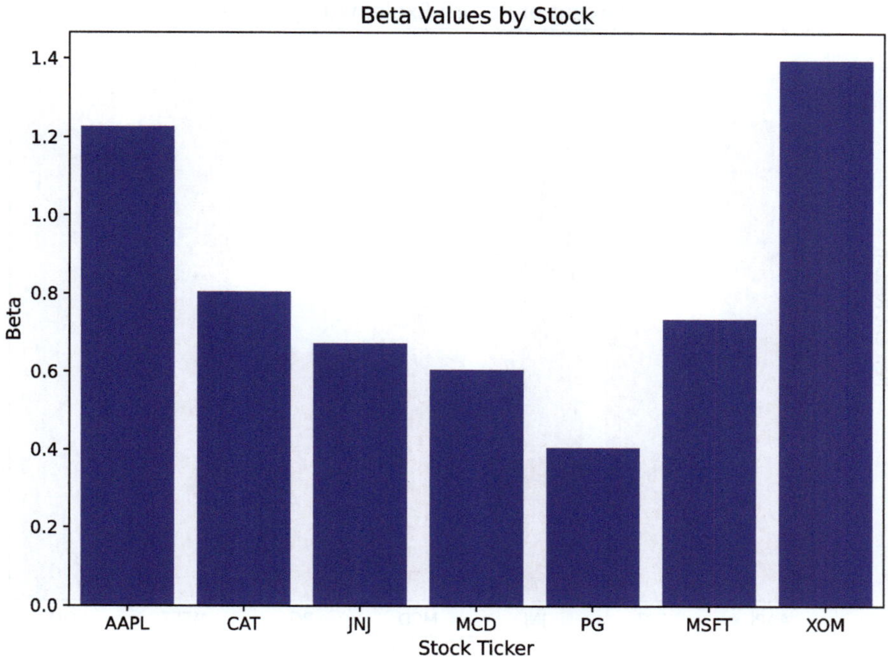

Fig. 4.12 Barplot of betas

4.19.6 Recommendations and Findings

From this analysis, we draw several conclusions. In taking the mean of each variable, Apple showed the best performance followed by Microsoft. However, when looking at the variability, Apple had the second highest variability, which demonstrates the risk-reward relationship. Looking at the other extreme, Exxon Mobil had the lowest average return in the data set and the highest standard deviation, making it a poor investment. Procter & Gamble had the third largest mean and a standard deviation that indicated a low risk. Having a low standard deviation and a high mean makes Procter & Gamble a sound investment.

From this analysis, we can use the following insights to formulate some investment strategies for management.

- It is probably a safe prediction that Procter & Gamble will not be affected as much as other stocks if the S&P 500 does not perform well. Therefore, investing in Procter & Gamble is a sound choice given that the return is relatively high.
- If Apple continues to have high variability and high returns, then it may be advantageous to buy Apple when the price is on the downswing, since the variability of Apple's stock is high and the long-term mean return is high.

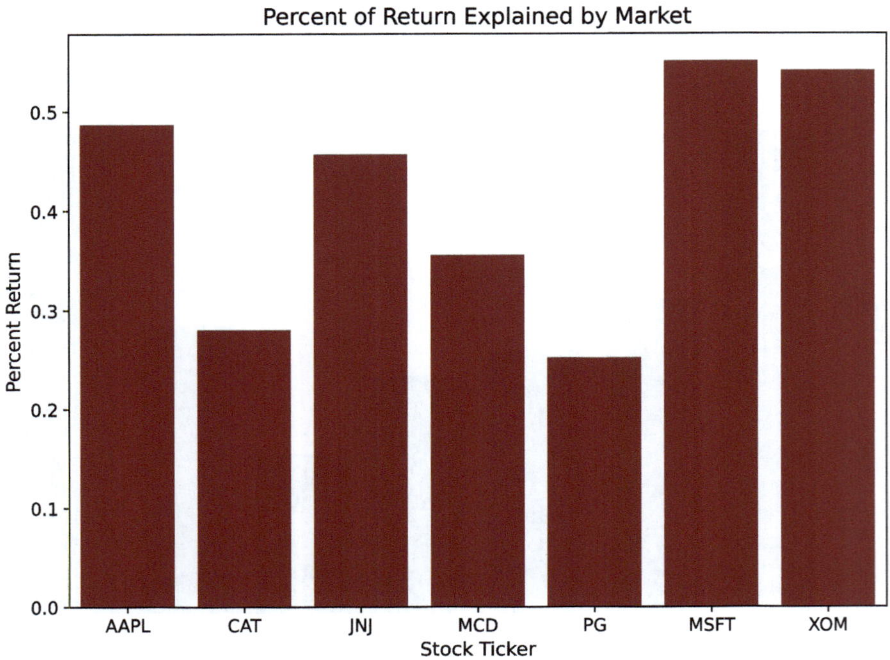

Fig. 4.13 Barplot of R^2 values

- Since Microsoft and Exxon Mobil are correlated with the market, if we are able to accurately predict the market's performance, it would be possible to accurately predict the returns of Microsoft and Exxon Mobil.

Despite the analysis done here, keep in mind that financial analysts warn that the past does not guarantee future performance.

4.19.7 Case Conclusion

This case study underscores the vital role of regression analysis in financial market analysis, particularly in calculating and interpreting stock betas. By utilizing simple linear regression, we were able to determine the beta values for a selection of stocks, which offer insights into their relative volatility and correlation with the market benchmark.

The analysis provided a clear depiction of each stock's performance relative to market movements, with some stocks like Apple showing high returns coupled with high risk, while others like Procter & Gamble demonstrated stable returns with lower risk. These insights are invaluable for formulating investment strategies and making informed decisions.

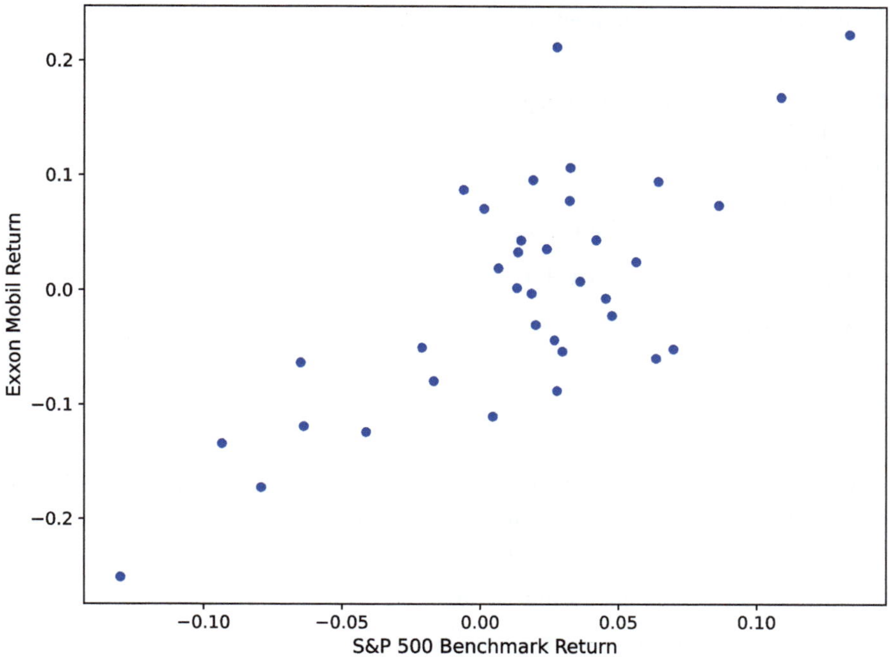

Fig. 4.14 Scatterplot of returns

In essence, the study highlights the power of combining regression analysis with Ravix to derive meaningful insights from financial data. As financial markets grow increasingly complex, the ability to analyze and interpret data efficiently becomes ever more crucial. The methods and findings presented here can be expanded upon to tackle larger and more intricate datasets, ultimately contributing to more nuanced and effective financial analyses.

Through this case study, we have reinforced the importance of analytics and the application of statistical methods in financial decision-making. The use of Ravix in this analysis has demonstrated its utility in simplifying and enhancing the process of regression analysis and visualization, paving the way for more robust and data-driven investment strategies.

Problems

1. Social Media Influencers F-Statistic

Using the social media influencer data from Table 3.2, calculate the F-statistic by answering the following questions. Assume we are trying to predict the number of clicks based on the amount of pay.

Without the use of a computer:

a. Find $\hat{y}_i$ for each x_i.
b. Find the residual corresponding to each y.
c. Calculate the SST.
d. Calculate the SSR.
e. Calculate the SSE.
f. Calculate the F-value for the regression.

While using a computer (with Python):

g. Verify your results from parts a through f.

2. Social Media Influencers t-Statistic
Using the data from Table 3.2, answer the following questions.
Without the use of a computer:

a. Calculate the mean squared error (MSE) and the residual standard error (RSE).
b. Calculate the standard error of the slope estimate ($\hat{\beta}_1$).
c. Calculate the t-value for the slope estimate ($\hat{\beta}_1$).

While using a computer (with Python):

d. Verify your results from parts a through c.

3. Social Media Influencers Hypothesis Tests
Using the data from Table 3.2, answer the following questions:

a. Specify the null and alternative hypotheses for the t-test.
b. Find the p-value for the t-test using Python.
c. Compare the p-value to the level of significance ($\alpha = 0.05$) and make the decision.
d. Specify the null and alternative hypotheses for the F-test.
e. Find the p-value for the F-test using Python.
f. How do the p-values of the F and t-tests differ?
g. Interpret the results of both tests.

4. Accounting Analytics F-Statistic
Using the data from the previous chapter case study, calculate the F-statistic by answering the questions. Assume you are predicting the itemized deductions using the adjusted gross income. The data are given in Table 3.1.
Without the use of a computer:

a. Find $\hat{y}_i$ for each x_i.
b. Find the residual corresponding to each y.
c. Calculate the SST.
d. Calculate the SSR.
e. Calculate the SSE.
f. Calculate the F-value for the regression.

While using a computer (with Python):

g. Verify your results from parts a through f.

5. Accounting Analytics t-Statistic
Using the data from Table 3.1, answer the following questions.
Without the use of a computer:

a. Calculate the mean squared error (MSE) and the residual standard error (RSE).
b. Calculate the standard error of the slope estimate ($\hat{\beta}_1$).
c. Calculate the t-value for the slope estimate ($\hat{\beta}_1$).

While using a computer (with Python):

d. Verify your results from parts a through c.

6. Accounting Analytics Critical Region Method for t
Using the data from Table 3.1, answer the following questions:

a. Specify the null and alternative hypotheses for the t-test.
b. Find the critical values for the t-test using Python with $\alpha = 0.05$.
c. Compare the critical value to the test statistic for t and make the decision.
d. Interpret the results of the t-test.

7. Sales Calls F-Statistic
Using Abdullah's calls and sales data from Table 2.4, answer the following questions:

a. Find $\hat{y}_i$ for each x_i.
b. Find the residual corresponding to each y_i.
c. Calculate the SST.
d. Calculate the SSR.
e. Calculate the SSE.
f. Calculate the F-value for the regression.

8. Sales Calls t-Statistic
Using the data from Table 2.4, answer the following questions:

a. Calculate the mean squared error (MSE) and the residual standard error (RSE).
b. Calculate the standard error of the slope estimate ($\hat{\beta}_1$).
c. Calculate the t-value for the slope estimate ($\hat{\beta}_1$).

9. Automotive Stocks F-Statistic
Using the automotive stock data from Table 3.3 to predict price from earnings, answer the following questions:

a. Find $\hat{y}_i$ for each x_i.
b. Find the residual corresponding to each y.
c. Calculate the SST.
d. Calculate the SSR.
e. Calculate the SSE.
f. Calculate the F-value for the regression.

10. Automotive Stocks Test Statistic for t

Using the data from Table 3.3, answer the following questions:

a. Calculate the mean squared error (MSE) and the residual standard error (RSE).
b. Calculate the standard error of the slope estimate ($\hat{\beta}_1$).
c. Calculate the t-value for the slope estimate ($\hat{\beta}_1$).

11. Automotive Stocks F-Tests

Using the data from Table 3.3, answer the following questions:

a. Find the critical value for the F-test using Python. Specify the function, arguments, and output.
b. Compare the critical value to the test statistic for F and make the decision.
c. Find the p-value for the F-test using Python. Specify the function, arguments, and output.
d. Compare the p-value to the level of significance ($\alpha = 0.05$) and make the decision.

Chapter 5
Multiple Regression

Prediction is very difficult, especially about the future.

—Niels Bohr

5.1 Introduction

In this chapter, we build upon the coverage of regression analysis by considering situations involving two or more predictor variables. For instance, while the weight of a person may be predicted using their height, we could use both the height and age of that person to predict their weight. Using more than one predictor variable to predict a response is called multiple regression analysis, which enables us to consider more predictor variables and thus obtain better estimates than those possible with simple linear regression.

Since the calculation of multiple regression coefficients follows the same principles as simple linear regression, it is essential to understand the simple case before moving on to multiple linear regression. As seen in Chap. 3, calculating coefficients and model statistics aids in the understanding of simple linear regression. In multiple regression, most of the calculations build on the knowledge gained from working with simple linear regression, but the calculations become more tedious due to the number of predictor variables involved. Therefore, we limit the calculations done by hand or calculator when working with multiple regression, and instead rely upon Python and its functions for such calculations.

We begin by specifying the multiple regression model and equation. Then, we discuss the limitations of using R^2 in multiple regression (referred to as multiple R^2). Next, we introduce a measure of model fit that is more appropriate for multiple regression (adjusted R^2). We use a website marketing application to illustrate these concepts while keeping the number of predictor variables limited to two for simplicity. In the final discussion of the chapter, we analyze a housing dataset in a case study, providing detailed analyses for the reader.

D. P. McGibney, *Applied Linear Regression for Business Analytics with Python*,
International Series in Operations Research & Management Science 378,
https://doi.org/10.1007/978-3-032-23806-1_5

5.2 Multiple Regression Model

In the previous chapters, we discussed in detail the simple linear regression model and the utility of having a mathematical model. In multiple regression, the aim is to capture the effect of multiple predictor variables in predicting a response variable. The equation that describes how the response variable Y is related to p predictor variables X_1, X_2, ..., X_p and an error term is:

$$Y = \beta_0 + \beta_1 X_1 + \beta_2 X_2 + \cdots + \beta_p X_p + \varepsilon, \tag{5.1}$$

where $\beta_0, \beta_1, \beta_2, \ldots, \beta_p$ are the parameters and ε is a random variable representing the error in the model. If only β_0 and β_1 are nonzero and the remaining coefficients $\beta_2, \beta_3, \ldots, \beta_p$ are zero, then the model reduces to the simple linear regression model introduced in the previous chapter:

$$Y = \beta_0 + \beta_1 X_1 + \varepsilon. \tag{5.2}$$

As is the case for both the simple linear regression model and the multiple regression model, the assumptions for the linear regression model must be adhered to. A detailed discussion of the assumptions for linear regression and how to deal with violations of the linear model is given in Chap. 8.

5.3 Multiple Regression Equation

The mathematical model in Eq. (5.1) is known as a multiple linear regression model. When we use and fit the multiple linear regression model with a real dataset, then the corresponding multiple linear regression equation is:

$$\hat{Y} = \hat{\beta}_0 + \hat{\beta}_1 X_1 + \hat{\beta}_2 X_2 + \cdots + \hat{\beta}_p X_p. \tag{5.3}$$

The sample statistics or coefficients $\hat{\beta}_0, \hat{\beta}_1, \hat{\beta}_2, \ldots, \hat{\beta}_p$ can be calculated using a given dataset and the set of matrix equations. These sample statistics are point estimators of the corresponding parameters: $\beta_0, \beta_1, \beta_2, \ldots, \beta_p$.

5.4 Website Marketing Application: Modeling

In this example, a newly formed startup is developing a website that provides news and entertainment. They notice that their monthly revenue (in thousands of dollars) is dependent on the number of ads and the amount of money (in thousands of dollars) spent on additional marketing. The data are shown in Table 5.1.

Investigate the relationship between the response and predictors by doing the following:

Table 5.1 Website marketing data

Ads	Marketing	Revenue
4	14	118
5	8	113
6	13	149
4	14	129
5	10	131
5	5	126
5	11	129
5	9	125
6	12	150
5	11	124
5	11	126
6	11	159
5	12	130
4	9	114
6	8	136
3	9	80
6	7	137
5	8	128
6	9	149
5	10	136

(a) Specify the multiple linear regression model which results from predicting revenue by the number of ads and the amount spent on marketing.
(b) Fit the model using Python.
(c) Explicitly state the regression equation.

Solution

(a) Since it is useful to predict the revenue based on the number of ads and the amount spent on marketing, it follows that the revenue generated is an appropriate response variable. The multiple linear regression model can be written as

$$Y = \beta_0 + \beta_1 X_1 + \beta_2 X_2 + \varepsilon,$$

where

- Y is the amount of revenue generated,
- X_1 is the number of ads,
- X_2 is the amount spent on additional marketing, and
- ε is the error.

The model, therefore, has parameters β_0, β_1, and β_2. Fitting the data to a model generates approximations of the parameters, which are referred to as $\hat{\beta}_0$, $\hat{\beta}_1$, and $\hat{\beta}_2$.

(b) The regression analysis for this problem can be efficiently performed using Ravix in Python. First, we use Pandas to construct a dataframe, which allows us to organize and manipulate the data efficiently. Then, we create lists to store the revenue, ads, and marketing spend data. These lists are then combined into a dictionary, which is subsequently converted into a dataframe using the `DataFrame` function.

```
import pandas as pd

Revenue = [118, 113, 149, 129, 131, 126, 129, 125, 150, 124,
           126, 159, 130, 114, 136, 80, 137, 128, 149, 136]
Ads = [4, 5, 6, 4, 5, 5, 5, 5, 6, 5, 5, 6, 5, 4, 6, 3, 6, 5, 6,
       5]
Marketing = [14, 8, 13, 14, 10, 5, 11, 9, 12, 11, 11, 11, 12, 9,
             8, 9, 7, 8, 9, 10]

df = pd.DataFrame({"Revenue": Revenue, "Ads": Ads,
                   "Marketing": Marketing})
```

After loading the dataset into Python, we use the `ols` function from the Ravix package to create a regression object. This object will contain the fitted model based on the specified formula and dataset. In the code below, the object is named `reg`. To examine the details of the created regression model, we use the `summary` function. This function generates a comprehensive summary of the model's statistics and coefficients, providing insights into the model's performance.

```
from ravix import ols
reg = ols("Revenue ~ Ads + Marketing", data = df)
reg.summary()
```

```
# Summary of OLS Regression Analysis:
# =========================================================
#
# Coefficients:
# ---------------------------------------------------------
#               Estimate  Std. Error  t-value    p-value
# Intercept       12.551      12.531    1.002     0.3306
# Ads             18.614      1.8930    9.833   1.98e-08 ***
# Marketing       2.2783     0.67139    3.393     0.0035 **
#
# Model Statistics:
# ---------------------------------------------------------
# Residual Std. Error: 6.7467
# R-squared:         0.8558            AIC: 135.8696
# Adj. R-squared: 0.8389               BIC: 138.8568
# F-statistic: 50.4534 on 2 and 17 DF, p-value: 7.09e-08
# =========================================================
```

The output from the `summary` function is detailed and includes various statistics such as coefficients, standard errors, and p-values. By examining the coefficients section, we can reconstruct the regression equation.

(c) Upon inspection of the summary output, the resulting multiple linear regression equation is

$$\hat{Y} = 12.55 + 18.61X_1 + 2.28X_2.$$

To enhance clarity, the multiple linear regression equation is often expressed using the actual variable names, which simplifies interpretation:

$$\text{Revenue} = 12.55 + 18.61(\text{Ads}) + 2.28(\text{Marketing}).$$

5.5 Significance Testing: t

In multiple regression, a separate t-test is conducted for each predictor variable to assess individual significance. These t-tests are essential for determining whether each coefficient is significantly different from zero, providing insight into the contribution of each predictor. A coefficient that is not significantly different from zero suggests that the corresponding predictor does not meaningfully contribute to the model. While the t-test is a powerful tool, it's important to recognize that statistical significance alone does not always imply practical significance.

The hypotheses for the t-test of the ith variable is

$$H_0 : \beta_i = 0$$

$$H_1 : \beta_i \neq 0.$$

The corresponding test statistic can be calculated by dividing the coefficient estimate ($\hat{\beta}_i$) by the corresponding standard error of the coefficient ($s_{\hat{\beta}_i}$):

$$t_i = \frac{\hat{\beta}_i}{s_{\hat{\beta}_i}}. \tag{5.4}$$

In general, a t-test statistic can be calculated for every coefficient in the multiple regression model. The t-value corresponds to a p-value that can be calculated using the `cdf` function in Python as shown in Sect. 4.7. For multiple regression models, we use a t distribution with $n - p - 1$ degrees of freedom. We reject the null hypothesis and deem the coefficient estimate ($\hat{\beta}_i$) statistically significant when the p-value is less than α. As discussed in Chap. 4, a two-tailed test is most relevant for a coefficient test since the p-value takes into consideration extreme values on both sides of the t distribution. Alternatively, the test can be concluded by comparing t_i with the critical values ($t_i \leq -t_{\alpha/2}$ or $t_i \geq t_{\alpha/2}$).

Upon first inspection, it may seem intuitive to remove predictors based solely on their significance in the t-test. However, decisions about including or excluding variables should take into account the broader context of the model and the specific goals of the analysis, rather than relying purely on p-values.

5.6 Coefficient Interpretation

In multiple regression analysis, the intercept ($\hat{\beta}_0$) is interpreted as the expected value of y when all other variables are equal to zero. The remaining regression coefficients ($\hat{\beta}_i$ for all i) represent the expected change in the response variable associated with a one-unit increase in the corresponding predictor, holding all other predictor variables constant. Coefficients that are not statistically significant provide insufficient evidence of a meaningful relationship and should be interpreted with caution.

5.7 Website Marketing Application: Individual Significance Tests

From the regression summary, a coefficient summary is constructed and shown below in Table 5.2. This table displays several statistics that indicate important details about the fitted regression coefficients. Most importantly, we see the estimated values of $\hat{\beta}_0$, $\hat{\beta}_1$, and $\hat{\beta}_2$, as given in the second column. The corresponding standard errors of $\hat{\beta}_0$, $\hat{\beta}_1$, and $\hat{\beta}_2$ are given in the third column, with the resulting t-values and p-values in the fourth and fifth columns, respectively.

Table 5.2 Website marketing regression

	Estimate	Std. error	t-value	p-value
(Intercept)	12.550623	12.5313599	1.001537	0.3306108
Ads	18.614323	1.8930117	9.833179	0.0000000
Marketing	2.278313	0.6713902	3.393426	0.0034566

Using the estimate and the standard error of the estimate from Table 5.2, do the following:

(a) Verify the calculation of the t-value using the estimate and the standard error.
(b) Verify the p-value for each coefficient using the t-value and the `cdf` function.
(c) Conclude the t-test for each coefficient.
(d) Interpret the coefficients.

Solution

(a) For the website marketing application, the t-values can be calculated using Eq. (5.4):

$$t_0 = \frac{\hat{\beta}_0}{s_{\hat{\beta}_0}} = \frac{12.551}{12.531} = 1.002$$

$$t_1 = \frac{\hat{\beta}_1}{s_{\hat{\beta}_1}} = \frac{18.614}{1.893} = 9.833$$

$$t_2 = \frac{\hat{\beta}_2}{s_{\hat{\beta}_2}} = \frac{2.278}{0.6714} = 3.393.$$

(b) Since the t distribution is symmetric, extreme t values can exist on both the right and left side of the distribution. Thus, we use a two-tailed test, which considers both extreme values. In this case, each p-value is two-tailed and can be calculated using the formula given here. Recall that the negative of the absolute t-value is taken in order to return the extreme area on the left.

```
from scipy.stats import t
2*t.cdf(-t, n-p-1)
```

From the code above and noting that the degrees of freedom are $n - p - 1 = 20 - 2 - 1 = 17$, the p-values corresponding to t_0, t_1, and t_2 are given by

```
2*t.cdf(-1.002, 17)
2*t.cdf(-9.833, 17)
2*t.cdf(-3.393, 17)
```

The code above yields near the values listed in the summary output, although slight discrepancies occur due to rounding. Note that, in this case, each p-value is two-tailed with $n - p - 1 = 20 - 2 - 1 = 17$ degrees of freedom since there are 20 observations and 2 predictors.

(c) Since the p-value of the intercept is larger than $\alpha = 0.05$, the null hypothesis cannot be rejected and is, therefore, not statistically significant. The slope estimates of $\hat{\beta}_1$ and $\hat{\beta}_2$ are both statistically significant since both of the corresponding p-values are less than $\alpha = 0.05$.

(d) Since the intercept is not statistically significant, it should not be interpreted. The coefficient corresponding to the "Ads" variable is $\hat{\beta}_1 = 18.61$. Therefore, the coefficient can be interpreted as:

For each additional ad, an increase of $18,610 in revenue occurs when money on additional marketing is held constant.

In a similar way, $\hat{\beta}_2 = 2.28$ can be interpreted as:

The revenue increases by \$2280 for every \$1000 spent via other marketing means assuming the number of ads is held constant.

5.8 Significance Testing: *F*

In simple linear regression, the F and t-tests provide the same conclusion since the null and alternative hypotheses are the same. In fact, the resulting p-values of the tests are also equivalent despite using different statistics. In multiple regression, however, these tests serve different purposes. We use the F-test to determine if the model is significant. Once the F-test deems the model significant, we should use the t-test to determine whether each individual predictor variable is significant.

The hypotheses for an F-test are as follows:

$$H_0 : \beta_1 = \beta_2 = \ldots = \beta_p = 0$$

$$H_1 : \text{At least one coefficient is nonzero.}$$

The null hypothesis of the F test states that there is no linear relation between the predictor variables and the response variable. The alternative hypothesis states that there is a linear relation between at least one predictor variable and the response variable.

The F-test takes into account the entire regression model when determining significance. The test statistic is given by the ratio of the mean squares due to regression (MSR) and the mean squared error (MSE):

$$F = \frac{MSR}{MSE}. \tag{5.5}$$

Alternatively, in terms of the multiple R^2, the F-statistic can be expressed as

$$F = \frac{R^2}{1 - R^2}\left(\frac{n - p - 1}{p}\right). \tag{5.6}$$

MSE measures the average of the squares of the errors between the observed and predicted values. It is a key metric used to evaluate the goodness-of-fit for regression models.

The rejection rule stipulates that one should reject H_0 if F is at or more extreme than the critical value for a given level of α. An equivalent method entails verifying if the p-value of the test statistic is at or below α. In particular, reject H_0 if p-value $\leq \alpha$ or if $F \geq F_\alpha$, where the critical value F_α is based on an F distribution with p degrees of freedom in the numerator and $n - p - 1$ degrees of freedom in the denominator.

5.9 Multiple R^2

For multiple regression, many of the components are similar to simple linear regression but take into consideration multiple predictor variables. Particularly, while we discussed the coefficient of determination (R^2) between two variables, we would like to find the coefficient of determination for the multiple predictor case (multiple R^2). Multiple R^2 is given by

$$R^2 = SSR/SST. \tag{5.7}$$

In the website marketing application, both variables were significant in predicting the revenue. In practice, however, a new predictor variable will not always aid in the prediction of the response variable. In some of the simple linear regression examples, the R^2 value was very low, indicating that the predictor variable did not adequately explain the variation in the response. In multiple regression, including an additional predictor variable, even new variables that are not statistically significant, will always result in a higher multiple R^2. This concept can be explained intuitively by considering that predictor variables with a low multiple R^2 still contribute to reducing the error in fitting the model.

Mathematically, the prediction errors become smaller, thus reducing the sum of squares due to error, SSE. Because

$$SSR = SST - SSE, \tag{5.8}$$

when SSE becomes smaller, SSR becomes larger, causing

$$R^2 = \frac{SSR}{SST} \tag{5.9}$$

to increase. Even though increasing the SSR causes the R^2 to increase, the addition of predictor variables does not always increase the true explained variation. Therefore, it is better to avoid using the multiple R^2 as a means of model evaluation for multiple regression.

Also consider the case in which over a hundred predictor variables exist, but each predictor variable consists of random noise with no predictive power. If a response variable were introduced with less than a hundred observations, then a model could be fit to predict the response variable with the random noise predictors. In fact, since the number of observations would be limited, the random noise could be easily fit to predict the response. The error or SSE would be at or near zero, implying a great fit, but the model fit would be misleading. The problem of overfitting is a common concern in machine learning.

5.10 Adjusted R^2

To address this limitation of R^2, we introduce a similar statistic without the pitfalls mentioned previously. The adjusted multiple coefficient of determination is known as the adjusted R^2 and denoted as R_a^2 which compensates for the number of predictor variables in the model and the number of observations, making it a more reliable measure of model fit particularly when multiple predictors are involved.

Unlike R^2, which considers the ratio of SSR and SST, the adjusted R^2 incorporates the mean sum of squares for both the numerator and the denominator, adjusting for the number of predictors and the sample size.

The R_a^2 amplifies the unexplained variation ratio $(1 - R^2)$ by the ratio of total degrees of freedom $(n - 1)$ with the degrees of freedom due to error $(n - p - 1)$. The result is the equation

$$R_a^2 = 1 - (1 - R^2)\frac{n-1}{n-p-1}, \tag{5.10}$$

which raises the unexplained variation by taking p, the number of predictor variables, into account.

The adjusted R^2 is most useful when comparing regression models fit to the same response variable that differ in the number of predictors. Like R^2, it measures the proportion of variation explained by the model, but it adjusts for model complexity, making it a more appropriate criterion in multiple regression settings.

5.11 Website Marketing Application: Multiple R^2 and Adjusted R^2

In the website marketing application, the SSR is 4593.13 and the SST is 5366.95. Recalling that there are 20 observations and two predictors, do the following:

(a) Calculate the multiple R^2.
(b) Calculate the adjusted R^2.
(c) Interpret the multiple R^2 and the adjusted R^2

Solution

(a) The multiple R^2 is given by Eq. (5.9):

$$\begin{aligned} R^2 &= \frac{SSR}{SST} \\ &= \frac{4593.13}{5366.95} \\ &= 0.8558. \end{aligned}$$

(b) From Eq. (5.10),

$$\begin{aligned} R_a^2 &= 1 - (1 - R^2)\frac{n-1}{n-p-1} \\ &= 1 - (1 - 0.8558)\frac{20-1}{20-2-1} \\ &= 0.8389, \end{aligned}$$

which agrees with the summary output from the modeling application above.

(c) The multiple R^2 indicates that 85.58% of the variation in Y is explained by the two predictor variables. The adjusted R^2 indicates that 83.89% of the variation in Y is explained by the two predictor variables taking into account the number of observations and the number of predictors.

5.12 Correlations in Multiple Regression

A correlation matrix shows the linear correlation between each pair of variables under consideration in a multiple regression model. While correlations between the predictor and response variable serve to predict the response variable, correlations among predictor variables may be problematic for modeling the regression. For instance, if predictor variables X_1 and X_2 have a correlation coefficient of 0.99, then there would be little benefit to using both X_1 and X_2 to predict the response variable, Y. In fact, using both X_1 and X_2 could be a detriment to the predictive power of the model since noise is likely the reason for the difference between the predictors. We expect some amount of random noise to be present in all random variables, whether they be a response or a predictor. Therefore, if X_1 contains most of the same information about Y that X_2 contains, using a multiple regression model with both predictor variables would compound the effects of the noise.

The effect of compounding noise as mentioned above is referred to as multicollinearity which occurs when two predictor variables have a high linear correlation. If two predictor variables in the regression model are highly correlated with each other, it is important to investigate further.

While we can inspect correlations among predictor variables, visualizing these relationships with a correlation plot often provides a clearer and more intuitive summary.

5.12.1 The `plot_cor` Function

To find the correlation coefficients among a dataframe `df`, you can use the built-in `corr` function in pandas, which calculates these values for you:

```
df.corr()
```

This method produces a correlation matrix containing pairwise correlations among all variables. While this feature is useful, a visual representation often makes it easier to identify strong relationships and patterns. A correlation plot provides a visually appealing depiction of the correlation coefficients calculated by running the `plot_cor` function.

To import and run `plot_cor` on a dataframe, simply pass the dataframe into the function:

```
from ravix import plot_cor
plot_cor(df)
```

The result is a matrix color-coded based on the correlation coefficient values. An example of this is given in Fig. 5.2.

While `plot_cor` has many of the standard Ravix arguments such as `xlab`, `ylab`, and `title`; it also has a `style` argument with preset formatting options.

```
plot_cor(df, style = 2)
```

This representation will not only help users identify potential multicollinearity but is also useful in finding correlations between the response variable and predictors. As with any correlation analysis, these values reflect linear relationships and should be interpreted alongside diagnostic plots and model results.

5.13 Variance Inflation Factors

When predictor variables are highly correlated, multicollinearity exists. One common way to assess multicollinearity is by calculating Variance Inflation Factor (VIF) values. For two variables, the correlation coefficient suffices to measure the degree of multicollinearity. However, with more than two predictors, one should first find the multiple R^2 value for each predictor variable with respect to the other predictor variables.

If R_x^2 is the multiple R^2 for a predictor X then the VIF for X is calculated as

$$VIF = \frac{1}{1 - R_x^2}. \tag{5.11}$$

Although there are no strict thresholds for measuring multicollinearity, the following guidelines can be used:

- A VIF of 1 shows no multicollinearity.
- A $VIF \geq 4$ indicates multicollinearity that may cause problems.
- A $VIF > 10$ indicates multicollinearity that will cause serious issues.

To address high VIF values, several strategies can be employed, including removing redundant variables either manually or through variable selection methods (as discussed in Chap. 9). Transforming variables, covered in Chap. 7, can also reduce VIF. Additionally, methods like regularization or principal component analysis can help reduce VIF, which are beyond the scope of this text.

5.13.1 The `vif` *Function*

Calculating VIF values in Ravix can be done using the `vif` function. After importing Ravix, you can use the function by providing a formula and a dataframe.

```
from ravix import vif
vif("Y ~ X1 + X2 + X3", data = df)
```

The `vif` function isolates the predictor variables based on the provided formula. If a dataframe containing the predictor variables is already available, you can call the function without specifying a formula. Additionally, the `vif` function can generate a bar plot of the VIF values if the `plot` parameter is set to `True`, providing a visual representation of multicollinearity.

```
vif("Y ~ X1 + X2 + X3", data = df, plot = True)
```

This flexibility and the plotting feature make the `vif` function a convenient tool for assessing multicollinearity in your regression models.

5.14 Case Study: Real Estate

5.14.1 Problem Statement

Real estate is a popular investment among individuals across a wide range of income levels. Because of this widespread interest, analysis of real estate data can be particularly interesting.

The board of directors for an investment firm has sought your expertise in analyzing their data. You are tasked with analyzing a dataset with 150 cities in a large state. You will need to fit a regression model, analyze the model, interpret the coefficients, and test a claim made by a local politician. The directors of the investment firm need your input so that they can determine which city they should invest in. First, you will need to understand the dataset using some basic Python commands.

5.14.2 Data Description

The data consists of 150 cities in a large state. The state of interest remains unknown as the investment firm directors would like to retain some secrecy. The firm leaders requested that the model predict price as a function of the remaining variables in the dataset.

The following variables describe each city in the dataset, using their corresponding column names from the data:

- `price in usd` represents the average price of a home.
- `rooms` represents the average number of rooms.
- `income` represents the average income.
- `Tax Rate` represents the property tax rate.
- `%Commercial` represents the percentage of commercial property.

As a first step, we load the data file into Python using the `get_data` function from Ravix and display the first five observations.

```
from ravix import get_data
df = get_data("house_prices.csv")
df.head()
```

```
#    price in usd  rooms  income  Tax Rate  %Commercial
# 0        107496  4.494   61168     1.937        21.26
# 1        102012  4.235   56311     3.975        20.22
# 2        118100  4.415   64746     1.557        13.45
# 3        102439  4.526   56074     2.771        20.52
# 4         97648  4.043   47908     4.275        12.93
```

Although Python allows column names with spaces and special characters (such as "%"), renaming them is recommended to avoid errors and improve readability. The variable names can easily be renamed by reassigning `df.columns`:

```
df.columns = ["Price", "Rooms", "Income", "TaxRate",
              "Commercial"]
```

The `plot` function in Ravix can be used to create scatterplots of two variables; however, it can also be used to quickly create a matrix of scatterplots. If the first argument of the `plot` function is a formula, then a matrix of scatterplots is created.

```
from ravix import plot
plot("Price ~ Rooms + Income + TaxRate + Commercial", data = df)
```

The scatterplots in Fig. 5.1 prove particularly interesting with regard to the response variable. Upon a cursory glance, the average price is positively correlated with "Rooms" and "Income" and negatively correlated with "Tax Rate" and "Commercial." Using the `corr` function from Pandas, the correlations can be verified.

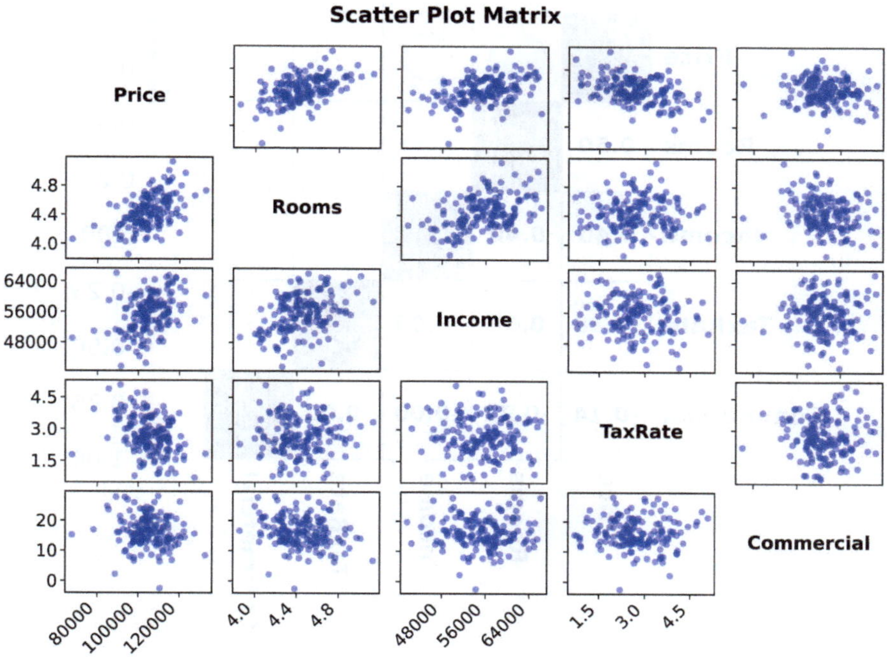

Fig. 5.1 Scatterplot matrix of housing data

```
df.corr()
```

```
#                Price     Rooms    Income   TaxRate  \
# Price       1.000000  0.501434  0.404822 -0.377880
# Rooms       0.501434  1.000000  0.402225  0.021021
# Income      0.404822  0.402225  1.000000 -0.078082
# TaxRate    -0.377880  0.021021 -0.078082  1.000000
# Commercial -0.136474 -0.217199 -0.000559 -0.002307
#
#             Commercial
# Price        -0.136474
# Rooms        -0.217199
# Income       -0.000559
# TaxRate      -0.002307
# Commercial    1.000000
```

While this information verifies our conclusions from the scatterplots, the correlation matrix can be difficult to read and understand. Using the `plot_cor` function from Ravix, the matrix may be color-coded and much more easily read. Simply pass the dataframe of interest into the function:

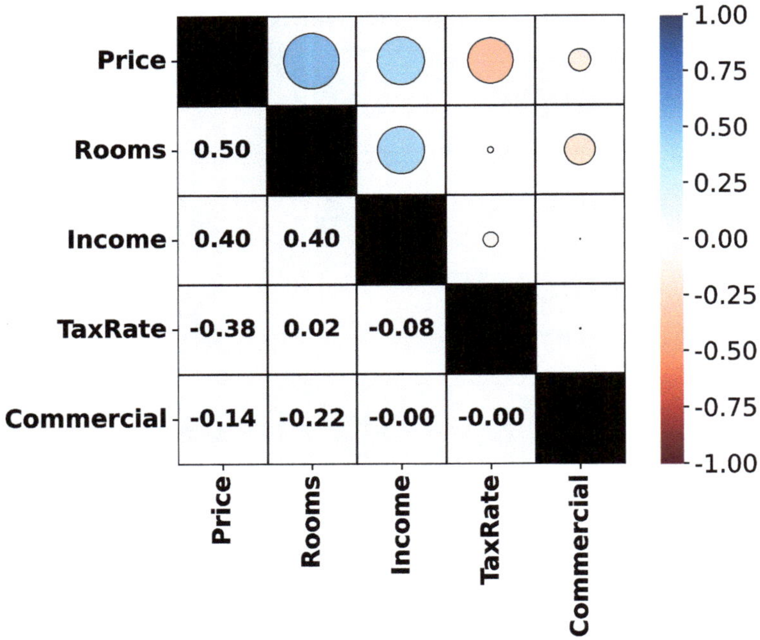

Fig. 5.2 Correlation plot of housing data

```
from ravix import plot_cor
plot_cor(df)
```

Figure 5.2 shows the correlations in an easy to read format. Note that some of the correlations between the response and other variables are relatively weak (particularly "Price" with "Commercial"). A t-test can better accomplish testing for individual significance between a predictor and response variable.

5.14.3 Calculating VIF Values

Correlation coefficients alone do not provide a clear indication of multicollinearity. As discussed earlier in this chapter, VIF values serve as a more reliable measure. Here, we pass the equation and the dataframe to `vif` allowing Ravix to calculate the values. Additionally, we set the `plot` argument to `True` to return a plot.

```
from ravix import vif
vif("Price ~ Rooms + Income + TaxRate + Commercial", data = df,
    plot = True)
```

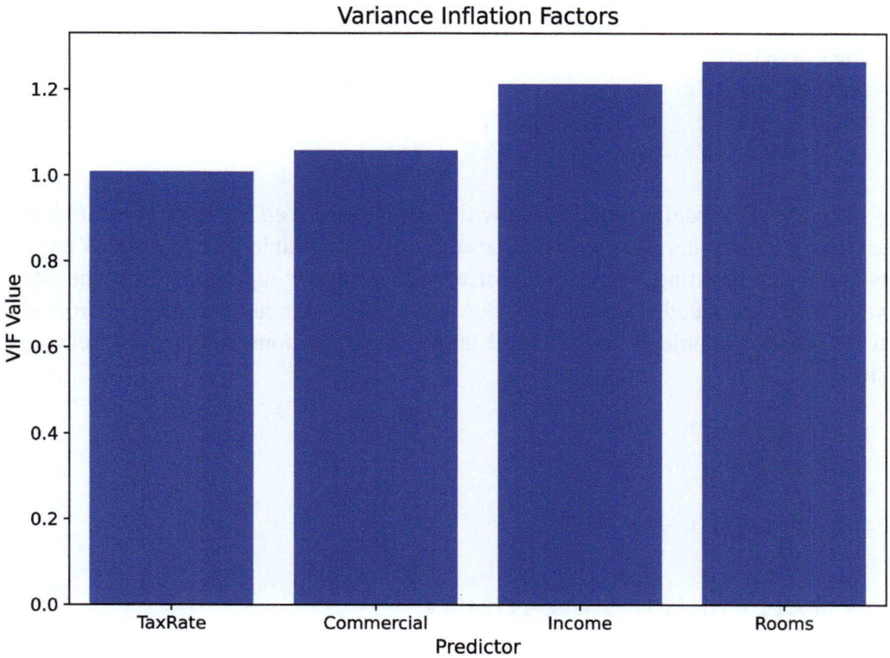

Fig. 5.3 Variance inflation factors barplot

```
#        TaxRate  Commercial     Income      Rooms
# VIF   1.009601    1.059649   1.215393   1.268351
```

Observing the barplot in Fig. 5.3, we note the VIF values are near 1, indicating very low multicollinearity.

5.14.4 Simple Linear Regression Models

If one were only familiar with simple linear regression, a simple linear regression model could be fit for each predictor variable. In particular, a model could be generated and fit for each $i = 1, 2, 3, 4$ where

$$Y = \beta_0 + \beta_i X_i + \varepsilon.$$

Here, Y is the average price (Price), X_1 is the average number of rooms (Rooms), X_2 is the average income (Income), X_3 is the tax rate (Tax Rate), X_4 is the percent of commercial real estate (Commercial), ε is the error, and the unknown parameter values are $\beta_0, \beta_1, \beta_2, \beta_3, \beta_4$. The models are fit one at a time using the `ols` function, as shown as follows:

```
from ravix import ols
reg1 = ols("Price ~ Rooms", data = df)
reg2 = ols("Price ~ Income", data = df)
reg3 = ols("Price ~ TaxRate", data = df)
reg4 = ols("Price ~ Commercial", data = df)
```

While each model is fit individually (as `reg1`, `reg2`, `reg3`, `reg4`), you could also automate this by iterating over the variables using a simple `for` loop. This can be useful if you're fitting many models or want to minimize repetitive code. The slope values are calculated from the previous code chunk and can be accessed from the fitted regression objects (`reg1`, `reg2`, `reg3`, `reg4`) as done in the stock beta case study:

```
reg1.params["Rooms"]
reg2.params["Income"]
reg3.params["TaxRate"]
reg4.params["Commercial"]
```

```
# np.float64(20468.440941184974)

# np.float64(0.757386551008151)

# np.float64(-3878.652249791035)

# np.float64(-248.5122542787285)
```

The `params` attribute of each regression model can be indexed by the corresponding coefficient name using square brackets. However, when using platforms like Google Colab, only the last result of a code block is displayed. To view the output of each regression, run the lines one at a time or use the print function to ensure all output is displayed.

The p-values of the coefficients are available in the `pvalues` attribute. Indexing this attribute by the coefficient name using square brackets yields the corresponding p-value.

```
reg1.pvalues["Rooms"]
reg2.pvalues["Income"]
reg3.pvalues["TaxRate"]
reg4.pvalues["Commercial"]
```

```
# np.float64(6.317987156558687e-11)

# np.float64(2.772950751902651 5e-07)

# np.float64(1.866896600145971 3e-06)

# np.float64(0.09585640150573431)
```

From the output above, all of the slope values are significant in each simple linear regression model at $\alpha = 0.1$. Furthermore, the R^2 values are given using the `rsquared` attribute:

```
reg1.rsquared
reg2.rsquared
reg3.rsquared
reg4.rsquared
```

```
# np.float64(0.2514357787663343)

# np.float64(0.1638811913738396)

# np.float64(0.1427933699948042)

# np.float64(0.01862528217695092)
```

and the R_a^2 values are given using the `rsquared_adj` attribute:

```
reg1.rsquared_adj
reg2.rsquared_adj
reg3.rsquared_adj
reg4.rsquared_adj
```

```
# np.float64(0.2463779124066473)

# np.float64(0.15823173996420337)

# np.float64(0.13700143330557 99)

# np.float64(0.011994371921389746)
```

which are slightly less than the R^2 values, since R_a^2 take into account the low number of observations.

5.14.5 Multiple Regression Model

As discussed previously, the directors of the investment firm would like us to create a model to predict the average price based on all other variables in the data. Therefore, our model will be a multiple regression model with four predictor variables:

$$Y = \beta_0 + \beta_1 X_1 + \beta_2 X_2 + \beta_3 X_3 + \beta_4 X_4 + \varepsilon,$$

where Y, X_1, X_2, X_3, X_4 are the same values from the simple linear regression models. However, the error (ε) and the unknown parameter values $\beta_0, \beta_1, \beta_2, \beta_3, \beta_4$ may potentially be different values.

The model can be fit from the dataframe using the `ols` command from Ravix:

```
from ravix import ols
reg = ols("Price ~ Rooms + Income + TaxRate + Commercial",
          data = df)
reg.summary()
```

```
# Summary of OLS Regression Analysis:
# ======================================================
#
# Coefficients:
# ------------------------------------------------------
#                 Estimate  Std. Error  t-value    p-value
# Intercept        20083.5     12355.8    1.625     0.1062
# Rooms            16937.6      2859.4    5.924   2.19e-08 ***
# Income           0.39098     0.12829    3.048     0.0027 **
# TaxRate          -3801.8      641.47   -5.927   2.16e-08 ***
# Commercial       -85.745      116.59   -0.735     0.4633
#
# Model Statistics:
# ------------------------------------------------------
# Residual Std. Error: 7131.77
# R-squared:       0.4391           AIC: 3092.29
# Adj. R-squared: 0.4236            BIC: 3107.34
# F-statistic: 28.3728 on 4 and 145 DF, p-value: < 2e-16
# ======================================================
```

In the summary, the values of the coefficients prove significantly different from those of the simple linear regression models. In fact, the coefficient for "Commercial" is not significant, even though the coefficient was significant in the simple linear regression case. Comparing the R^2 values from simple linear regression, even the highest value of 0.2514358 is not higher than the R^2 value from the multiple regression model, indicating a better model fit using multiple variables. Earlier in this chapter, the reader was warned about comparing models using R^2 alone. Therefore, we compare R_a^2 and note that, when taking into consideration the number of observations and the number of variables, the variation in Y is better explained using a combination of variables than using a single variable.

The `reg` object was created with a formula that specified four variables. For convenience, the `reg` object could have been equivalently defined using a coding shorthand as shown as follows:

```
reg = ols("Price ~ .", data = df)
```

Note the "." on the right side of the formula. Recall that the right side of the formula is where we denoted our predictor variables previously. Having "." on the right side of the formula tells Python to include all of the variables except the response variable in the dataframe as predictor variables in the model.

The regression equation can therefore be expressed as

$$\hat{Y} = 20083.5355 + 16937.6420X_1 + 0.3910X_2 - 3801.8121X_3 - 85.7450X_4,$$

where the variables are defined above. Additionally, the regression equation can be expressed as

$$\begin{aligned}\text{Price} = {} & 20083.5355 + 16937.6420 \times \text{Rooms} + 0.3910 \times \text{Income} \\ & -3801.8121 \times \text{TaxRate} - 85.7450 \times \text{Commercial}.\end{aligned}$$

5.14.5.1 Model Interpretation

The summary function provides a wealth of information necessary for our analysis. First, the F-value corresponds to a small p-value, indicating that the overall model is statistically significant and merits further investigation. While the multiple R^2 value of 0.4391 shows that approximately 43.91% of the variation in the response variable is explained by the predictors, it does not account for the number of predictors used in the model. In the context of multiple regression, where additional variables can inflate R^2 without truly improving the model, the adjusted R^2 (R_a^2) is a more reliable measure.

In this case, the adjusted R^2 value of 0.4236 should be preferred, as it corrects for the number of predictors and provides a more accurate estimate of the variation explained by the model. This adjusted value suggests that, after accounting for the number of predictors, about 42.36% of the variation in the response variable is explained by the model.

5.14.5.2 Coefficient Interpretation

Each coefficient estimate is tested when fitting a regression model in Python. The coefficients can be interpreted as follows:

- **Intercept**—the intercept coefficient p-value is 0.106, which is larger than 0.05 (p-value $> \alpha$). The conclusion of the hypothesis test results in an insignificant intercept. There is not enough evidence to support that the intercept is nonzero for this model.
- **Rooms**—the coefficient for rooms has a small p-value, much smaller than 0.05 (p-value $< \alpha$). The conclusion holds that the coefficient is not 0. The coefficient can be interpreted as: "For every room, an increase in price of $16,937.64 is expected." This interpretation assumes that all of the other variables remain unchanged or constant.
- **Income**—the coefficient for income has a small p-value, also smaller than 0.05 (p-value $< \alpha$). The conclusion holds that the coefficient is not 0. The coefficient may be interpreted as: "For every dollar of income, an increase in price of $0.391

is expected." Like the "Rooms" coefficient, this interpretation assumes that all of the other variables remain unchanged or constant.

- **Tax Rate**—the coefficient for tax rate has a small p-value, much smaller than 0.05 (p-value $< \alpha$). The conclusion holds that the coefficient is not 0. We can further interpret the coefficient as: "For every percentage increase in the tax rate, a decrease in price of $3801.81 is expected." This interpretation assumes that all of the other variables remain unchanged or constant.
- **Commercial**— coefficient for price has a p-value larger than 0.05 (p-value $> \alpha$). Therefore, the commercial variable is not nonzero for this model, and the coefficient for commercial should not be interpreted.

5.14.5.3 Coefficient Confidence Interval

As another task given by the directors of the investment firm: we must test the claim of a local politician. The politician states that, for every percent decrease in the real estate tax, the property value increases by $5000. Can this claim be refuted?

From the tax rate coefficient of the fitted model, the estimate for the increase in property value for a 1% decrease is $3801.81. While the estimate in Python should be trusted over a random claim, a simple hypothesis can be conducted. We may set up the null and alternative hypotheses as follows:

$$H_0 : \beta_3 = -\$5000$$

$$H_1 : \beta_3 \neq -\$5000.$$

Note that the null hypothesis states that the coefficient is negative, which implies home values decrease in value by $5000 for every 1% increase in tax. This is equivalent to saying that home values increase by $5000 for every 1% decrease in tax.

This test is easy to conclude using the confidence interval for β_3. Getting a default summary in Ravix will calculate 95% confidence intervals for each coefficient, but specifying the `out` argument as `"confint"` or `"ci"` focuses on the coefficients and the coefficient confidence intervals.

```
reg.summary(out = "confint")
```

```
# Confidence Intervals:
# ============================================================
#                 Estimate 95% CI Lower 95% CI Upper
# Intercept        20083.5     -4337.22      44504.3
# Rooms            16937.6      11286.2        22589
# Income          0.390977      0.13742     0.644535
# TaxRate         -3801.81     -5069.66     -2533.97
# Commercial       -85.745     -316.177      144.687
# ============================================================
```

Since −\$5000 falls within the 95% confidence interval, the claim by the politician cannot be rejected at the 5% level of significance.

Was the politician correct in his statement? Our data indicate that the best estimate is \$3801.81, while the politician claims the value as \$5000. A significant difference between the estimate and the claim seems clear. In fact, if we change the level of significance to $\alpha = 0.1$, we attain a different result. We produce the summary again while specifying `alpha = 0.1`, which is equivalent to specifying a confidence level of 0.9 as shown in the Python code:

```
reg.summary(out = "confint", alpha = 0.1)
```

```
# Confidence Intervals:
# ===========================================================
#                 Estimate 90% CI Lower 90% CI Upper
# Intercept      20083.5     -370.647     40537.7
# Rooms          16937.6      12204.2     21671.1
# Income        0.390977     0.178604     0.60335
# TaxRate       -3801.81     -4863.73     -2739.9
# Commercial     -85.745     -278.749     107.259
# ===========================================================
```

The above represents a scenario where an estimate was made possibly with the intent to deceive by positing a claim that the value is close to one end of the confidence interval. However, as demonstrated above, the claim is difficult to disprove.

5.14.6 Case Conclusion

In this analysis, we examined a housing dataset by creating and interpreting a scatterplot matrix and a correlation plot for the relevant variables. We initially fit simple linear regression models, which, while statistically significant, explained only a limited portion of the variation in home prices.

To enhance our predictive power, we specified and fit a multiple regression model that incorporated all the relevant variables. This multiple regression model demonstrated a significant improvement in explaining the variation in home prices compared to the simple models. By interpreting the coefficients of the multiple regression model and conducting additional analysis using confidence intervals, we effectively addressed the politician's claim regarding the impact of tax rates on property values.

This analysis highlights the practical utility of various functions in Python's Ravix package, such as `ols`, `vif`, `plot`, and `plot_cor`, showcasing their effectiveness in multiple regression analysis. These functions were instrumental in fitting the model, summarizing the results, and visualizing the data, making the analysis more comprehensible and robust.

Multiple regression analysis proves to be a powerful tool for understanding complex relationships between variables in real estate and investment contexts. By identifying key drivers behind property prices, businesses and investors can make more informed decisions to optimize their strategies and enhance profitability. This case study demonstrates the value of multiple regression techniques in deriving actionable insights from data, thereby underscoring their importance in real-world applications.

Problems

1. Automotive Tire Sales Simple and Multiple Regression
For this example, we use a subset of the tire sales dataset. The dataset has several variables and is used to calculate the sales of tires at 500 different stores. The variables of the dataset are as follows:

- **Sales**—the number of tires sold at each store (in thousands)
- **CompPrice**—the local competitor's price for the same or similar tire in USD
- **Income**—the average annual income in thousands of USD within a 5 mile radius of the store
- **Ads**—the number of ads from a particular store
- **Cars**—the average number of cars in a household within a 5 mile radius
- **Price**—the sales price of the tire offered in USD
- **SellerType**—one of three possible categories: supercenter, automotive, or dealer
- **District**—one of four possible categories denoting the district of the store: north, south, east, or west
- **Age**—the average age of customers from each store

A subset of the data can be accessed using the Python code as follows:

```
from ravix import get_data
df = get_data("tire_sales.csv")
sub = df[['Sales', 'CompPrice', 'Income', 'Ads', 'Cars',
          'Price', 'Age']]
```

The data subset sub provides the variables required for the questions below. Using the sub dataframe, do the following:

(a) Get a description of the dataframe.
(b) For each variable in the dataframe, fit a simple linear regression model to predict Sales.
(c) In which of these models is there a statistically significant relationship between predictor and response? **Hint: the p-value is available within the params for each model fit.**

(d) Fit a multiple regression model predicting `Sales` as a function of all of the predictors. For which predictors can you reject the null hypothesis $H_0 : \beta_j = 0$?
(e) How do your results from the simple linear regression models compare to those from the multiple regression model?

2. Automotive Tire Sales Correlations
A subset of the data can be accessed using the Python code as follows:

```
from ravix import get_data
df = get_data("tire_sales.csv")
sub = df[['Sales', 'CompPrice', 'Income', 'Ads', 'Cars',
          'Price', 'Age']]
```

The data subset `sub` provides the variables required for the questions below. Using the `sub` dataframe, do the following:

(a) Find a correlation plot between the variables within the data subset (`Sales`, `CompPrice`, `Income`, `Ads`, `Population`, `Price`).
(b) From the correlation plot, which predictor variable is sales most correlated with?
(c) Generate a scatterplot matrix from the variables within the data subset (`Sales`, `CompPrice`, `Income`, `Advertising`, `Population`, `Price`).
(d) From the scatterplot matrix, which predictor variable is sales most correlated with? Compare your answer with your result from part b.

3. Automotive Tire Sales Modeling
Using the data from `tire_sales.csv`, answer the following questions.

(a) Fit a linear model to predict the response variable (`Sales`) from the variables: `CompPrice`, `Ads`, `Cars`, and `Price`.
(b) Display a summary of the model given from part a and clearly state the regression equation.

4. Automotive Tire Sales by Store Type
Using the data from `tire_sales.csv`, answer the following questions:

(a) Construct boxplots of the number of sales by store type.
(b) It is believed that sales are significantly different across the different store types. Can you refute this claim using an F test with a significance level of $\alpha = 0.05$? Clearly state the null and alternative hypotheses.

5. Automotive Tire Sales by District
Using the data from `tire_sales.csv`, answer the following questions:

(a) Construct boxplots of the number of sales by district.
(b) It is believed that sales are significantly different across the different districts. Can you refute this claim using an F test with a significance level of $\alpha = 0.05$? Clearly state the null and alternative hypotheses.

Chapter 6
Estimation Intervals and Analysis of Variance

An approximate answer to the right problem is worth a good deal more than an exact answer to an approximate problem.

—John Tukey

6.1 Introduction

In some cases, a point estimate alone proves insufficient and requires an interval approximation. Hence, we extend our discussion to include confidence intervals for the mean response and prediction intervals. Chapters 2 and 3 introduced the fundamental concepts of sum of squares, along with the components of explained and unexplained variance. Here, we explore these fundamental concepts more deeply while introducing additional analysis techniques.

We begin this chapter with a discussion of confidence intervals of the mean response for simple linear regression and then cover prediction intervals for individual values. These estimation intervals provide different types of estimates, which we in turn explore with a housing price application. We then discuss analysis of variance for regression and expand on the house price application. In the final discussion, we present a case study from human resources and solve it in detail. The case study involves finding both prediction and confidence intervals to find interval estimates for both a mean response and an individual response. This case study makes use of previously learned Python functions with the aforementioned concepts and shows the relevant source code.

6.2 Expected Value

The expected value or expectation refers to the mean value. As the name implies, the expected value provides the long-term value that we would expect in a given scenario. For instance, if one were to flip a fair coin many times, one would expect

D. P. McGibney, *Applied Linear Regression for Business Analytics with Python*, International Series in Operations Research & Management Science 378,
https://doi.org/10.1007/978-3-032-23806-1_6

an equal number of heads and tails. Therefore, the expected number of heads from many flips of a fair coin would be one half of the flips. It follows that the expectation of one flip is 50% heads. Mathematically, if we let X represent the number of heads, then the expected value for the number of heads is written as $E[X]$.

In regression analysis, we use the conditional expectation to find the expected value of Y given X, written as $E[Y|X]$. In fact, making predictions using fitted regression models can be thought of as calculating conditional expectations from known data. Previously, we fit a regression model to predict the house price using size, which is an estimate of the expected value of house price given the size.

The expected value of the simple linear regression model is written as

$$E[Y|X] = \beta_0 + \beta_1 X, \tag{6.1}$$

and the expected value of the multiple regression model, is written as

$$E[Y|X_1, X_2, \ldots, X_p] = \beta_0 + \beta_1 X_1 + \beta_2 X_2 + \ldots + \beta_p X_p. \tag{6.2}$$

Notice that Eqs. (6.1) and (6.2) do not include the error term (ε) since the error has a mean of 0 ($E[\varepsilon] = 0$) as stated in the assumptions of linear regression. Although the true value of $E[Y|X_1, X_2, \ldots, X_p]$ is unknown, it can be estimated by $\hat{Y}$ which represents the predicted value of Y.

6.3 Confidence Interval of a Mean Response

As mentioned in the previous section, the expected value of Y given X is synonymous with the mean of Y given X. Instead of estimating the expected value with a point estimate, which provides a single numerical approximation, we may prefer an interval estimate. A c% confidence interval of a mean response provides an interval estimate of the expected value such that the true mean at a value of X is within the range c% of the time. A confidence interval can be constructed about this mean and, since its estimate is $\hat{Y}$ corresponding to a value of X, the best estimate of Y is $\hat{Y}$. In order to find this interval, we must approximate the standard error.

Assuming the simple linear regression case and using lower case x to denote a particular X value of interest, the standard error of the mean is calculated to be

$$SE_{CI} = s\sqrt{\frac{1}{n} + \frac{(x - \bar{x})^2}{\sum_{i=1}^{n}(x_i - \bar{x})^2}}. \tag{6.3}$$

Using this standard error, a $(1 - \alpha) \times 100\%$ confidence interval for $E[Y|X]$, the mean response of Y for a specified value of X, is given by the interval:

$$\hat{y} \pm t_{\alpha/2} SE_{CI}, \tag{6.4}$$

where $t_{\alpha/2}$ is the critical value with $n - 2$ degrees of freedom for the simple linear regression case. For a more detailed discussion on critical values, see Chap. 4.

6.3.1 *The* `intervals` *Function for Confidence Intervals*

The `intervals` function in Ravix can calculate the confidence interval of a mean response for a regression model as formulated in Formula 6.4. It has two required arguments and the following syntax:

```
from ravix import intervals
intervals(model, newX, interval = "confidence", level = 0.95)
```

The first argument (`model`) is the regression model and the second argument (`newX`) is a dataframe containing the new observations for which you want to calculate the intervals. This dataframe must have the same structure as the data used to fit the model, including all relevant predictor variables. The `interval` argument is optional and defaults to `"confidence"`, which calculates confidence intervals. If set to `"prediction"`, the function will calculate prediction intervals, which provide a range for individual future observations. Further discussion on prediction intervals is given in Sect. 6.5.1. The `level` argument specifies the confidence level, with the default set to 0.95.

In the application that follows, the `intervals` function will be used to calculate a 95% confidence interval for a given set of observations.

6.4 House Prices Application: Confidence Interval

Charles is interested in purchasing a house in Dallas and wants to purchase a house that is roughly 2500 square feet. He collects the size and sale price of 12 houses in a particular Dallas neighborhood. The data shown here reflect the size (in square feet) and the sale price (in thousands of dollars). Construct a 95% confidence interval about the predicted mean house price for all houses with a size of 2500 square feet.

Use the data from Table 6.1 to find the following:

(a) Calculate the mean of X ($\bar{x}$).
(b) Calculate SS_{xx} and SS_{xy}.

Table 6.1 House price data

Variable												
Size (X)	3500	2400	4900	1900	1200	1600	1450	1550	1600	1750	2850	3800
Price (Y)	588	490	675	425	350	412	385	405	420	418	509	550

(c) Find the least squares line.
(d) Calculate the $RMSE$.
(e) Predict the house price of a house with 2500 square feet.
(f) Calculate the critical value $t_{\alpha/2}$ for $\alpha = 0.05$.
(g) Find the standard error (SE_{CI}).
(h) Calculate the upper and lower bounds of the confidence interval.
(i) Provide an interpretation of the confidence interval.

Solution

(a) The mean is easily calculated to be $\bar{x} = 2375$. This can be performed manually or using Python with NumPy as shown as follows:

```
import numpy as np
X = [3500, 2400, 4900, 1900, 1200, 1600, 1450, 1550, 1600, 1750,
    2850, 3800]
np.mean(X)

    # np.float64(2375.0)
```

(b) From formula (2.5) in Chap. 2, we have

$$SS_{xx} = \sum_{i=1}^{n} x_i^2 - \frac{\left(\sum_{i=1}^{n} x_i\right)^2}{n}$$

$$= 82{,}320{,}000 - \frac{(28{,}500)^2}{12}$$

$$= 14{,}632{,}500$$

```
X = np.array(X)
SSxx = np.sum(X**2) - np.sum(X)**2/12
SSxx

    # np.float64(14632500.0)
```

To find the SS_{xy} we use Eq. (3.2):

$$SS_{xy} = \sum_{i=1}^{n} x_i y_i - \frac{\left(\sum_{i=1}^{n} x_i\right)\left(\sum_{i=1}^{n} y_i\right)}{n}$$

$$= 14558350 - \frac{(28500)(5627)}{12}$$

$$= 1{,}194{,}225.$$

```
Y = np.array([588, 490, 675, 425, 350, 412, 385, 405, 420, 418,
              509, 550])
SSxy = np.sum(X*Y) - np.sum(X)*np.sum(Y)/12
SSxy

    # np.float64(1194225.0)
```

(c) Using Eqs. (3.11) and (3.12), we calculate the regression equation. Specifically:

$$\begin{aligned}\hat{\beta}_1 &= \frac{SS_{xy}}{SS_{xx}} \\ &= \frac{1{,}194{,}225}{14{,}632{,}500} \\ &= 0.0816146\end{aligned}$$

$$\begin{aligned}\hat{\beta}_0 &= \bar{y} - \hat{\beta}_1\bar{x} \\ &= 468.9167 - (0.0816146)(2375) \\ &= 275.082.\end{aligned}$$

From the slope and intercept estimates we have:

$$\hat{Y} = 275.082 + 0.0816146X.$$

A Pandas dataframe can be created from the lists `X` and `Y` by importing Pandas, defining the lists, and using the `DataFrame` function from a dictionary.

```
import pandas as pd
X = [3500, 2400, 4900, 1900, 1200, 1600, 1450, 1550, 1600, 1750,
     2850, 3800]
df = pd.DataFrame({"X": X, "Y": Y})
```

We saw in the stock beta case study from Chap. 4 that the coefficients can be calculated from the regression summary. Here, the regression coefficients are within the model object `reg`, and we specify `out = "confint"` to get the confidence intervals of the coefficients.

```
from ravix import ols
reg = ols("Y ~ X", data = df)
reg.summary(out = "confint")
```

```
# Confidence Intervals:
# ============================================================
#                 Estimate 95% CI Lower 95% CI Upper
# Intercept     275.082       247.652       302.513
# X           0.0816146     0.0711416     0.0920875
# ============================================================
```

(d) From formula (4.4), we have

$$
\begin{aligned}
s = RMSE &= \sqrt{\frac{SSE}{n-2}} \\
&= \sqrt{\frac{3232.773}{10}} \\
&= 17.9799
\end{aligned}
$$

```
SSE = np.sum(reg.resid**2)
SSE
RMSE = np.sqrt(SSE/(12 - 2))
RMSE
```

```
# np.float64(3232.772766102852)
```

```
# np.float64(17.979913142456645)
```

(e) Given that $\hat{y}$ is calculated from $x = 2500$ square feet, the regression equation from part c enables us to calculate

$$\hat{y} = 275.08209465 + 0.08161456(2500) = 479.1185.$$

This is done in Python by creating a dataframe consisting of the single predictor value labeled as `X`. Then, this dataframe, `x`, is used as the second argument in the `predict` function.

```
x = pd.DataFrame({"X": [2500]})
from ravix import predict
predict(reg, x)
```

```
# 0    479.118486
# dtype: float64
```

(f) The `ppf` function in SciPy allows us to calculate the critical value ($t_{\alpha/2}$) as discussed in Sect. 4.7. Since the critical value places 2.5% in each tail, we want the t value where 97.5% of the area is to the left. Therefore, we use 0.975 as the first argument of the `ppf` function. The second argument is the degrees of freedom $n - 2$. Using the command line in Python, we have:

```
from scipy.stats import t
t.ppf(0.975,10)
```

```
# np.float64(2.2281388519649385)
```

which can be written as $t_{\alpha/2} = t_{0.025} = 2.228$ with 10 degrees of freedom.

(g) The standard error for $x = 2500$ is calculated using Eq. (6.3):

$$SE_{CI} = s\sqrt{\frac{1}{n} + \frac{(x-\bar{x})^2}{SS_{xx}}}$$

$$= 17.9799\sqrt{\frac{1}{12} + \frac{(2500-2375)^2}{14{,}632{,}500}}$$

$$= 5.223528.$$

Calculating this value in Python is done using the basic operators and the `sqrt` function from NumPy.

```
SE_ci = 17.9799 * np.sqrt(1/12 + (2500-2375)**2/14632500)
SE_ci
```

```
# np.float64(5.223498620712397)
```

(h) Using the standard error, the confidence interval is

$$\hat{y} \pm t_{\alpha/2} SE_{CI},$$

which gives a lower bound of

$$479.1185 - 2.228139(5.223499) = 467.4798,$$

and an upper bound of

$$479.1185 + 2.228139(5.223499) = 490.7572.$$

We can find these bounds using Ravix by using the `intervals` function and specifying the dataframe created from part e and the `interval` argument specified as `"confidence"`.

```
from ravix import intervals
intervals(reg, x, interval = "confidence")
```

```
#    Prediction  Lower Bound  Upper Bound
# 0  479.118486   467.479798   490.757175
```

This output matches the lower and upper bounds of our calculation specified above. These values are 467.4798 and 490.7572, respectively.

(i) We are 95% confident that the price of a randomly selected house with a size of 2500 square feet will fall between \$467,480 and \$490,757.

6.5 Prediction Intervals

In many cases, a prediction interval for a particular value is more useful. Prediction intervals are calculated when we wish to know the range under which $(1-\alpha) \times 100\%$ of the predictions lie. Unlike confidence intervals for the mean, prediction intervals account for individual variability and apply to a single observation.

Note the subtle difference between the standard error of the confidence interval in Eq. (6.3) and the standard error of the prediction interval:

$$SE_{PI} = s\sqrt{1 + \frac{1}{n} + \frac{(x - \bar{x})^2}{\sum_{i=1}^{n}(x_i - \bar{x})^2}}. \tag{6.5}$$

Similar to that of the confidence interval, a $(1 - \alpha) \times 100\%$ prediction interval for an individual response about $\hat{y}$, for a specified value of $X = x$, is given by

$$\hat{y} \pm t_{\alpha/2} SE_{PI} \tag{6.6}$$

where x is the given value of the predictor variable, n is the number of observations, and $t_{\alpha/2}$ is the critical value with $n - 2$ degrees of freedom.

Confidence intervals describe uncertainty about a mean response and therefore have a smaller standard error. Prediction intervals, like confidence intervals, are centered at the predicted value $\hat{y}$, but they widen to account for both estimation uncertainty and the additional variability of individual outcomes.

6.5.1 *The* `intervals` *Function for Prediction Intervals*

The `intervals` function can also calculate prediction intervals that provide a range for individual future observations of y given new data. Unlike confidence intervals, which estimate the mean response, prediction intervals account for both the uncertainty in the model and the variability in individual outcomes. This function has the following syntax:

```
from ravix import intervals
intervals(model, newX, interval = "prediction", alpha = 0.05)
```

Similar to how we calculate confidence intervals with the intervals function, the first argument (`model`) is the fitted regression model, and the second argument (`newX`) is a dataframe containing the new observations for which the prediction intervals are to be computed. If you wish to calculate a prediction interval, the `interval` argument must be specified as "`prediction`"; otherwise, the function will default to calculating a confidence interval. The default level of significance is 0.05.

In the application below, we will compare the results of prediction intervals and confidence intervals, highlighting the differences in their interpretation and use.

6.6 House Price Application: Prediction Interval

Construct a 95% prediction interval about the predicted house price for all houses with a size of 2500 square feet.

Use the data from Table 6.1 to find the following:

(a) Calculate the upper and lower bounds of the prediction interval.
(b) Provide an interpretation of the prediction interval.
(c) Explicitly state the difference between the interpretation of the prediction interval and the confidence interval.
(d) Indicate which interval has a larger range.

Solution

(a) Recall from the previous application that $SS_{xx} = 14632500, s = 17.9799, t_{0.025} = 2.228139$, and $\hat{y} = 479.1185$. Using the aforementioned calculations, we find the standard error of the prediction interval at $x = 2500$:

$$
\begin{aligned}
SE_{PI} &= s\sqrt{1 + \frac{1}{n} + \frac{(x - \bar{x})^2}{SS_{xx}}} \\
&= 17.9799\sqrt{1 + \frac{1}{12} + \frac{(2500 - 2375)^2}{14{,}632{,}500}} \\
&= 18.72329.
\end{aligned}
$$

In Python, the calculation can also be easily found doing the following:

```
import numpy as np
SE_pi = 17.9799*np.sqrt(1 + 1/12 + (2500-2375)**2/14632500)
SE_pi

# np.float64(18.723294097209077)
```

Using the standard error, the prediction interval is

$$\hat{y} \pm t_{\alpha/2} SE_{PI},$$

which gives a lower bound of

$$479.1185 - 2.228139(18.72329) = 437.4004,$$

and an upper bound of

$$479.1185 + 2.228139(18.72329) = 520.8366.$$

The dataframe is constructed using Pandas:

```
import pandas as pd
x = pd.DataFrame({"X": [2500]})
```

We can find these bounds using Python by using the `intervals` function and specifying the dataframe `x` and the `interval` argument specified as a `"prediction"` interval.

```
from ravix import intervals
intervals(reg, x, interval = "prediction")
```

```
#     Prediction  Lower Bound  Upper Bound
# 0   479.118486   437.400357   520.836616
```

This output matches the lower and upper bounds of our calculation specified above. These values are 437.4004 and 520.8366, respectively.

(b) The interpretation is such that we are 95% confident that house price for a random house with a size of 2500 square feet is between \$437,400 and \$520,837.
(c) The result from this application represents the interval of a particular house price, whereas the previous application represents that of a mean house price.
(d) The prediction interval is a larger range interval than the confidence interval.

6.7 Confidence Intervals Versus Prediction Intervals

In the previous applications, we calculated both confidence and prediction intervals. These intervals are often used in regression analysis to quantify uncertainty in predictions, but they serve different purposes. A confidence interval provides a

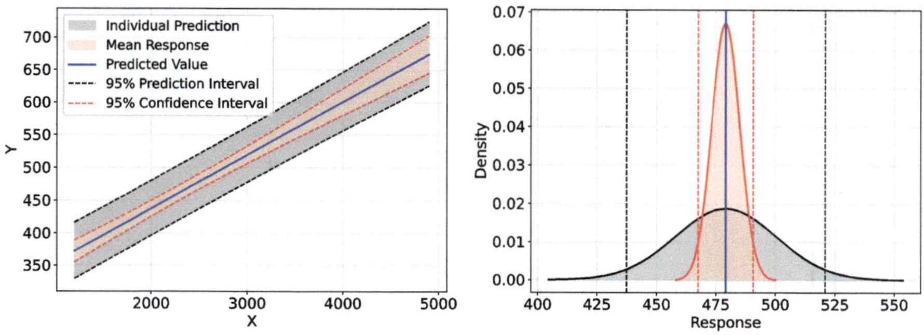

Fig. 6.1 Confidence and prediction interval comparison

range for the mean response, $E[y|x]$, while a prediction interval gives a range for individual future values of y. Understanding the difference between the two is crucial for correctly interpreting model results.

In both cases, the predicted value of y for a given x is $\hat{y}$, which serves as the point estimate around which both intervals are centered. However, because the mean response has less variability than individual outcomes, the standard error of $E[Y|X]$ is smaller than the standard error of y. Prediction intervals account for the additional variability in individual outcomes and are therefore always wider than confidence intervals.

Figure 6.1 illustrates these differences. The left-side graph shows a regression line in blue with the 95% confidence interval as dashed red lines. The wider dashed black lines represent the 95% prediction interval, reflecting the greater uncertainty when predicting individual outcomes. On the right side, the confidence and prediction distributions at $x = 2500$ are compared. Both distributions are centered around the predicted value, $\hat{y} = 479$, but the prediction interval is visibly wider, as seen by both the spread of the distribution and the dashed lines.

6.8 Analysis of Variance

Analysis of Variance (ANOVA), developed by statistician and evolutionary biologist Ronald Fisher, is a collection of statistical models used to analyze the differences among group means and their associated procedures (such as "variation" among and between groups).

By analyzing the variation, the least squares regression equation becomes a natural result. The components of the variation in regression modeling are represented by the sum of squares of the total, regression, and error: SST, SSR, and SSE, respectively. Recall from Chap. 3 the relationship between the sum of squares in Eq. (3.10):

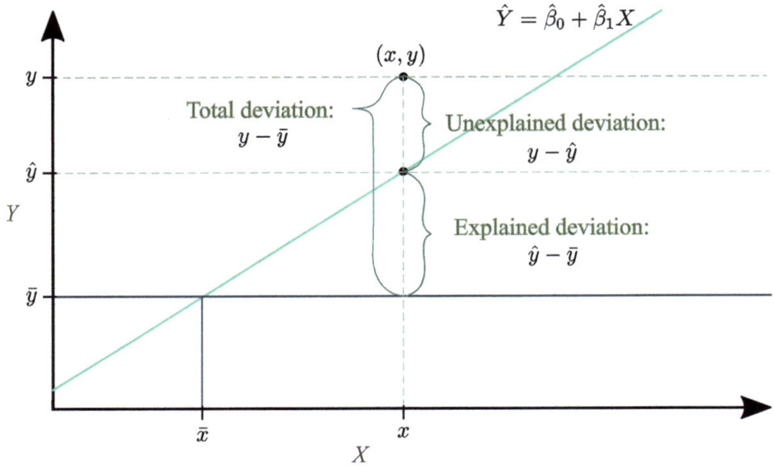

Fig. 6.2 Explained vs. Unexplained deviation

$$SST = SSR + SSE.$$

This equation holds not only for the simple linear regression case but also for the multiple regression case. Figure 6.2 provides an intuitive depiction between the total, unexplained, and explained deviations. These deviations represent the components of the corresponding sum of squares values.

6.8.1 *Mean of Squares Due to Regression*

While the SSR is variation, it is preferred that this variation be as large as possible since SSR represents the amount of variation "explained" by the regression model. It follows that the larger values of the Mean of Squares due to Regression (MSR) indicate large amounts of explained variation. The MSR represents the mean variation explained by each predictor. From the SSR and the degrees of freedom, the Mean of Squares due to Regression (MSR) can be calculated as

$$MSR = SSR/p. \tag{6.7}$$

The degrees of freedom for the SSR correspond to the number of predictors (p) in the model. For the simple linear regression model, there is one degree of freedom since we have only one predictor variable (X). In the case of a model with $p = 10$ predictors, there are 10 degrees of freedom. The MSR, the SSR, and corresponding degrees of freedom are shown in the ANOVA table (Table 6.2).

Table 6.2 The ANOVA table

Variation source	DF	SS	MS	F	p-value
Regression	p	SSR	$MSR = \frac{SSR}{p}$	$F = \frac{MSR}{MSE}$	p-value (F)
Error	$n - p - 1$	SSE	$MSE = \frac{SSE}{n-p-1}$		
Total	$n - 1$	SST			

6.8.2 *Mean Squared Error*

The degrees of freedom for the SSE correspond to the number of observations (n) minus the predictors (p) minus 1 or

$$n - p - 1.$$

In the simple linear regression case, there are $n - 2$ degrees of freedom. For the model with 20 observations and 7 predictors, the degrees of freedom of the SSE would be $20 - 7 - 1 = 12$.

The previous chapters were devoted to simple linear regression and resulted in $n - 2$ used as the degrees of freedom, with the MSE found by dividing by $n - 2$. Generally, the MSE can be found by

$$MSE = \frac{SSE}{n - p - 1}. \tag{6.8}$$

Both SSE and MSE are considered measures of error; therefore, we desire them to be as small as possible. The MSE, SSE, and corresponding degrees of freedom are shown in the ANOVA table (Table 6.2).

6.8.3 *The F Statistic*

Taking the ratio of the MSR to MSE results in the F statistic:

$$F = \frac{MSR}{MSE}. \tag{6.9}$$

Since the degrees of freedom of the MSR is p and the degrees of freedom of the MSE is $n - p - 1$, the F statistic has p degrees of freedom in its numerator and $n - p - 1$ degrees of freedom in its denominator. A larger F statistic generally corresponds to lower p-values, which indicates that the model is statistically significant.

Table 6.3 The ANOVA table for simple linear regression

Variation source	DF	SS	MS	F	p-value
Regression	1	SSR	$MSR = \frac{SSR}{1}$	$F = \frac{MSR}{MSE}$	p-value (F)
Error	$n-2$	SSE	$MSE = \frac{SSE}{n-2}$		
Total	$n-1$	SST			

6.9 ANOVA Table

From the ANOVA components in the previous section, a table can be created to summarize the F statistic. Specifically, the MSR, MSE, and their underlying components can be shown in an ANOVA table (Table 6.2).

Considering the simple linear regression case, the ANOVA table can be simplified such that $p = 1$. Filling in $p = 1$ for Table 6.2 results in Table 6.3. In this table,

$$\begin{aligned} MSR &= \frac{SSR}{1} \\ &= SSR \end{aligned}$$

and

$$\begin{aligned} MSE &= \frac{SSE}{n-p-1} \\ &= \frac{SSE}{n-2} \end{aligned}$$

as discussed in Chap. 4.

Using the `summary` method from `reg` and the `out` argument set as "`anova`", the components of ANOVA are easily returned from the fitted regression model.

```
reg.summary(out = "anova")
```

We demonstrate the utility of this function in the following application.

6.10 House Price Application: ANOVA Table

Calculate the values in the ANOVA table for the data from Table 6.1. In particular, do the following to construct the ANOVA table.

(a) Calculate the SSR.
(b) Calculate the SST.

(c) Calculate the MSR.
(d) Calculate the F statistic.
(e) Use Python to find the p-value of the F statistic.
(f) Use the calculated information from the previous parts to construct the ANOVA table for the data.
(g) Verify your table using the `summary` function from Ravix.

Solution

(a) Using Eq. (3.16), the SSR is

$$SSR = \sum_{i=1}^{n} (\hat{y}_i - \bar{y})^2$$
$$= 97{,}466.14$$

While we could easily find the $\hat{y}$ values using some simple calculation, these values are given to us within the regression object `reg` accessed as `reg.fittedvalues`. Using the vector already calculated for us, we use the `sum` and `mean` functions and pay careful attention to the orders of operation by adding in extra parentheses.

```
import numpy as np
SSR = np.sum((reg.fittedvalues - np.mean(Y))**2)
SSR
```

```
# np.float64(97466.14390056366)
```

(b) Equation (3.15) is

$$SST = \sum_{i=1}^{n} (y_i - \bar{y})^2$$
$$= 100{,}698.9$$

Using the `sum` and `mean` functions from NumPy, the calculation is done using the formula here.

```
SST = np.sum((Y - np.mean(Y))**2)
SST
```

```
# np.float64(100698.91666666666)
```

Equivalently, the SST can be found with Eq. (3.18):

$$SST = SSR + SSE,$$

which yields the same result (using $SSE = 3232.773$).

Table 6.4 The ANOVA table for the house price data

Variation source	DF	SS	MS	F	p-value
Regression	1	97466	97466	301.49	8.506e-09
Error	10	3232.8	323.28		
Total	11	100698.9			

(c) For the MSR, we have $MSR = SSR/1 = 97466.14$, since $p = 1$.
(d) Combining Eqs. (6.9) and (6.8), the F statistic is

$$F = \frac{MSR}{SSE/(n-2)} = \frac{97466.14}{3232.773/10} = 301.4939$$

(e) The cdf function from SciPy allows us to calculate the p-value of the F statistic as discussed in Sect. 4.16. Using the function with $F = 301.4939$ and degrees of freedom in the numerator $\upsilon_1 = 1$ and $\upsilon_2 = 10$ in the denominator, we obtain:

```
from scipy.stats import t
1 - t.cdf(301.4939, 1, 10)

    # np.float64(0.0010919907327687772)
```

(f) Filling in the calculated values, we obtain Table 6.4.
(g) First, we construct the dataframe from the lists using Pandas.

```
import pandas as pd
X = [3500, 2400, 4900, 1900, 1200, 1600, 1450, 1550, 1600, 1750,
     2850, 3800]
Y = [588, 490, 675, 425, 350, 412, 385, 405, 420, 418, 509, 550]
df = pd.DataFrame({"X": X, "Y": Y})
```

Once the dataframe (df) is available, we use ols from Ravix to fit the model and then use the summary method.

```
from ravix import ols
reg = ols("Y ~ X", data = df)
reg.summary(out = "anova")

 #               df      Sum Sq Mean Sq        F   p-value
 # Regression     1       97466   97466  301.494  8.51e-09
 # Residual      10      3232.8  323.28
 # Total         11   1.007e+05
```

While the labels differ slightly, the information in the Python table is consistent with the table created in part f. We see a slight rounding error in the p-value, but this error is negligible and doesn't affect the test results with $\alpha = 0.05$.

6.11 Generalized *F* Statistic

Arguably, the most important F-test for multiple linear regression consists of the test for overall significance. However, this test can be modified from the hypothesis tests in Chap. 4 to assess differences between models. Recall from Chap. 4 the null and alternative hypotheses for an F-test:

$$H_0 : \beta_1 = \beta_2 = \ldots = \beta_p = 0$$

$$H_1 : \text{At least one coefficient is nonzero.}$$

The null and alternative hypotheses can be thought of in terms of the models that result. For example, the null hypothesis with $\beta_1 = \beta_2 = \ldots = \beta_p = 0$ would give us the following reduced model:

$$H_0 : Y = \beta_0 + \varepsilon.$$

The alternative hypothesis would result in the full model:

$$H_1 : Y = \beta_0 + \beta_1 X_1 + \beta_2 X_2 + \cdots + \beta_p X_p + \varepsilon.$$

The F statistic was previously calculated to be

$$F = \frac{SSR/p}{MSE}.$$

Recall that the SSR is the difference between the sum of squares total (SST) and the sum of squared error in the full model (SSE). The F statistic can therefore be written as

$$F = \frac{(SST - SSE)/p}{MSE}. \tag{6.10}$$

The test statistic, as calculated in Eq. (6.10), represents the difference between the SST and the SSE. The SST is the amount of error that would result from assuming the null hypothesis that $\beta_1 = \beta_2 = \ldots = \beta_p = 0$, whereas the SSE is the error that results from using the alternative hypothesis.

Now consider the reduced model (null hypothesis model) to contain one less predictor variable than the full model (alternative hypothesis model). If the full model contains p predictors, then the statistical test to compare the models with and without X_p would be:

$$H_0 : Y = \beta_0 + \beta_1 X_1 + \beta_2 X_2 + \cdots + \beta_{p-1} X_{p-1} + \varepsilon$$

$$H_1 : Y = \beta_0 + \beta_1 X_1 + \beta_2 X_2 + \cdots + \beta_p X_p + \varepsilon.$$

The numerator of the F statistic, for the test mentioned here, is the difference between the sum of squared error in the reduced model and the sum of the squared error of the full model divided by the number of predictors. Here, we denote the sum of the squared error in the reduced model to be $SSE(X_1, X_2, \ldots, X_{p-1})$. Since the difference in the number of predictors is equal to one, the degrees of freedom in the numerator are equal to one, resulting in an F statistic of

$$F = \frac{SSE(X_1, X_2, \ldots, X_{p-1}) - SSE}{MSE}. \tag{6.11}$$

In general, it is possible to generalize the F statistic to compare models with any number of predictor variables as long as the reduced model is a subset of the full model. For simplicity, the Eq. (6.11) only considers the comparison of models with a difference of one predictor variable.

6.12 Case Study: Employee Retention Modeling

6.12.1 Problem Statement

Managers at modern companies work hard to ensure that their employees are satisfied with their current jobs. Several corporate problems stem from employees' lack of satisfaction in the workplace, including costly restaffing fees or a lacking end product due to poor quality of work or productivity. Companies hence investigate when a disproportionate number of employees churn from their current roles.

Angelo, a hiring manager at Royalty Cruise Lines, carries out the task of predicting future employee retention in new applicants. He collects data from several current employees and, based on the data, creates a regression model. In particular, he prepares to predict the number of years an employee will stay with Royalty using the number of years that they worked at a previous job, education level, starting salary, and a dummy variable (a more thorough coverage of dummy variables is provided in Chap. 7) that states if they would be hired in branch A of the company. Angelo has a particular interest in a few employees who will be working with him on an important project.

Angelo has to decide among seven job candidates. If he would like the employee with the longest retention time, whom should he choose? For this role, Angelo would prefer that the candidate remains at his job for at least five years. Calculate a 95% prediction interval for this candidate's retention.

Angelo also faces the challenge of whether to hire applicants from competing staffing and recruiting companies Gravity AP or Ascend. The applicants from these companies generally have different characteristics due to differing recruiting strategies.

6.12.2 Data Description

The data consists of 100 former employees followed throughout their careers at Royalty. The variables for each employee are as follows:

- **Retention**—The number of years that the employee worked at Royalty.
- **BranchA**—Denotes whether employee worked in branch A during their work career (1 if they worked in branch A; 0 otherwise).
- **Education**—The employee's highest education level (in years).
- **Experience**—The number of years of previous experience before joining Royalty.

Here, we load the file and use the `head` and `summary` functions to obtain an overview of the data.

```
from ravix import get_data
df = get_data("hr_retention.csv")
df.describe()
```

```
#          Retention     BranchA   Education  Experience
# count  100.000000  100.000000  100.000000  100.000000
# mean     6.401660    0.500000   16.050000    9.863345
# std      4.700394    0.502519    1.855241    2.896466
# min      0.195685    0.000000   12.000000    4.959452
# 25%      2.384130    0.000000   15.000000    7.399694
# 50%      5.560948    0.500000   16.000000    9.388705
# 75%      9.838773    1.000000   17.000000   11.760050
# max     21.377025    1.000000   21.000000   17.975974
```

The employees that Angelo recently hired have the characteristics given in Table 6.5.

The dataset can be arranged into a dataframe by creating lists for each variable using the square brackets and then using the `DataFrame` function from Pandas.

Table 6.5 Recent hires data

BranchA	Experience	Education
1	0	12
0	10	16
0	0	16
1	10	16
0	15	18
1	10	16
1	10	18

```
import pandas as pd
BranchA = [1, 0, 0, 1, 0, 1, 1]
Exp = [0, 10, 0, 10, 15, 10, 10]
Edu = [12, 16, 16, 16, 18, 16, 18]
df_new = pd.DataFrame({"BranchA": BranchA, "Experience": Exp,
                       "Education": Edu})
```

Running the above code provides the information from the new hires into the dataframe `df_new`.

6.12.3 *Multiple Regression Model*

As discussed previously, Angelo's firm would like us to predict the average retention based on all other variables in the data. Therefore, our model will be a multiple regression model:

$$Y = \beta_0 + \beta_1 X_1 + \beta_2 X_2 + \beta_3 X_3 + \varepsilon,$$

where Y, X_1, X_2, and X_3 variables are Retention, BranchA, Education, and Experience, respectively. The remaining component in the model, ε, refers to the error.

The regression model fit is given as follows:

```
from ravix import ols
reg = ols("Retention ~ BranchA + Education + Experience",
          data = df)
reg.summary()
```

```
# Summary of OLS Regression Analysis:
# ==========================================================
#
# Coefficients:
# ----------------------------------------------------------
#                  Estimate  Std. Error  t-value    p-value
# Intercept         -7.9360      1.8278   -4.342   3.50e-05 ***
# BranchA            1.4332     0.40929    3.502     0.0007 ***
# Education      -0.017395     0.11021   -0.158     0.8749
# Experience         1.4093    0.071649   19.669    < 2e-16 ***
#
# Model Statistics:
# ----------------------------------------------------------
# Residual Std. Error: 2.0143
# R-squared:        0.8219          AIC: 427.7586
# Adj. R-squared: 0.8164            BIC: 438.1792
# F-statistic: 147.6965 on 3 and 96 DF, p-value: < 2e-16
# ==========================================================
```

The regression equation, as determined by the summary above, is given by:

$$\hat{Y} = -7.93599 + 1.43316X_1 - 0.01740X_2 + 1.40928X_3.$$

Substituting in for the full variable names, we have:

$$Retention = -7.93599 + 1.43316 \times BranchA - 0.01740 \times Education$$
$$+1.40928 \times Experience.$$

6.12.3.1 Model Interpretation

Since the F statistic has a very low p-value in the summary, the model proves statistically significant. The high values of the R^2 and R_a^2 indicate the model is fit well. The most significant predictor variable based on the p-values consists of `Experience`, followed by `BranchA`. The `Education` variable does not appear to be significant since its p-value is well above 0.05.

6.12.3.2 Coefficient Interpretation

Fitting a regression model tests each coefficient estimate against a null value of zero. The coefficients can be interpreted as follows:

- **Intercept**—The intercept represents the retention when all the variables are 0. The p-value is significant since it falls below 0.05. In this case, the retention is below 0, which does not make sense and therefore should not be interpreted. To circumvent this issue, refer to the discussion in Sect. 7.11.

- **BranchA**—Since the p-value indicates significance, the interpretation is:
  The retention will increase by 1.433 if they are to work in Branch A, holding experience and education constant.
- **Education**—Since the p-value indicates the variable is insignificant, the coefficient for education should not be interpreted.
- **Experience**—The p-value indicates this variable is significant. The interpretation is:
  For every 1 year of previous experience, the retention is expected to increase by 1.409 years, holding Branch A and education constant.

6.12.3.3 ANOVA Table

Within the ANOVA table, we should note several numerical computations which are the fundamental building blocks of the linear regression equation. We easily find these values with the `summary` function in Ravix.

```
reg.summary(out = "anova1")
```

```
#              df  Sum Sq Mean Sq          F   p-value
# BranchA       1  200.54  200.54    49.4271  3.00e-10
# Education     1  27.534  27.534     6.7862  0.010647
# Experience    1  1569.7  1569.7   386.8763   < 2e-16
# Residual     96  389.51  4.0574
```

Note that the order of the variables matters. The F statistic for `BranchA` is calculated using the sum of squares of 200.54. This sum of squares represents the variation explained by using `BranchA` as a predictor. The sum of squares of `Education` represents the additional variation explained by including `Education` after `BranchA`.

The structure of the partial F statistic is

$$F = \frac{SSE(X_1, X_2) - SSE}{MSE}.$$

The $SSE(X_1, X_2) = 1569.69 + 389.51$ represents the sum of squared error since 1569.69 is the variation explained through including experience in the model and 389.51 is the SSE that would still be included. The F statistic for a model without experience would be:

```
F = (1569.69)/(389.51/(100 - 4))
F
```

```
# 386.8712998382583
```

Notice that this F value corresponds to the F value for experience in the Python output above.

From the output above, we can calculate the SST by summing up all of the sum of squares values. Specifically, we have:

```
SST = 200.54 + 27.53 + 1569.69 + 389.51
SST
```

```
# 2187.27
```

One should further note that the value of R^2 can be calculated using these sum of squares values:

$$R^2 = \frac{SSR}{SST} = 1 - \frac{SSE}{SST}.$$

This calculation can easily be done in Python:

```
Rsquare = 1 - 389.51/SST
Rsquare
```

```
# 0.8219195618282151
```

Furthermore, R_a^2 can be calculated as

$$R_a^2 = 1 - (1 - R^2)\frac{n-1}{n-p-1}$$

which is given by

```
AdjRsquare = 1 - (1 - Rsquare)*(100 - 1)/(100 - 3 - 1)
AdjRsquare
```

```
# 0.8163545481353468
```

The values of R^2 and R_a^2 can be verified in the regression summary output.

6.12.4 Predictions

Predict the retention for an employee who will be assigned to branch A with no work experience and 12 years of education:

$$\begin{aligned}\text{Retention} &= -7.93599 + 1.43316 \times 1 - 0.01740 \times 12 + 1.40928 \times 0 \\ &= -6.7116.\end{aligned}$$

Making predictions with Ravix is straightforward using the `predict` command. The first argument is the regression object, and the second argument is a dataframe. Note, the variable names in the dataframe must reflect those of the predictor variables used within the `ols` function.

```
df_temp = pd.DataFrame({"BranchA": [1], "Experience": [0],
                        "Education": [12]})
from ravix import predict
predict(reg, df_temp)
```

```
# 0   -6.711566
# dtype: float64
```

Since negative retention is not feasible, this estimate does not make sense. Based on the model above, we would predict the retention to be 0. Further note that this prediction represents an extrapolation since our original dataset has minimum years of experience of almost 5 years.

The dataframe of new hires defined previously, `df_new`, can be predicted in its entirety.

```
predict(reg, df_new)
```

```
# 0    -6.711566
# 1     5.878533
# 2    -8.214309
# 3     7.311696
# 4    12.890165
# 5     7.311696
# 6     7.276906
# dtype: float64
```

Based on this model, Applicant 5, who has 15 years of experience, is predicted to have the highest retention in the group.

6.12.4.1 Prediction Interval

The `intervals` function can also be extended to compute the limits of the prediction interval by specifying the `interval` option. Calculate 95 and 99% prediction intervals for applicant 5. The 95% prediction interval serves as the default, given by the `predict` function:

```
applicant5 = df_new.iloc[[4], :]

from ravix import intervals
intervals(reg, applicant5, interval = "prediction")
```

```
#     Prediction  Lower Bound  Upper Bound
# 4    17.170202     12.93342    21.406985
```

The `intervals` function may be easily modified to return a prediction interval with a different α using the `alpha` option within the `intervals` function. A 99% prediction interval for Applicant 5 can be found through

```
intervals(reg, applicant5, interval = "prediction", alpha = 0.01)
```

```
#     Prediction  Lower Bound  Upper Bound
# 4    17.170202    11.560923    22.779482
```

Observing a 99% prediction interval for Applicant 5, we are very confident that he or she will remain at Royalty for more than 5 years as desired.

6.12.4.2 Confidence Intervals

Ascend mentions that their typical applicants have the properties given in Table 6.6, whereas applicants from Gravity AP have the properties given in Table 6.7.

Find 95% confidence intervals for each set of applicants. Creating a dataframe from Table 6.6 and plugging the resulting `df_ascend` dataframe into the intervals function provides a necessary input for the `intervals` function.

```
df_ascend = pd.DataFrame({'BranchA': [1], 'Experience': [10],
                          'Education': [20]})

intervals(reg, df_ascend, interval = "confidence")
```

```
#     Prediction  Lower Bound  Upper Bound
# 0    21.508909     19.40258    23.615237
```

Table 6.6 Ascend data

BranchA	Experience	Education
1	10	20

Table 6.7 Gravity AP data

BranchA	Experience	Education
1	7	16

Similarly, a dataframe for Gravity AP candidates is created and used as input to find the corresponding confidence intervals.

```
df_gravity = pd.DataFrame({'BranchA': [1], 'Experience': [7],
                           'Education': [16]})

intervals(reg, df_gravity, interval = "confidence")
```

```
#    Prediction  Lower Bound  Upper Bound
# 0   15.923957    13.622335    18.225579
```

The observed confidence intervals and predictions indicate that hiring Ascend applicants would result in higher retention values.

6.12.5 Case Conclusion

This case study applied multiple regression analysis using the Ravix package in Python to predict employee retention. We explored the relationships between retention and variables such as previous experience, education level, and the employment branch. The analysis demonstrated how to interpret model coefficients, calculate prediction and confidence intervals, and utilize the ANOVA table to assess model significance.

This analysis underscores the versatility of Python's Ravix package, particularly through the use of the functions `ols`, `predict`, `intervals`, and `get_data`. These tools were essential in building the regression model, providing detailed summaries, generating predictions, calculating intervals, and efficiently handling data. By leveraging these functions, the analysis became more intuitive, allowing for clear insights and robust decision-making.

Through this analysis, we identified the candidate with the highest predicted retention and calculated a prediction interval confirming that their expected retention exceeds the desired threshold with high confidence. Additionally, we compared potential hires from two staffing companies, Ascend and Gravity AP, finding that Ascend candidates are likely to have higher retention rates based on their calculated confidence intervals.

This study underscores the practical applications of regression analysis in human resources, particularly in making informed decisions on employee selection based on statistical evidence. While the analysis utilized a limited set of Python commands, it built on concepts from previous chapters, reinforcing the importance of understanding and correctly applying statistical methods in predictive modeling.

Table 6.8 Coral gables house price data

Size	Price
2700	650
3500	875
2600	670
5300	1200
4200	981

Problems

Coral Gables Housing Confidence Interval

Jigar is interested in purchasing a house in Coral Gables and wants to get the largest total living area for the price. He collects the size and sale price of five houses in a particular Coral Gables neighborhood. The data shown in Table 6.8 reflect the size (in square feet) and the sale price (in thousands of USD).

Without the use of a computer:

(a) Find the regression equation that predicts price as a function of the house size.
(b) Calculate the 95% critical value $t_{\alpha/2}$ for $\alpha = 0.05$ (use Python here).
(c) Use $\bar{x} = 3660$, $SS_{xx} = 5052000$, and $RMSE = 26.71$ to find the standard error for a 95% confidence interval (SE_{CI}) for a house price with a size of 2500 square feet.
(d) Calculate the upper and lower bounds of the 95% confidence interval for a house price with a size of 2500 square feet.
(e) Provide an interpretation of the confidence interval.

Coral Gables Housing Prediction Interval

Use the Coral Gables housing data from Table 6.8 and do the following without the use of a computer.

(a) Use $\bar{x} = 3660$, $SS_{xx} = 5052000$, and $RMSE = 26.71$ to find the standard error for a 95% prediction interval (SE_{PI}) for a house price with a size of 2500 square feet.
(b) Calculate the upper and lower bounds of the prediction interval for a house price with a size of 2500 square feet.
(c) Provide an interpretation of the prediction interval.
(d) Explicitly state the difference between the interpretation of the prediction interval and the confidence interval.
(e) Indicate which interval has a larger range.

Coral Gables Housing Intervals in Python

Use the Coral Gables housing data from Table 6.8 and do the following using Python.

(a) Print a summary of the regression equation that predicts price as a function of house size.
(b) Use the regression equation to predict the price when the size of the house is 2500 square feet.

(c) Find a 95% confidence interval for the price when the size of the house is 2500 square feet.
(d) Find a 95% prediction interval for the price when the size of the house is 2500 square feet.

Coral Gables Housing ANOVA
Using the Coral Gables Housing data from Table 6.8, do the following calculations without the use of a computer.

(a) Calculate the SSR.
(b) Calculate the SST.
(c) Calculate the MSR.
(d) Calculate the F statistic.

Coral Gables Housing ANOVA in Python
Using the Coral Gables Housing data from Table 6.8, do the following using Python.

(a) Find the p-value of the F statistic calculated in the previous problem.
(b) Use Python to construct the ANOVA table.

Streaming Service Confidence Interval
A popular video streaming service relies on paid subscribers. In Table 6.9, we display the revenue (in millions of $) and the number of paid memberships (in millions) for the years 2007–2022.

Without the use of a computer:

(a) Find the regression equation that predicts revenue as a function of the number of memberships.

Table 6.9 Streaming service data

Year	Memberships (X)	Revenue (Y)
2007	1.19	339
2008	5.78	499
2009	6.93	600
2010	7.08	687
2011	9.63	836
2012	12.28	1079
2013	11.10	1604
2014	7.24	1807
2015	23.18	2189
2016	27.03	2752
2017	34.78	3395
2018	38.67	4417
2019	56.84	5845
2020	71.32	7890
2021	88.98	10,081
2022	101.44	12,498

(b) Use the regression equation to predict the revenue when the memberships is 100 million.
(c) Calculate the critical value $t_{\alpha/2}$ for $\alpha = 0.05$ (use Python here).
(d) Find the standard error for a 95% confidence interval (SE_{CI}) for revenue when there are 100 million memberships.
(e) Calculate the upper and lower bounds of the 95% confidence interval centered about 100 million memberships.
(f) Provide an interpretation of the confidence interval.

Streaming Service Prediction Interval
Use the streaming service data from Table 6.9 and do the following without the use of a computer.

(a) Find the standard error for a 95% prediction interval (SE_{PI}) for revenue when there are 100 million memberships.
(b) Calculate the upper and lower bounds of the prediction interval for revenue when there are 100 million memberships.
(c) Provide an interpretation of the prediction interval in part b.
(d) Explicitly state the difference between the interpretation of the prediction interval and the confidence interval.
(e) Indicate which interval has a larger range.

Streaming Service Intervals in Python
Use the streaming service data from Table 6.9 and do the following using Python.

(a) Print a summary of the regression equation that predicts revenue as a function of the number of memberships.
(b) Use the regression equation to predict the revenue when the memberships is 100 million.
(c) Find a 95% confidence interval for the revenue when the memberships is 100 million.
(d) Find a 95% prediction interval for the revenue when the memberships is 100 million.
(e) Find a 95% confidence interval for the slope coefficient.

Streaming Service ANOVA
Use the streaming service data from Table 6.9 and do the following calculations without the use of a computer.

(a) Calculate the SSR.
(b) Calculate the SST.
(c) Calculate the MSR.
(d) Calculate the SSE.
(e) Calculate the F statistic.

Streaming Service ANOVA Calculations in Python
Use the streaming service data from Table 6.9 and do the following using Python.

(a) Calculate the SSR.
(b) Calculate the SST.
(c) Calculate the MSR.
(d) Calculate the SSE.
(e) Calculate the F statistic.

Streaming Service ANOVA in Python
Use the streaming service data from Table 6.9 and do the following using Python.

(a) Find the p-value of the F statistic calculated in the previous problem.
(b) Use the anova function in Python to construct the ANOVA table.

Chapter 7
Predictor Variable Transformations

If you torture the data enough, nature will always confess.

— Ronald Coase

7.1 Introduction

In regression analysis, it is often useful to modify predictor variables to better reflect their relationship with the response variable. This chapter explores transformations of predictor variables. One popular transformation involves dummy variables, which allow the effects of categorical variables to be incorporated into the model.

Dummy variables can be used in regression analysis as both predictor and response variables, but we will limit our discussion to predictor variables. Using dummy variables as the response variable is referred to as classification, which will remain outside of the scope of this book. In previous chapters, we assumed linear models with untransformed predictor variables. Here, we introduce nonlinear transformations of predictor variables, thereby making the model linear.

First, we will discuss categorical variables containing two different categories, and then broaden the discussion to include more than two. Applications using salary data are used to reinforce the concepts. Second, we introduce quadratic relationships which can be modeled using nonlinear transformations. Third, interactions between predictor variables are explained, followed by a discussion of the general linear model. The final discussion consists of a case study that predicts the number of likes of videos based on the characteristics and sentiment of the comments.

7.2 Categorical Variables

Not all data is numeric or quantitative in nature. For instance, a categorical variable refers to a non-numeric variable that can take on one or more possible groupings or categories. Therefore, each observation can be assigned to a particular group.

D. P. McGibney, *Applied Linear Regression for Business Analytics with Python*,
International Series in Operations Research & Management Science 378,
https://doi.org/10.1007/978-3-032-23806-1_7

One example of a categorical variable is business type, which can be broken down into two categories: for-profit and not-for-profit. Methods of payment used at a convenience store can also be considered a categorical variable.

It is often the case that categorical data has a relationship to a numeric response variable, which can help predict the response variable. Also, while categorical variables may seem problematic because of their non-numeric characteristic, it is possible to convert categorical variables into quantitative ones. For example, X might represent business type, for a specific company. We can let 0 indicate a not-for-profit business, and 1 indicate a for-profit business. For a particular observation of $X = x$, we specify

$$x = \begin{cases} 0, & \text{if not-for-profit} \\ 1, & \text{if for-profit.} \end{cases}$$

In this case, X is called a dummy or indicator variable. When dummy variables are used in a regression model, the dummy variable coefficients are interpreted in relation to a base level. Here, the dummy variable X denotes the for-profit category and the base level is not-for-profit, since for-profit corresponds to 1 and not-for-profit corresponds to 0. For example, if the coefficient for X is 20, then the for-profit businesses contribute 20 more to the response variable than the not-for-profit businesses. Since the comparison is in reference to the not-for-profit level, we refer to it as the base level.

From a survey, we may observe a binary variable with categories Yes and No. For a particular observation of $X = x$, we define

$$x = \begin{cases} 0, & \text{if No} \\ 1, & \text{if Yes.} \end{cases}$$

Close attention should be taken in defining dummy variables since their definitions will dictate their interpretations. For example, a dummy variable representing the "Yes" category will have a coefficient in a fitted regression model. This coefficient is typically more intuitive to interpret than a dummy coefficient coded for "No."

Dummy variables provide a convenient and easy way to account for categorical data, as we demonstrate in the next application.

7.3 Employee Salary Application: Dummy Variables

Suppose the management of North Grumman Inc. wants to test whether their employees' annual salaries relate to certain variables. The education level, highest degree achieved, years of experience, and the annual salary (in thousands of dollars) for each of the sampled 28 employees are shown in the data table above (Table 7.1). The degree variable contains three levels: a Bachelor of Arts (B.A.) degree, a Master

Table 7.1 Salary data

Education	Degree	Experience	Salary
Graduate	MSBA	2	85
Graduate	MS	3	74
Graduate	MS	8	100
Graduate	MSBA	0	78
Undergraduate	BA	2	52
Undergraduate	BA	0	60
Graduate	MSBA	0	85
Graduate	MSBA	0	74
Graduate	MSBA	5	124
Graduate	MS	0	73
Graduate	MSBA	2	82
Undergraduate	BA	3	55
Undergraduate	BA	10	115
Undergraduate	BA	0	43
Graduate	MSBA	2	97
Graduate	MS	0	74
Graduate	MS	0	72
Graduate	MSBA	0	80
Undergraduate	BA	5	75
Undergraduate	BA	0	62
Graduate	MSBA	2	90
Graduate	MSBA	3	88
Graduate	MS	8	72
Graduate	MS	0	65
Graduate	MSBA	2	85
Undergraduate	BA	0	54
Undergraduate	BA	20	103
Graduate	MSBA	0	78

of Science (M.S.) degree, and a Master of Science in Business Analytics (M.S.B.A.) degree.

Use the data from Table 7.1 to do the following:

(a) Specify the model to be fit using "Education" and "Experience" variables to predict salary.
(b) Manually code a dummy variable denoting whether or not an employee has a graduate degree. Use 0 to denote if a candidate has an undergraduate degree, and use 1 to denote if a candidate has a graduate degree.
(c) Fit the linear regression model.
(d) Explicitly state the linear regression equation.
(e) Interpret the intercept coefficient.
(f) Interpret the variable coefficients.
(g) Discuss the model fit.

Solution

(a) The multiple linear regression model can be written as

$$Y = \beta_0 + \beta_1 X_1 + \beta_2 X_2 + \varepsilon,$$

where

- Y is the annual salary (in thousands),
- X_1 is the years of experience,
- X_2 is a graduate degree dummy variable:

$$X_2 = \begin{cases} 1 \text{ if graduate degree,} \\ 0 \text{ otherwise,} \end{cases}$$

- ε is the error.

(b) Creating a list of zeros and ones can be done with square brackets. These values follow from Table 7.1.

```
degree_dummy = [1, 1, 1, 1, 0, 0, 1, 1, 1, 1, 1, 0, 0, 0, 1, 1,
                 1, 1, 0, 0, 1, 1, 1, 1, 1, 0, 0, 1]
```

(c) Lists are created to represent the experience and salary for each employee. Then, Pandas is imported and used to put the variables in a dataframe.

```
experience = [2, 3, 8, 0, 2, 0, 0, 0, 5, 0, 2, 3, 10, 0, 2, 0,
               0, 0, 5, 0, 2, 3, 8, 0, 2, 0, 20, 0]
salary = [85, 74, 100, 78, 52, 60, 85, 74, 124, 73, 82, 55, 115,
           43, 97, 74, 72, 80, 75, 62, 90, 88, 72, 65, 85, 54,
           103, 78]

import pandas as pd
df = pd.DataFrame({"experience": experience, "salary": salary,
                   "degree_dummy": degree_dummy})
```

To fit the regression model, Ravix is imported and the fit function is used. After the model is created, the summary function is invoked to print a model summary.

```
from ravix import ols
reg = ols("salary ~ experience + degree_dummy", data = df)
reg.summary()
```

```
# Summary of OLS Regression Analysis:
# ======================================================
#
# Coefficients:
# ------------------------------------------------------
#                  Estimate  Std. Error  t-value    p-value
# Intercept          55.850      4.9308   11.327   2.45e-11 ***
# experience         2.9089     0.57713    5.040   3.36e-05 ***
# degree_dummy       21.433      5.3113    4.035     0.0005 ***
#
# Model Statistics:
# ------------------------------------------------------
# Residual Std. Error: 12.6333
# R-squared:          0.5696           AIC: 224.3224
# Adj. R-squared: 0.5352               BIC: 228.319
# F-statistic: 16.5433 on 2 and 25 DF, p-value: 2.65e-05
# ======================================================
```

(d) From the summary coefficients, we have:

$$\hat{Y} = 55.8495 + 2.9089X_1 + 21.4332X_2.$$

(e) We first note that the intercept coefficient is significant from the p-value, indicating that the interpretation is valid. The expected salary is predicted to be \$55,850 when all predictor variables (`experience` and `degree_dummy`) are 0. Note that the dummy variable (X_2) is 0 when an employee has an undergraduate degree (base level). Thus, an employee with an undergraduate degree and no experience is predicted to make \$55,850.

(f) Since both variable coefficient p-values indicate statistical significance, it is appropriate to interpret the variable coefficients.

- For every additional year of experience, the annual salary is expected to increase by \$2,909, assuming the degree variable remains constant.
- Since 1 designates that an employee has a graduate degree, the coefficient indicates an increase of \$21,433 in annual salary if someone has a graduate degree compared to someone at the base level, assuming the experience variable remains constant.

(g) The model fit proves reasonably well given that the R^2 is 0.5696, indicating the regression model explains approximately 57% of the variation in the annual salary.

7.4 Employee Salary Application: Dummy Variables 2

In this application, we will demonstrate how to modify the model interpretation by changing the dummy variable. Use the data from Table 7.1 to do the following:

(a) Create a new dummy variable that is 1 if someone has an undergraduate degree, and 0 if someone has a graduate degree.
(b) Fit a linear regression using this new dummy variable and experience to predict salary.
(c) Interpret the intercept coefficient.
(d) Interpret the variable coefficients.
(e) Discuss the differences between this model and the first application model.

Solution

(a) Here, we would like to change the meaning of the dummy variable to indicate someone has an undergraduate degree:

$$X_2 = \begin{cases} 1 \text{ if undergraduate degree,} \\ 0 \text{ otherwise.} \end{cases}$$

One approach to creating the new `degree_dummy` would be to manually create the vector as done in the previous example. However, the dummy variable can be coded using the `where` function from NumPy or even more simply by subtracting the `degree_dummy` from 1. The head is shown for the reader to verify the correct representation.

```
import numpy as np
df['degree_dummy2'] = 1 - np.array(degree_dummy)
df['degree_dummy2'].head()
```

```
# 0    0
# 1    0
# 2    0
# 3    0
# 4    1
# Name: degree_dummy2, dtype: int64
```

(b) The regression is fit and summarized using the commands below:

```
from ravix import ols
reg2 = ols("salary ~ experience + degree_dummy2", data = df)
reg2.summary()
```

```
# Summary of OLS Regression Analysis:
# =========================================================
#
# Coefficients:
# -------------------------------------------------------
#                    Estimate  Std. Error  t-value    p-value
# Intercept           77.283      3.1086   24.861   < 2e-16 ***
# experience          2.9089     0.57713    5.040  3.36e-05 ***
# degree_dummy2      -21.433      5.3113   -4.035    0.0005 ***
#
# Model Statistics:
# -------------------------------------------------------
# Residual Std. Error: 12.6333
# R-squared:         0.5696           AIC: 224.3224
# Adj. R-squared: 0.5352              BIC: 228.319
# F-statistic: 16.5433 on 2 and 25 DF, p-value: 2.65e-05
# =========================================================
```

(c) As in the previous problem, all three coefficient p-values indicate statistical significance; hence, it is appropriate to interpret the intercept and both variable coefficients. The expected salary is $77,283 when all predictor variables are 0. Note that the dummy variable (X_2) is 0 when an employee has a graduate degree. An employee with a graduate degree and no experience is predicted to make $77,283.

(d) The variable coefficient interpretations are:

- The `experience` coefficient interpretation is the same as in the previous application.
- Since 1 designates that an employee has an undergraduate degree, the coefficient indicates a decrease of $21,433 in annual salary if someone has an undergraduate degree compared to someone with a graduate degree, assuming `experience` remains constant. Note that this coefficient is the negative of the dummy coefficient from the model fit in part c of the previous application.

(e) The model in this application proves a similar overall fit to the previous application model as shown by the R_a^2 and R^2. The sign of the dummy coefficient is now the negative of the previous dummy coefficient. In addition, the intercept has changed to reflect the base level of employees with a graduate degree. Despite these differences, both models show the expected salary for an employee with no experience is $55,850 for those with an undergraduate degree and $77,283 for those with a graduate degree.

7.5 Multilevel Categorical Variables

A categorical variable with k levels is encoded using $k - 1$ dummy variables, ensuring the information is fully captured without redundancy. For example, a variable with levels A, B, and C could be represented by the dummy variables x_1 and x_2. If x_1 and x_2 are assigned as

$$x_1 = \begin{cases} 1 \text{ if level B} \\ 0 \text{ otherwise,} \end{cases}$$

$$x_2 = \begin{cases} 1 \text{ if level C} \\ 0 \text{ otherwise,} \end{cases}$$

then (x_1, x_2) will have values of $(0, 0)$ when level A occurs (base level), $(1, 0)$ when level B occurs, and $(0, 1)$ when level C occurs. Even though level A doesn't have a dummy variable, the absence of the other two dummy variables indicates level A.

Notice that since there were three levels, it was necessary to have two dummy variables. Furthermore, two dummies retain the information of all three levels. More generally, a categorical variable with k levels requires $k - 1$ dummy variables. For example, a variable with 10 levels would require 9 dummy variables to fully represent its information.

The following application demonstrates how to code dummy variables with more than two categories.

7.6 Employee Salary Application: Dummy Variables with Multiple Levels

Continuing with the salary data (Table 7.1) from the previous two applications, use the data to do the following:

(a) Code the "Degree" variable as a vector of text strings.
(b) Fit a linear regression using degree and experience to predict salary.
(c) Explicitly state the linear regression equation.
(d) Interpret the intercept coefficient.
(e) Interpret the variable coefficients.
(f) Discuss the differences between this model and the previous application model.

Solution

(a) A text string can be denoted using double quotes or single quotes. Square brackets are used to create a list of text strings with values of `MSBA`, `MS`, or `BA`.

```
degree = ["MSBA", "MS", "MS", "MSBA", "BA", "BA", "MSBA",
          "MSBA", "MSBA", "MS", "MSBA", "BA", "BA", "BA",
          "MSBA", "MS", "MS", "MSBA", "BA", "BA", "MSBA",
          "MSBA", "MS", "MS", "MSBA", "BA", "BA", "MSBA"]
```

(b) The `ols` function formula denotes that `salary` is the response with `experience` and `degree` as predictors. The `ols` function automatically converts the `degree` variable to dummy variables.

```
from ravix import ols
reg3 = ols("salary ~ experience + degree", data = df)
reg3.summary()
```

```
# Summary of OLS Regression Analysis:
# =======================================================
#
# Coefficients:
# -------------------------------------------------------
#                  Estimate  Std. Error  t-value    p-value
# Intercept          55.093      4.3599   12.636   4.26e-12 ***
# experience         3.0791     0.51286    6.004   3.38e-06 ***
# degree_MS          12.264      5.6886    2.156     0.0413 *
# degree_MSBA        27.455      5.1433    5.338   1.77e-05 ***
#
# Model Statistics:
# -------------------------------------------------------
# Residual Std. Error: 11.1498
# R-squared:       0.6782          AIC: 218.1841
# Adj. R-squared: 0.6379           BIC: 223.5129
# F-statistic: 16.8573 on 3 and 24 DF, p-value: 4.17e-06
# =======================================================
```

(c) From the summary coefficients, we have:

$$\hat{Y} = 55.0931 + 3.0791X_1 + 12.2638X_2 + 27.4550X_3,$$

where

- X_1 is the years of experience,
- X_2 is a dummy denoting if an employee has an M.S. degree, and
- X_3 is a dummy denoting if an employee has an M.S.B.A. degree.

(d) The p-value of the intercept indicates statistical significance, which means the intercept can be interpreted: *An employee with no experience who has a B.A. degree is expected to earn $55,093.*

(e) All variable coefficient p-values indicate statistical significance; hence, it is appropriate to interpret all three variable coefficients.

- For every additional year of experience, the annual salary is expected to increase by $3,079, assuming all other variables remain constant.
- If someone has an M.S. degree, then their salary is expected to be $12,264 higher than someone with a B.A. degree, assuming all other variables remain constant.
- If someone has an M.S.B.A. degree, then their salary is expected to be $27,455 higher than someone with a B.A. degree, assuming all other variables remain constant.

(f) Since the number of predictor variables increased in this application, the R^2 value will certainly increase. Recall that R^2 should not be used to compare models with different numbers of predictors. Therefore, we should use R_a^2 rather than R^2 when comparing models. The model in the previous application had an R_a^2 of 0.5352, whereas the current model's R_a^2 of 0.6379 demonstrates a better fit.

7.7 Coding Dummy Variables

In the first application, the dummy variable for noting if an employee had a graduate degree (`degree_dummy`) was created by replacing "Graduate" with 1 and "Undergraduate" with 0. In the third application, we took advantage of the `ols` function's ability to automatically transform the vector of text strings (`degree`) to dummy variables. When Ravix automatically created the dummy variable, two variables could have been created, a dummy for graduate degree or a dummy for undergraduate degree. However, only one dummy variable was needed, and so Ravix dropped the first alphabetical column. In many instances, manually coding a dummy variable provides more flexibility. This can be done in a few different ways in Python.

To examine the different methods for creating dummy variables, the dataframe from Table 7.1 is created from three lists.

```
import pandas as pd
experience = [2, 3, 8, 0, 2, 0, 0, 0, 5, 0, 2, 3, 10, 0, 2, 0,
              0, 0, 5, 0, 2, 3, 8, 0, 2, 0, 20, 0]
salary = [85, 74, 100, 78, 52, 60, 85, 74, 124, 73, 82, 55, 115,
          43, 97, 74, 72, 80, 75, 62, 90, 88, 72, 65, 85, 54,
          103, 78]
degree = ["MSBA", "MS", "MS", "MSBA", "BA", "BA", "MSBA",
          "MSBA", "MSBA", "MS", "MSBA", "BA", "BA", "BA",
          "MSBA", "MS", "MS", "MSBA", "BA", "BA", "MSBA",
          "MSBA", "MS", "MS", "MSBA", "BA", "BA", "MSBA"]
df = pd.DataFrame({"experience": experience, "salary": salary,
                   "degree": degree})
```

Creating dummies can be done in many different ways.

- A dummy for the MSBA category within the degree can be done with a simple equivalency test.

```
MSBA_dummy = df.degree == "MSBA"
MSBA_dummy.head()
```

```
# 0     True
# 1    False
# 2    False
# 3     True
# 4    False
# Name: degree, dtype: bool
```

- The where function from Numpy is also convenient for creating dummies. In this case, the equivalency test is the first argument. If the argument is true, then we get a one, otherwise zero. The first five values of the dummy are given using indexing.

```
import numpy as np
MSBA_dummy2 = np.where(df.degree == "MSBA", 1, 0)
MSBA_dummy2[:5]
```

```
# array([1, 0, 0, 1, 0])
```

- The get_dummies function from Pandas allows users to easily convert all categorical columns to dummies.

```
df_dummies = pd.get_dummies(df)
df_dummies.head()
```

```
#    experience  salary  degree_BA  degree_MS  degree_MSBA
# 0           2      85      False      False         True
# 1           3      74      False       True        False
# 2           8     100      False       True        False
# 3           0      78      False      False         True
# 4           2      52       True      False        False
```

Note that the original column of string values is not available in the newly created dataframe. Since there are three levels in the original degree column, three dummies were created. Each dummy is a vector of booleans, which can be either True or False. If we were to convert these boolean columns to integers, then False would be zero and True would be 1. In fact, when using arithmetic operations on these booleans, they are automatically treated as zeros and ones.

Regression modeling only requires that we include two dummies for three category levels. Therefore, it is often the case that one will want to drop one of the dummy variables. The `drop_first` option allows us to drop the dummy that occurs first alphabetically.

```
df_dummies2 = pd.get_dummies(df, drop_first = True)
df_dummies2.head()
```

```
#    experience  salary  degree_MS  degree_MSBA
# 0           2      85      False         True
# 1           3      74       True        False
# 2           8     100       True        False
# 3           0      78      False         True
# 4           2      52      False        False
```

Notice that the output here does not have `degree_BA` since `drop_first = True`.

If the `degree_MSBA` column should be dropped, then we could use the `drop` function to manually drop it. Since we want to include the `degree_BA` dummy, we omit the `drop_first` argument or set it to `False`, which is the default. This will make `degree_MSBA` the base level and use the dummies `degree_BA` and `degree_MS`.

Rather than using "=" to reassign the `df_dummies` dataframe, the `inplace = True` argument is specified in the `drop` function.

```
df_dummies = pd.get_dummies(df)
df_dummies.drop(columns = ["degree_MSBA"], inplace = True)
df_dummies.head()
```

```
#    experience  salary  degree_BA  degree_MS
# 0           2      85      False      False
# 1           3      74      False       True
# 2           8     100      False       True
# 3           0      78      False      False
# 4           2      52       True      False
```

In many applications, manually coding dummy variables is not necessary because it is automatically done using Ravix. However, when we are particular about which dummy variables we want coded, there are some great options available, particularly the `get_dummies` function or a manual manipulation.

7.8 Employee Salary Application: Dummy Variable Coding

With the salary dataset, do the following:

(a) Create a new pandas dataframe including the dummies but with the "MS" dummy excluded.
(b) Fit a linear regression using all of the predictors in the dataframe to predict salary.
(c) Interpret the intercept coefficient.
(d) Interpret the variable coefficients.
(e) Discuss the differences between this model and the previous application model.

Solution

(a) Here, we use the `get_dummies` function from Pandas to include all of the dummies. Then, we drop the `degree_MS` dummy variable from the dataframe with the `drop` function. The head is printed as confirmation of the desired result.

```
df_dummies = pd.get_dummies(df)
df_dummies.drop(columns = ["degree_MS"], inplace = True)
df_dummies.head()
```

```
#    experience  salary  degree_BA  degree_MSBA
# 0           2      85      False         True
# 1           3      74      False        False
# 2           8     100      False        False
# 3           0      78      False         True
# 4           2      52       True        False
```

(b) The model is fit using the `ols` function from Ravix. A summary is produced using the `summary` function.

```
from ravix import ols
reg = ols("salary ~ experience + degree_BA + degree_MSBA",
          data = df_dummies)
reg.summary()
```

```
# Summary of OLS Regression Analysis:
# =========================================================
#
# Coefficients:
# ---------------------------------------------------------
#                  Estimate  Std. Error  t-value    p-value
# Intercept          67.357      4.4382   15.177   8.41e-14 ***
# experience         3.0791     0.51286    6.004   3.38e-06 ***
# degree_BA         -12.264      5.6886   -2.156     0.0413 *
# degree_MSBA        15.191      5.3392    2.845     0.0089 **
#
# Model Statistics:
# ---------------------------------------------------------
# Residual Std. Error: 11.1498
# R-squared:         0.6782          AIC: 218.1841
# Adj. R-squared: 0.6379             BIC: 223.5129
# F-statistic: 16.8573 on 3 and 24 DF, p-value: 4.17e-06
# =========================================================
```

(c) The p-value of the intercept indicates statistical significance, which means that the intercept can be interpreted. The interpretation holds that an employee with no experience with an M.S. degree is expected to earn $67,357.

(d) All variable coefficient p-values indicate statistical significance, so it is appropriate to interpret all three variable coefficients.

- The interpretation of the experience coefficient remains unchanged from that of the previous application.
- An individual with an M.S.B.A. degree is expected to earn $15,191 higher than someone with an M.S. degree, holding all other variables constant.
- An individual with a B.A. degree is expected to earn $12,264 less (or -$12,264 higher) than someone with an M.S. degree, holding all other variables constant.

(e) The model in this application proves a similar overall fit to the previous application model as shown by the R_a^2 and R^2. The dummy coefficients and the intercept have changed to reflect the base level of employees with an M.S. degree. Despite these differences, both models show that the expected salary for an employee with no experience is $55,093 for those with a B.A. degree, $67,357 for those with an M.S. degree, and $82,548 for those with an M.S.B.A. degree.

7.9 Modeling Curvilinear Relationships

When using linear regression, one makes the assumption that the data follows a linear relationship, but this scenario may not always be the case. Linear regression, despite the name, can handle nonlinear data by transforming the nonlinear components of the data.

This is demonstrated in Fig. 7.1, which shows a nonlinear relationship between X and Y. The relationship in each scatterplot is said to be curvilinear. When we use a simple linear regression model to fit the data, we do not properly capture the pattern as shown by the blue line. Therefore, it becomes necessary to fit the data to a quadratic model by introducing an X^2 term. This relationship is shown in Fig. 7.1 by the red curve. The quadratic model can be written as:

$$Y = \beta_0 + \beta_1 X + \beta_2 X^2 + \varepsilon, \tag{7.1}$$

where β_0, β_1, and β_2, are the coefficient parameters of the model and ε is the error term. Fitting this equation to a dataset provides the regression equation:

$$\hat{Y} = \hat{\beta}_0 + \hat{\beta}_1 X + \hat{\beta}_2 X^2, \tag{7.2}$$

where $\hat{\beta}_0$, $\hat{\beta}_1$, and $\hat{\beta}_2$ are the estimated coefficients.

7.10 Sales Performance Application: Quadratic Modeling

A manufacturer of quantum computing servers wants to investigate the relationship between the length of employment of their salespeople and the number of servers sold. The data in Table 7.2 lists the number of months each salesperson has been employed by the firm, "Employment," and the number of servers sold, "Sales," by 21 randomly selected salespeople.

If the response variable is Sales and the predictor variable is Employment, then do the following:

(a) Fit a simple linear regression line to the data and print the model summary.
(b) Plot the data in a scatterplot superimposing the simple linear regression line on the plot.
(c) Fit a quadratic regression model to the data and produce the corresponding summary.
(d) Plot the data in a scatterplot superimposing the quadratic model curve on the plot.
(e) Explain which relationship (simple linear or quadratic) more accurately reflects the relationship between "Sales" and "Employment."

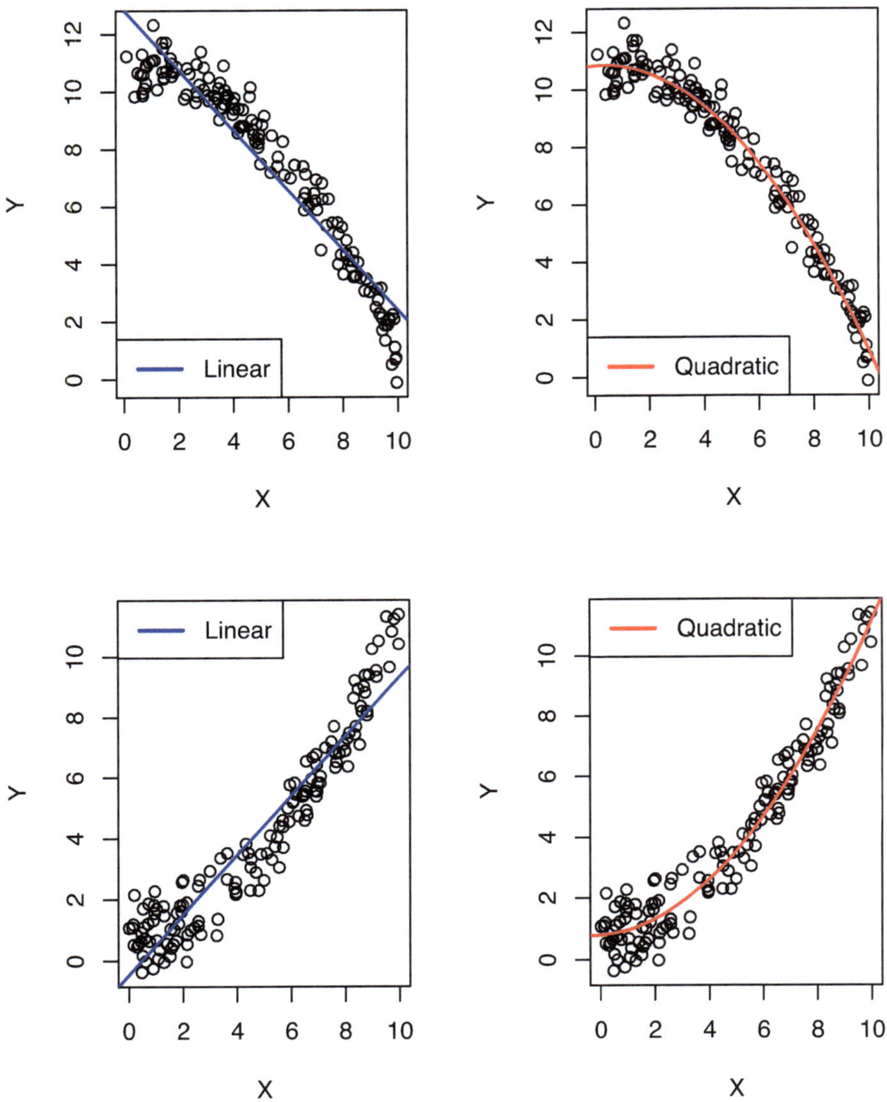

Fig. 7.1 Scatterplots of curvilinear data

Solution

(a) After importing Ravix, the `get_data` function is used to load the CSV file. The sales are then predicted using employment specifying the `data` argument as `df` in the `ols` function. A summary is printed using the `summary` function.

Table 7.2 Sales data

Sales	Employment
379	57
239	114
225	100
240	58
403	75
227	46
338	106
5	2
167	50
284	77
378	93
399	70
239	106
306	89
325	86
113	33
270	67
131	24
30	5
125	15
267	52

```python
from ravix import get_data, ols
df = get_data("employment.csv")
reg = ols("Sales ~ Employment", data = df)
reg.summary()
```

```
# Summary of OLS Regression Analysis:
# ======================================================
#
# Coefficients:
# ------------------------------------------------------
#                Estimate  Std. Error  t-value   p-value
# Intercept        88.151      38.492    2.290    0.0336 *
# Employment       2.4444     0.54121    4.517    0.0002 ***
#
# Model Statistics:
# ------------------------------------------------------
# Residual Std. Error: 81.4114
# R-squared:       0.5178          AIC: 246.2733
# Adj. R-squared: 0.4924           BIC: 248.3624
# F-statistic: 20.3992 on 1 and 19 DF, p-value: 0.000236
# ======================================================
```

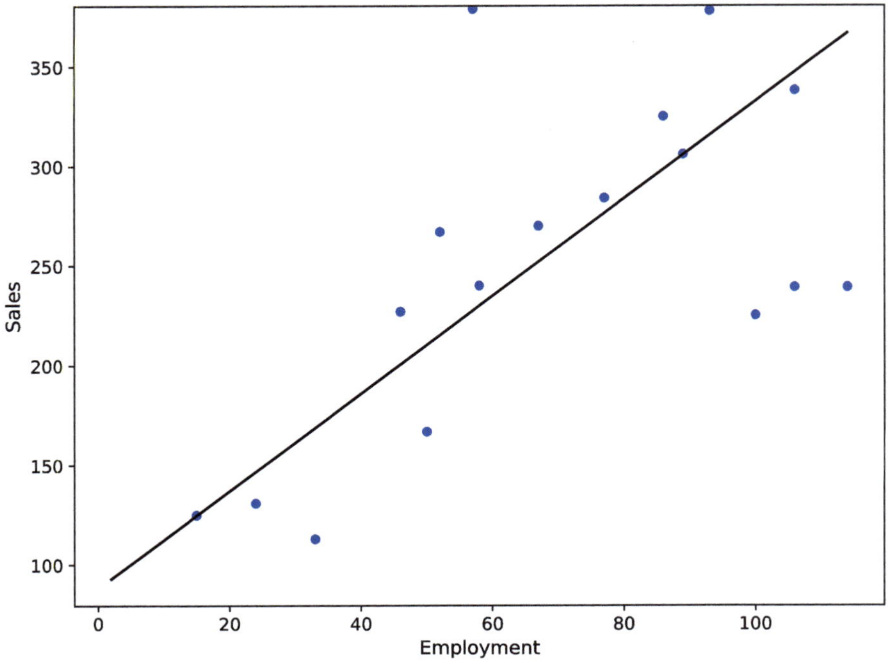

Fig. 7.2 Sales and employment: linear trend

(b) The `plot` function from Chap. 3 allows us to display a scatterplot and superimpose the line on the plot. After specifying the formula, the `reg` model created from part (a) is used to specify the plot given in Fig. 7.2.

```
from ravix import plot, abline
plot("Sales ~ Employment", data = df)
abline(reg)
```

(c) The `Employment**2` can be included in the formula to denote the squared employment as a predictor. The `ols` and `summary` functions can be used in a similar way to fit and summarize the quadratic model.

```
quad = ols("Sales ~ Employment + Employment**2", data = df)
quad.summary()
```

```
# Summary of OLS Regression Analysis:
# ======================================================
#
# Coefficients:
# ------------------------------------------------------
#                  Estimate  Std. Error  t-value    p-value
# Intercept         -26.445      39.290   -0.673     0.5094
# Employment         8.3236      1.4603    5.700   2.09e-05 ***
# Employment^2 -0.050677    0.012115   -4.183     0.0006 ***
#
# Model Statistics:
# ------------------------------------------------------
# Residual Std. Error: 59.5621
# R-squared:         0.7555          AIC: 234.0131
# Adj. R-squared: 0.7283             BIC: 237.1467
# F-statistic: 27.8034 on 2 and 18 DF, p-value: 3.13e-06
# ======================================================
```

(d) To plot the quadratic curve , simply specify the model created in the `model` argument of the `plot` . The other components of the syntax remain unchanged from the previous plot. The updated plot from the code below is given in Fig. 7.3.

```
plot("Sales ~ Employment", data = df)
abline(quad)
```

(e) The superimposed model fit line and the curve in the scatterplots both indicate strong fits. However, the relationship between the length of time employed and the number of sales is better captured using the quadratic model. The R_a^2 values in the summaries also show the quadratic model has a better fit since 0.7283 is greater than 0.4924.

7.11 Mean-Centering

In many cases, the intercept does not provide anything meaningful about the data, or worse, the intercept does not make sense. These scenarios can be attributed to the intercept being the expected value of the response when all of the predictor variables are zero. If a predictor variable being zero does not make sense, then the corresponding prediction will not make sense either.

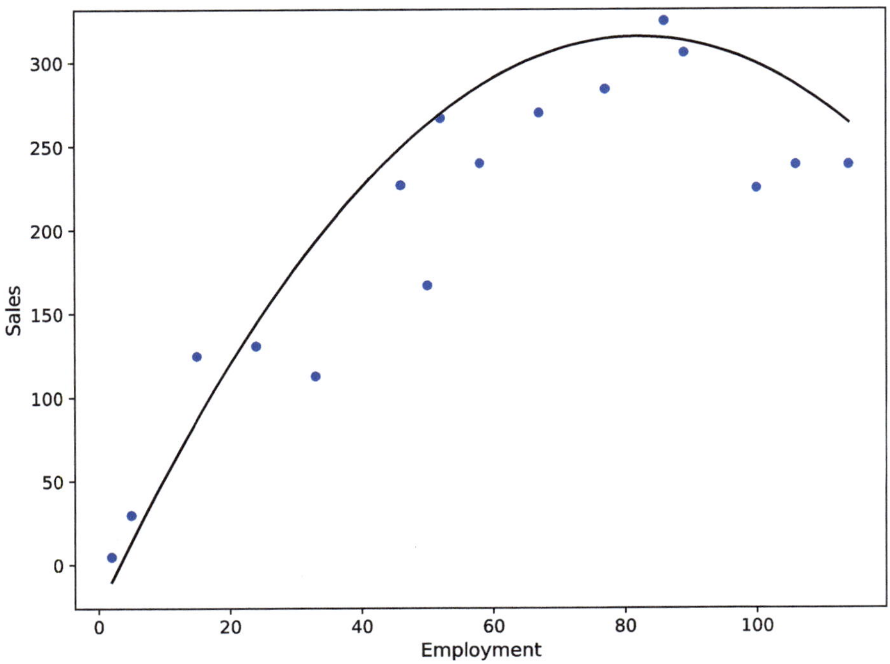

Fig. 7.3 Sales and employment: curvilinear trend

Mean-centering a predictor variable allows the variable to preserve its variability, but the value of the mean-centered variable is relative to the mean. To mean-center the variable X, simply subtract the mean from each observation. If x_i represents the ith value of X, then the ith transformed value is:

$$x_i - \bar{x}. \tag{7.3}$$

Note the mean of the transformed data will be zero, which allows for a convenient interpretation of the intercept. The application below provides an example of how mean-centering can be utilized.

7.12 Marketing Toys Application: Mean-Centering

A popular toy company sells two different toys: toy A and toy B. The company wishes to know how much to spend on marketing to achieve their sales goals. The data are in the file: `marketing_toys.csv` which can be loaded into Python using the `get_data` function. Units of `Sales` and `Marketing` are in thousands of USD. Additionally, we observe the first five observations:

```
from ravix import get_data
df = get_data("marketing_toys.csv")
df.head()
```

```
#   Toy    Sales  Marketing
# 0   A  855.844    174.081
# 1   A  728.745    175.027
# 2   B  887.742    180.757
# 3   B  604.876    168.615
# 4   B  703.407    162.877
```

After loading in the data, complete the following:

(a) Create a scatterplot of the marketing expenditures versus the sales revenue for all observations colored by toy.
(b) Fit a model and print a summary predicting sales as a function of marketing and the toy type.
(c) Interpret the coefficients.
(d) Repeat part a, but include a line depicting Toy A and another depicting Toy B based on the model summary from part b.
(e) Mean center the marketing variable and repeat part b.
(f) Interpret the intercept coefficient for the model in part d.
(g) Change the base level, produce a summary, and state the new meaning of the intercept.

Solution

(a) Here, we will designate Toy A as light blue and Toy B as light pink. Particularly, the `where` function from NumPy does an equivalency test and returns the corresponding color in the `colors` NumPy array. The overall scatterplot is given using the `plot` command. The `xlab` and `ylab` options are specified to produce a well-labeled plot. The `color` argument is set to the `colors` vector. The result is shown in Fig. 7.4.

the `colors` vector. The result is shown in Figure 7.4.

```
from ravix import plot
import numpy as np
colors = np.where(df.Toy == "A", "lightblue", "lightpink")
plot("Sales ~ Marketing", data = df,
     xlab = "Marketing (in $1000)",
     ylab = "Sales (in $1000)",
     color = colors)
```

(b) Using the full dataframe `df`, a fit can easily be made using the `ols` function:

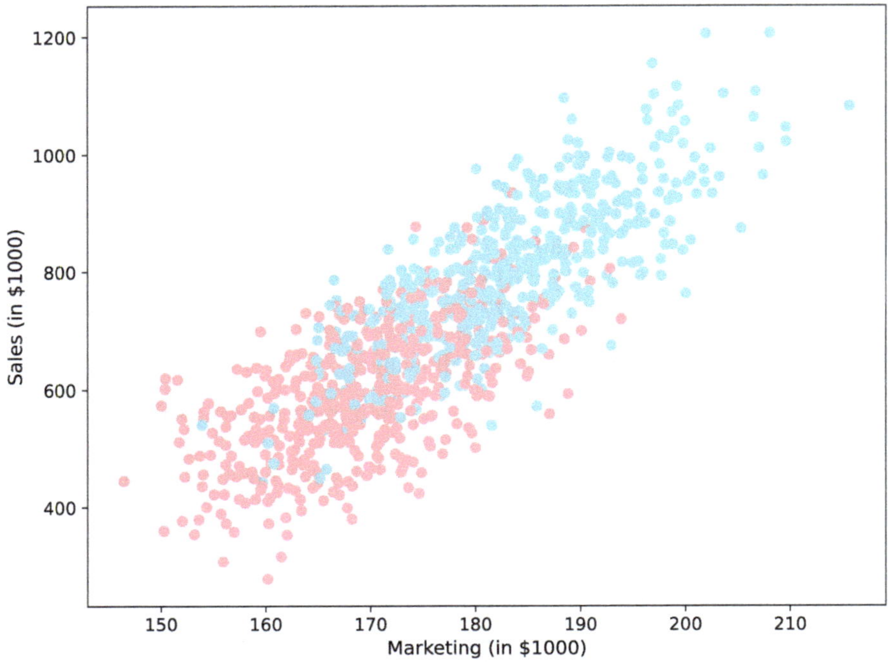

Fig. 7.4 Combined scatterplot of toys

```
from ravix import ols
reg1 = ols("Sales ~ Marketing + Toy", data = df)
reg1.summary()
```

```
# Summary of OLS Regression Analysis:
# ========================================================
#
# Coefficients:
# --------------------------------------------------------
#                Estimate  Std. Error  t-value    p-value
# Intercept       -907.55      52.749  -17.205   < 2e-16 ***
# Marketing        9.3253     0.28752   32.433   < 2e-16 ***
# Toy_B           -76.892      6.5574  -11.726   < 2e-16 ***
#
# Model Statistics:
# --------------------------------------------------------
# Residual Std. Error: 83.7402
# R-squared:      0.7158            AIC: 11696.31
# Adj. R-squared: 0.7152            BIC: 11711.03
# F-statistic: 1255.43 on 2 and 997 DF, p-value: < 2e-16
# ========================================================
```

(c) The interpretation of the coefficients is as follows:

- Toy B has on average \$76,892 less sales than toy A, or alternatively, toy A has sales on average \$76,892 more than toy B.
- For every \$1,000 increase in marketing expenditures, there is expected to be \$9,325 more sales.
- The intercept is the value when all variables are 0 (`Toy_B = 0` indicates toy A). Note that it would not make sense for sales to be negative, which implies the intercept should not be interpreted.

(d) From the model summary in part c, the slope for both toys is 9.3253. The intercept from toy A would be the intercept of -907.5484, ignoring the `Toy_B` dummy coefficient. The `abline` function can be used here to plot the regression line by simply stating the intercept (`a`) and slope (`b`) as arguments. Then use the intercept added with the coefficient of the `Toy_B` to get the intercept for `modelB`. The result is shown in Fig. 7.5.

The scatterplot arguments from part a remain unchanged. Figure 7.5 extends the original plot by adding regression lines using the `abline` function, specified after the `plot` call.

```
from ravix import abline
plot("Sales ~ Marketing", data = df,
     xlab = "Marketing (in $1000)",
     ylab = "Sales (in $1000)",
     color = colors)
abline(-907.5484, 9.3253, color = "darkblue")
abline(-907.5484 + (-76.8916), 9.3253, color = "darkred")
```

(e) Subtract the mean of the marketing expenditures from all the marketing values. The `describe` function displays this new variable and confirms the new mean is zero:

```
df['Mktg_MC'] = df.Marketing - np.mean(df.Marketing)
df.Mktg_MC.describe()
```

```
# count    1.000000e+03
# mean    -3.353762e-15
# std      1.140895e+01
# min     -2.980126e+01
# 25%     -8.292258e+00
# 50%     -8.772580e-01
# 75%      7.385492e+00
# max      3.938974e+01
# Name: Mktg_MC, dtype: float64
```

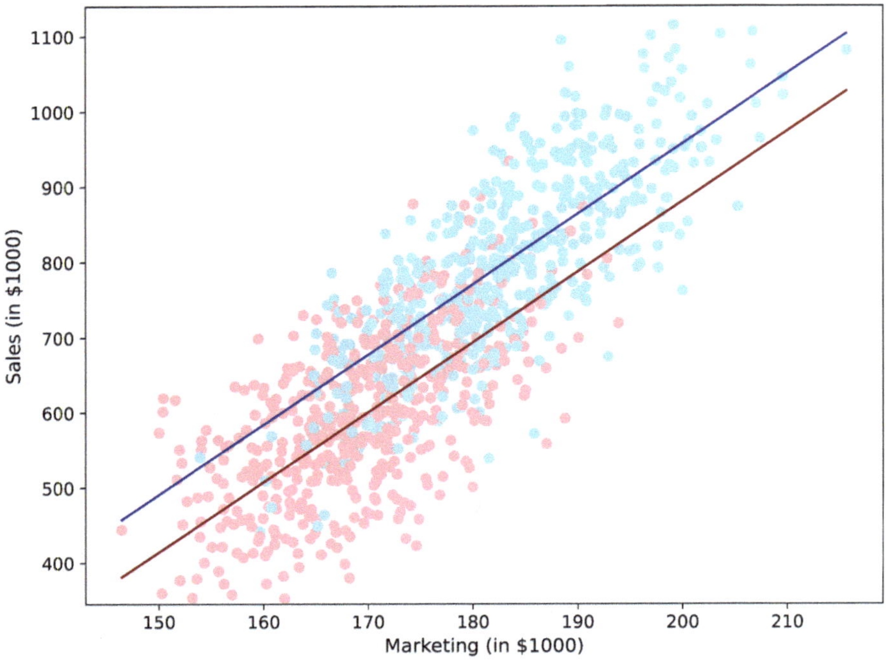

Fig. 7.5 Combined scatterplot of toys with lines 1

The regression is now fit using the mean-centered variable with Toy:

```
reg2 = ols("Sales ~ Mktg_MC + Toy", data = df)
reg2.summary()
```

```
# Summary of OLS Regression Analysis:
# ======================================================
#
# Coefficients:
# ------------------------------------------------------
#                  Estimate  Std. Error  t-value    p-value
# Intercept          736.26      4.2145  174.696   < 2e-16 ***
# Mktg_MC            9.3253     0.28752   32.433   < 2e-16 ***
# Toy_B             -76.892      6.5574  -11.726   < 2e-16 ***
#
# Model Statistics:
# ------------------------------------------------------
# Residual Std. Error: 83.7402
# R-squared:         0.7158            AIC: 11696.31
# Adj. R-squared: 0.7152               BIC: 11711.03
# F-statistic: 1255.43 on 2 and 997 DF, p-value: < 2e-16
# ======================================================
```

(f) In the summary from part e, a zero for both variables indicates that the sales from toy A without any marketing expenses will be $736,261.

(g) To change the base level, we redefine the Toy variable with B as the base level by manually creating the dummy variable (ToyA) and then fitting the regression model:

```
df['ToyA'] = np.where(df.Toy == "A", 1, 0)
reg3 = ols("Sales ~ Mktg_MC + ToyA", data = df)
reg3.summary()
```

```
# Summary of OLS Regression Analysis:
# =======================================================
#
# Coefficients:
# -------------------------------------------------------
#                 Estimate  Std. Error  t-value    p-value
# Intercept         659.37      4.2145  156.452   < 2e-16 ***
# Mktg_MC           9.3253     0.28752   32.433   < 2e-16 ***
# ToyA              76.892      6.5574   11.726   < 2e-16 ***
#
# Model Statistics:
# -------------------------------------------------------
# Residual Std. Error: 83.7402
# R-squared:       0.7158          AIC: 11696.31
# Adj. R-squared: 0.7152           BIC: 11711.03
# F-statistic: 1255.43 on 2 and 997 DF, p-value: < 2e-16
# =======================================================
```

In this case, a 0 for both variables indicates that the sales with mean marketing expenditures will be $659,370 for Toy B.

7.13 General Linear Regression Model

Models in which the parameters ($\beta_0, \beta_1, \ldots, \beta_p$) all correspond to variables with exponents of one are called linear models. However, as shown in the quadratic model, it is possible to square the predictor variable and use the resulting square as another predictor variable. This method allows us to model nonlinear functions using linear regression techniques. Since the aforementioned transformed models can be represented as linear models, they are a class of regression models that are referred to as general linear models.

The general linear model involving p predictor variables is

$$Y = \beta_0 + \beta_1 Z_1 + \beta_2 Z_2 + \ldots + \beta_p Z_p + \varepsilon. \tag{7.4}$$

Table 7.3 Ravix functions for linear and generalized linear models

	General linear model	Generalized linear models
Typical estimation method	Least squares	Maximum likelihood
Examples	Linear regression	Logistic, Poisson regression
Ravix function	`ols`	`logistic`, `poisson`

Each of the predictor variables Z_i is a function of $X_1, X_2, \ldots, X_k$ (the variables for which data have been collected).

The simplest case occurs when we have collected data for just one variable X_1 and want to estimate Y by using a straight-line relationship. In this case, $Z_1 = X_1$. The resulting model, the simple linear regression model, is:

$$Y = \beta_0 + \beta_1 X_1 + \varepsilon. \tag{7.5}$$

Within the framework of the general linear regression model, this model is called a simple first-order model with one predictor variable.

To account for a curvilinear relationship, we might set $Z_1 = X_1$ and $Z_2 = X_1^2$. The resulting quadratic model is

$$Y = \beta_0 + \beta_1 X_1 + \beta_2 X_1^2 + \varepsilon, \tag{7.6}$$

and is called a simple second-order model with one predictor variable.

If the original dataset consists of observations for Y and two predictor variables X_1 and X_2, we might develop a second-order model with two predictor variables:

$$Y = \beta_0 + \beta_1 X_1 + \beta_2 X_2 + \beta_3 X_1^2 + \beta_4 X_2^2 + \beta_5 X_1 X_2 + \varepsilon. \tag{7.7}$$

In this model, we include the variable $Z_5 = X_1 X_2$ to account for the potential effects of the two variables acting together. This type of effect is called an interaction.

Sometimes, the general linear regression model is referred to as the general linear model, which should not be confused with the Generalized Linear Model (GLM). The GLM serves as a flexible generalization of linear regression that allows for response variables that have non-normal distributions. The GLM extends linear regression by applying a link function to the response variable. These differences are summarized in Table 7.3.

7.14 Interactions

To simplify the discussion, we focus here on the case of two predictor variables with an interaction. The result is a first-order model with two predictor variables:

$$Y = \beta_0 + \beta_1 X_1 + \beta_2 X_2 + \beta_3 X_1 X_2 + \varepsilon. \tag{7.8}$$

If X_2 is a dummy variable, then the model changes for each outcome of X_2. When $X_2 = 0$, the model is:

$$Y = \beta_0 + \beta_1 X_1 + \varepsilon, \tag{7.9}$$

but when $X_2 = 1$, the model is:

$$\begin{aligned} Y &= \beta_0 + \beta_1 X_1 + \beta_2(1) + \beta_3 X_1(1) + \varepsilon \\ &= \beta_0 + \beta_1 X_1 + \beta_2 + \beta_3 X_1 + \varepsilon \\ &= (\beta_0 + \beta_2) + (\beta_1 + \beta_3) X_1 + \varepsilon. \end{aligned} \tag{7.10}$$

Notice the intercept for Eq. (7.10) is $\beta_0 + \beta_2$ and the slope is $\beta_1 + \beta_3$. By including an interaction with a dummy variable, the interaction model is equivalent to creating two separate models for each category the dummy variable represents.

7.15 Marketing Toys Application: Interactions

Using the marketing toys data from the previous application, perform the following steps:

(a) Fit a model with the mean-centered marketing expenditures, the toy type, and an interaction as predictors of the sales. Use toy B as the base level for the dummy variable and print a summary of the fitted model.
(b) Interpret the coefficient of the interaction.
(c) Use the summary to find the fitted models that would result if we were to fit models for type A and type B separately.
(d) Plot the mean-centered marketing variable on the x-axis and the sales on the y-axis, then plot lines representing both toy types.
(e) Fit a model using only the Toy A data to predict sales using marketing. How do the coefficient values differ from those in the model with the interaction?

Solution

(a) Using the mean-centered marketing variable and the interaction `Mktg_MC:Toy`, the model is fit:

```
from ravix import ols
reg = ols("Sales ~ Mktg_MC + ToyA + Mktg_MC:ToyA", data = df)
reg.summary()
```

```
# Summary of OLS Regression Analysis:
# ======================================================
#
# Coefficients:
# ------------------------------------------------------
#                  Estimate  Std. Error  t-value    p-value
# Intercept          650.37      4.7433  137.113   < 2e-16 ***
# Mktg_MC            7.9869     0.43829   18.223   < 2e-16 ***
# ToyA               79.269      6.5348   12.130   < 2e-16 ***
# Mktg_MC:ToyA       2.3232     0.57744    4.023  6.18e-05 ***
#
# Model Statistics:
# ------------------------------------------------------
# Residual Std. Error: 83.1097
# R-squared:       0.7203          AIC: 11682.19
# Adj. R-squared: 0.7195           BIC: 11701.82
# F-statistic: 855.0936 on 3 and 996 DF, p-value: < 2e-16
# ======================================================
```

(b) The interaction can be interpreted as:

For every dollar spent on marketing, there is an additional \$2.32 increase in sales for toy A as compared to toy B.

(c) From the summary, when `ToyA` is 0, the dummy variable and interaction coefficient can be disregarded because they are multiplied by 0. If X and Y denote mean-centered marketing and sales in thousands of USD, the resulting model for toy B is:

$$\hat{Y} = 650.3707 + 7.9869X.$$

The intercept for the toy A model can be found by adding the coefficient of the `ToyA` dummy variable to the intercept. The interaction term of the model can be added to the slope of the toy B model to get the slope of the toy A model. If X and Y denote mean-centered marketing and sales, respectively, the resulting model for toy A is:

$$\hat{Y} = (650.3707 + 79.2695) + (7.9869 + 2.3232)X$$

$$\hat{Y} = 729.6402 + 10.3101X.$$

(d) From the previous part, we note the slopes and intercepts to redefine `modelA` and `modelB` using the `abline` function.

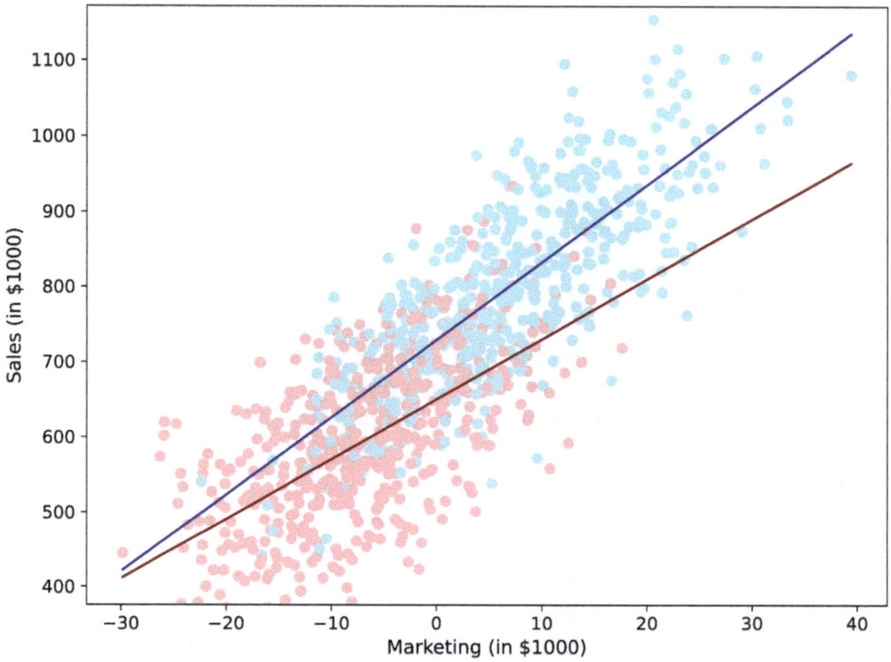

Fig. 7.6 Combined scatterplot of toys with lines 2

Altering the plot code from the previous plot to contain the mean-centered marketing expenditures on the x-axis, we have:

```
from ravix import plot, abline
plot("Sales ~ Mktg_MC", data = df, color = colors,
    xlab = "Marketing (in $1000)", ylab = "Sales (in $1000)")
abline(729.6402, 10.3101, color = "darkblue")
abline(650.3707, 7.9869, color = "darkred")
```

In the above code, we explicitly state the options of the `abline` function to match the intercept (a) and the slope (b) for each model. We also add a legend to clarify which line denotes each toy. The result is shown in Fig. 7.6. Note here the different slope and intercept values since the lines more accurately reflect both categories of the data. This effect is possible since an interaction is included in the model.

(e) Fitting a model for Toy A using the mean-centered marketing variable is done using the following code. A subset of the dataframe (`df_A`) corresponding to the Toy A portion of the dataset is done.

```
df_A = df[df['Toy'] == 'A'].copy()
df_A['Mktg_MC'] = df_A['Marketing'] - np.mean(df_A['Marketing'])
reg = ols("Sales ~ Mktg_MC", data = df_A)
reg.summary()
```

```
# Summary of OLS Regression Analysis:
# =======================================================
#
# Coefficients:
# -------------------------------------------------------
#               Estimate  Std. Error  t-value    p-value
# Intercept       798.96      3.6056  221.586   < 2e-16 ***
# Mktg_MC         10.310     0.36471   28.269   < 2e-16 ***
#
# Model Statistics:
# -------------------------------------------------------
# Residual Std. Error: 80.6247
# R-squared:       0.6161            AIC: 5810.74
# Adj. R-squared: 0.6153             BIC: 5819.17
# F-statistic: 799.1336 on 1 and 498 DF, p-value: < 2e-16
# =======================================================
```

For the toy A model, the slope is the same as the result from part c. Since marketing was centered based upon the marketing for Toy A alone, the intercept differs between the model given here and from part c.

7.16 Case Study: Video Engagement Modeling

7.16.1 Problem Statement

Online video platforms such as YouTube allow users to interact by leaving comments and liking content. These platforms provide a forum for individuals to express their views, share knowledge, and learn from others. However, while some videos receive little attention, others unexpectedly go viral. Producing highly engaging videos can be a profitable skill, as demonstrated by the revenue generated by influencers who consistently create attention-grabbing content.

Luke is the founder and CEO of a small video marketing and production company. He currently manages many of his clients' videos and distributes them on a popular online video platform. Luke measures the success of his videos by the number of "likes" that the videos get, since the number of likes has a direct relationship with the profit from ads that will be collected. After collecting some data, Luke is interested in identifying which promoter he should use, and the effects the sentiment and the age of a video have on the number of likes.

We will conduct a regression analysis to evaluate the relationships between the number of likes and key factors such as promoter, video age, and sentiment score. This analysis aims to provide Luke with actionable insights to increase his clients' revenue by identifying the most influential factors driving engagement.

7.16.2 Data Description

Luke collected a dataset comprising 157 videos with four key variables: the number of likes, the promoter responsible for the video, the video's age in days, and the average sentiment score derived from user comments. The number of likes serves as the response variable, while promoter, age, and sentiment serve as predictors.

- **Likes**—The number of likes a video received.
- **Promoter**—A categorical variable representing the promoter corresponding to the video.
- **Age**—The age of the video in days from the reference date of August 10, 2021.
- **Sentiment**—The average sentiment score of the comments from the video. The sentiment score, computed via a sentiment scoring algorithm, assigns values to comments based on tone, where negative scores indicate poor reception and positive scores reflect favorable feedback. By including sentiment as a predictor, we aim to understand whether user attitudes significantly affect the number of likes a video receives.

The data is contained within the `video_engagement.csv` file. Here, we explore the dataset by loading in and obtaining the `head` of the dataframe.

```
from ravix import get_data
df = get_data("video_engagement.csv")
df.head()
```

```
#      Likes Promoter  Age  Sentiment
# 0   240372        A  437   0.115399
# 1   100243        A  234   0.005290
# 2    19015        A    4   0.022850
# 3   188370        A  364  -0.003858
# 4   173882        A  358  -0.016700
```

From the `dtypes` attribute, the `Promoter` variable is identified as an object that cannot be used directly in the regression model. Therefore, we will need to process the variable accordingly.

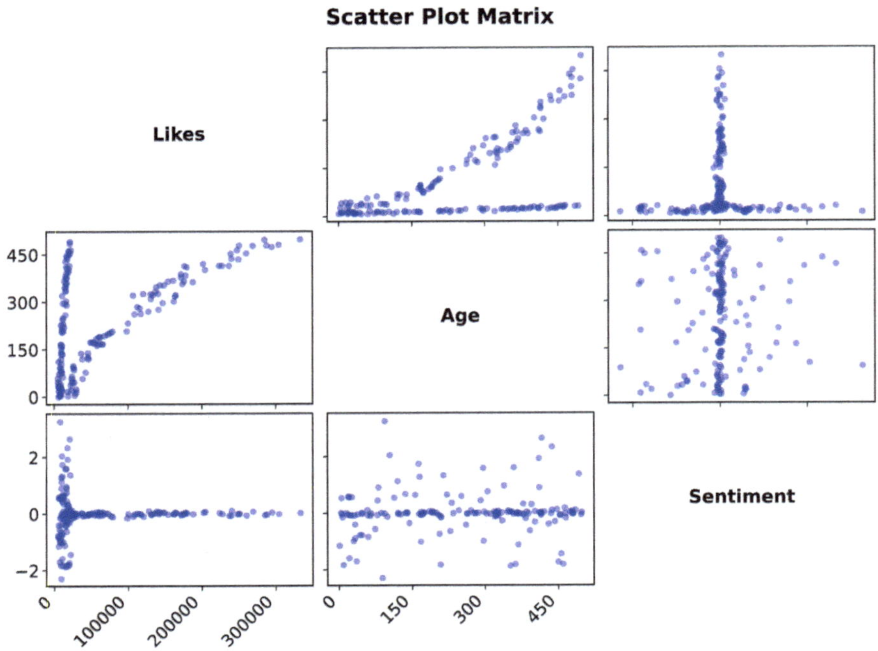

Fig. 7.7 Scatterplots of video engagement data

```
df.dtypes
```

```
# Likes          int64
# Promoter      object
# Age            int64
# Sentiment    float64
# dtype: object
```

Next, we visualize the data using the `plot` function. The resulting scatterplot matrix is shown in Fig. 7.7.

```
from ravix import plot
plot("Likes ~ Age + Sentiment", df)
```

Investigating the scatterplot matrix above yields some interesting findings. Most notably, the `Likes` variable has a particularly interesting relationship with the `Age` variable, warranting further analysis. A correlation matrix plot can be done using the `plot_cor` from Ravix. The resulting output is shown in Fig. 7.8.

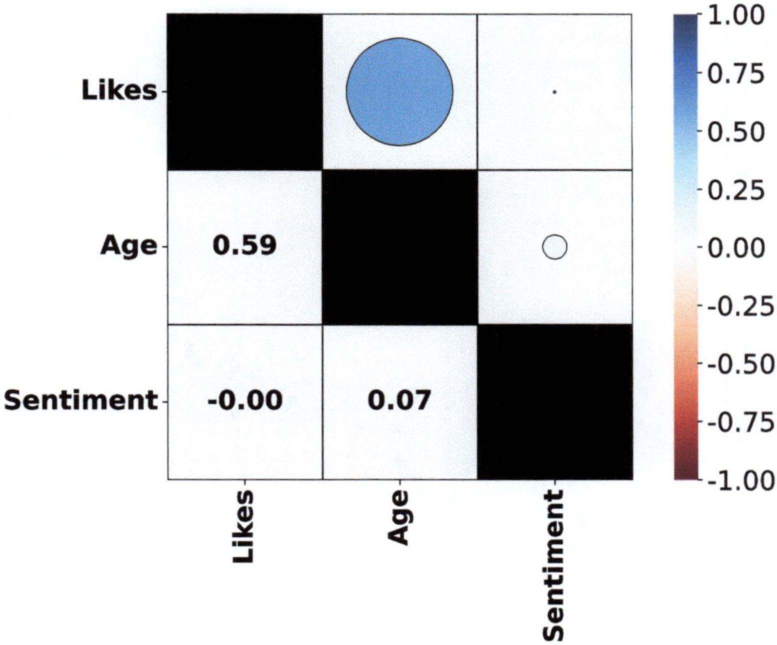

Fig. 7.8 Correlation plot of video engagement data

```
from ravix import plot_cor
plot_cor(df[["Likes", "Age", "Sentiment"]])
```

To explore the relationship between Likes and Age in more detail, we utilize the plot function to generate a scatterplot. Using NumPy's where function, we can differentiate between the promoters by assigning blue to promoter A and red to promoter B. The color parameter in plot allows us to apply this color coding based on the vector we created, visually distinguishing the two promoters in the plot. The resulting plot is shown in Fig. 7.9.

```
import numpy as np
colors = np.where(df.Promoter == "A", "blue", "red")
plot("Likes ~ Age", data = df, color = colors)
```

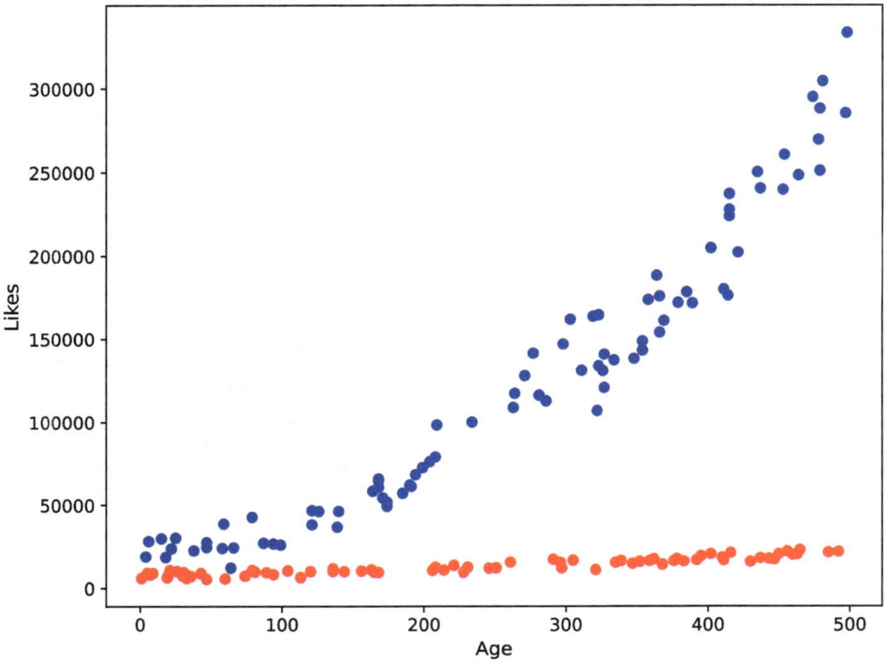

Fig. 7.9 Scatterplot of likes vs. age by promoter 1

7.16.3 *Promoter A Model*

Since the scatterplot matrix above has notably different characteristics for each promoter, we can segment the data by promoter, which will allow us to analyze the data from each promoter separately.

First, we create a dataframe (`df_A`) consisting of all observations from promoter A.

```
df_A = df[df.Promoter == "A"]
```

We fit a multiple regression model on the dataframe for promoter "A" using all of the other variables to predict the `Likes` variable.

```
from ravix import ols
reg = ols("Likes ~ Age + Sentiment", data = df_A)
reg.summary()
```

```
# Summary of OLS Regression Analysis:
# =========================================================
#
# Coefficients:
# ---------------------------------------------------------
#                    Estimate  Std. Error  t-value    p-value
# Intercept          -20326.6      5665.9   -3.588     0.0006 ***
# Age                  552.79      19.087   28.961    < 2e-16 ***
# Sentiment          -28295.6     54576.7   -0.518     0.6055
#
# Model Statistics:
# ---------------------------------------------------------
# Residual Std. Error: 25229.93
# R-squared:        0.9133            AIC: 1967.25
# Adj. R-squared: 0.9112              BIC: 1974.58
# F-statistic: 432.0125 on 2 and 82 DF, p-value: < 2e-16
# =========================================================
```

This summary has a high R_a^2 with the `Age` being significant and `Sentiment` being insignificant according to the p-values. Recalling from the scatterplot matrix in Fig. 7.9, the relationship between `Likes` and `Age` is curvilinear. Hence, we add another variable for the squared `Age` to the regression formula.

```
reg = ols("Likes ~ Age + Age**2 + Sentiment", data = df_A)
reg.summary()
```

```
# Summary of OLS Regression Analysis:
# =========================================================
#
# Coefficients:
# ---------------------------------------------------------
#                    Estimate  Std. Error  t-value    p-value
# Intercept           23074.6      5045.2    4.574  1.70e-05 ***
# Age                  28.309      45.920    0.616    0.5393
# Age^2                1.0447    0.088474   11.808   < 2e-16 ***
# Sentiment            831.26     33379.4    0.025    0.9802
#
# Model Statistics:
# ---------------------------------------------------------
# Residual Std. Error: 15388.57
# R-squared:        0.9681            AIC: 1884.16
# Adj. R-squared: 0.9670              BIC: 1893.93
# F-statistic: 820.6514 on 3 and 81 DF, p-value: < 2e-16
# =========================================================
```

From the summary, the R_a^2 increased by more than 0.05; hence, the model fit has improved. Next, we again analyze the p-values of the coefficients, with the largest p-value being that of the Sentiment variable. From this observation, we drop the Sentiment from the model and observe the resulting R_a^2.

```
reg = ols("Likes ~ Age**2 + Age", data = df_A)
reg.summary()
```

```
# Summary of OLS Regression Analysis:
# ======================================================
#
# Coefficients:
# ------------------------------------------------------
#                Estimate  Std. Error  t-value    p-value
# Intercept       23055.4      4955.3    4.653   1.24e-05 ***
# Age^2            1.0445    0.087693   11.911    < 2e-16 ***
# Age              28.445      45.314    0.628     0.5319
#
# Model Statistics:
# ------------------------------------------------------
# Residual Std. Error: 15294.50
# R-squared:       0.9681          AIC: 1882.16
# Adj. R-squared: 0.9674           BIC: 1889.49
# F-statistic: 1246.16 on 2 and 82 DF, p-value: < 2e-16
# ======================================================
```

While removing Sentiment only marginally increases the R_a^2, the model complexity decreases, which adds to its interpretability. Also, the p-value of the unsquared Age variable is insignificant according to its p-value; therefore, we remove this variable from the model as well.

```
regA = ols("Likes ~ Age**2", data = df_A)
regA.summary()
```

```
# Summary of OLS Regression Analysis:
# ==========================================================
#
# Coefficients:
# ----------------------------------------------------------
#                    Estimate  Std. Error  t-value    p-value
# Intercept           25713.8      2563.5   10.031   5.72e-16 ***
# Age^2                1.0978    0.021911   50.103    < 2e-16 ***
#
# Model Statistics:
# ----------------------------------------------------------
# Residual Std. Error: 15238.57
# R-squared:        0.9680            AIC: 1880.57
# Adj. R-squared: 0.9676              BIC: 1885.45
# F-statistic: 2510.26 on 1 and 83 DF, p-value: < 2e-16
# ==========================================================
```

The R_a^2 again increases by a miniscule amount. However, by removing the unsquared variable from the model, the interpretation of the coefficient in the fitted model for promoter A is simplified significantly. The interpretation is:

- *For every unit increase in the squared age, the number of likes is expected to increase by 1.098. In more practical terms, the positive quadratic term indicates accelerating growth in likes as videos age.*

  The interpretation of the intercept for promoter A videos is:

- *When the video is first released (the age is zero days), the expected number of likes is 25,710.*

7.16.4 Promoter B Model

A similar process can be done to get a model for promoter B. The resulting model is shown below in the summary here.

```
from ravix import ols
df_B = df[df.Promoter == "B"]
regB = ols("Likes ~ Age**2", data = df_B)
regB.summary()
```

```
# Summary of OLS Regression Analysis:
# ======================================================
#
# Coefficients:
# ------------------------------------------------------
#                 Estimate  Std. Error  t-value    p-value
# Intercept         8596.2      288.57   29.789   < 2e-16 ***
# Age^2           0.057260   0.0025987   22.034   < 2e-16 ***
#
# Model Statistics:
# ------------------------------------------------------
# Residual Std. Error: 1695.98
# R-squared:        0.8740           AIC: 1277.09
# Adj. R-squared: 0.8722             BIC: 1281.64
# F-statistic: 485.5156 on 1 and 70 DF, p-value: < 2e-16
# ======================================================
```

The interpretation of the Age-squared coefficient for the promoter B model is:

- *For every unit increase in the squared age, the number of likes is expected to increase by 0.05726.*

  The interpretation of the intercept for promoter B videos is:

- *When the video is first released (the age is zero days), the expected number of likes is 8,596.*

7.16.5 Combined Model

As shown earlier in this chapter, an interaction can be added to model the promoter A and B models jointly.

```
from ravix import ols
import numpy as np
df['PromoterA'] = np.where(df.Promoter == "A", 1, 0)
reg = ols("Likes ~ PromoterA + Age**2 + PromoterA:Age**2",
          data = df)
reg.summary()
```

```
# Summary of OLS Regression Analysis:
# ==========================================================
#
# Coefficients:
# ----------------------------------------------------------
#                      Estimate Std. Error  t-value   p-value
# Intercept              8596.2     1919.7    4.478  1.47e-05 ***
# PromoterA             17117.6     2699.5    6.341  2.44e-09 ***
# Age^2                0.057260   0.017287    3.312    0.0012 **
# PromoterA:Age^2        1.0405   0.023707   43.892   < 2e-16 ***
#
# Model Statistics:
# ----------------------------------------------------------
# Residual Std. Error: 11282.21
# R-squared:        0.9820          AIC: 3379.42
# Adj. R-squared: 0.9816            BIC: 3391.65
# F-statistic: 2781.32 on 3 and 153 DF, p-value: < 2e-16
# ==========================================================
```

From this model with interactions, the R_a^2 has improved due to the number of model observations increasing. The coefficient for `Age`-squared remains 1.0405 + 0.0573 = 1.098 for promoter A and 0.0573 for promoter B. The interpretation of the coefficient for the interaction is:

For every unit increase in the squared age, there is a 1.041 increase in likes for promoter A videos as compared to promoter B videos.

Note that the intercept for promoter B videos is also less than that of the intercept for promoter A videos since the promoter A dummy coefficient is positive.

```
from ravix import plot
colors = np.where(df.Promoter == "A", "blue", "red")
plot("Likes ~ Age", color = colors, data = df)
abline(regA)
abline(regB)
```

In Fig. 7.10, the promoter A and B videos are distinguished by blue and red colors, respectively. The `Likes` are plotted on the y-axis and the `Age` is plotted on the x-axis. The regression trends are plotted in black for both the promoter A model and the promoter B model.

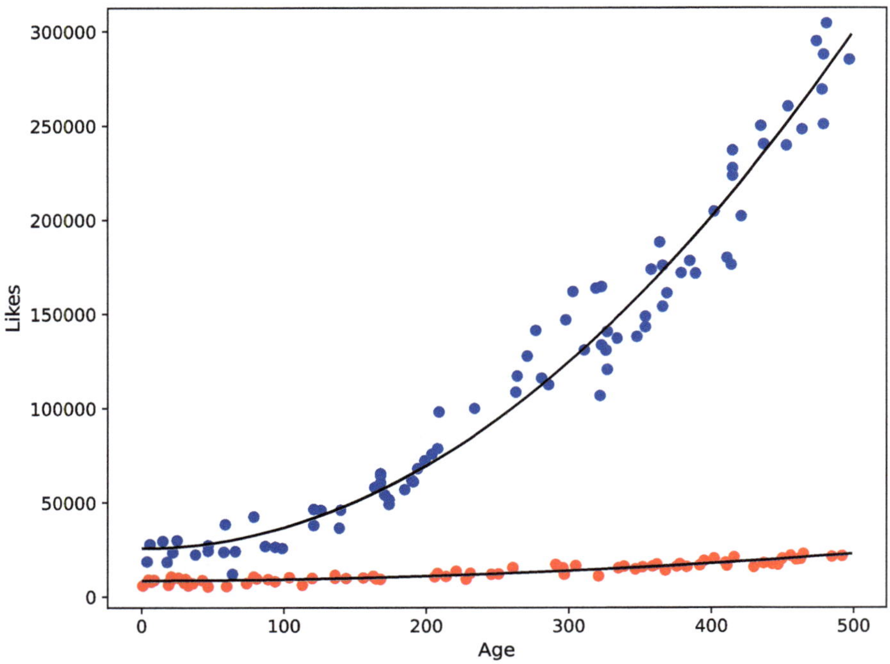

Fig. 7.10 Scatterplot of likes vs. age by promoter

7.16.6 Case Conclusion

This case study underscores the practical application of regression analysis in understanding the relationship between video engagement and demographic factors. By incorporating interaction terms and dummy variables, we examined how different promoters influence the number of likes based on the user's age, offering valuable insights into promotional effectiveness.

The analysis revealed that Promoter A consistently outperformed Promoter B, particularly in generating higher engagement among younger audiences. This distinction is crucial for advertisers like Luke, who seek to maximize visibility and impact on online video platforms. Promoter A's stronger trendline and higher baseline engagement suggest a more compelling promotional strategy that could deliver better returns for video campaigns.

This analysis highlights the versatility of Python's Ravix package, particularly through the use of both its plotting and modeling functions. The `plot` and `plot_cor` functions allowed for visualizing the relationships between variables, while the `ols` and `summary` functions were essential for building and interpreting the regression models. Additionally, functions like `where` and `head` from Pandas and NumPy streamlined data handling, ensuring a smooth workflow from data preparation to model execution.

By leveraging these tools, the analysis became intuitive and efficient, providing clear insights that informed Luke's decision-making process. The use of Ravix in this context highlights how data-driven insights can guide digital marketing strategies to maximize engagement.

Through this case study, we reinforce the value of data analytics in digital marketing. The use of Ravix has simplified the process of examining multiple factors, making it easier to build customized models that account for complex relationships. As businesses seek to optimize campaign performance, these tools will continue to play a pivotal role in making informed, data-driven decisions.

Problems

1. Job Changes Data with Dummy Variables
Using the `job_changes.csv` file from Ravix, answer the following questions:

a. Fit a simple linear regression model to predict the annual salary in thousands of USD (`Salary`) as a function of the number of job changes (`Jobs`). Print a summary of the model.
b. What percent of the variation in `Salary` can be explained by the model in the previous part?
c. Create a barplot of the three education levels in the `Education` variable by passing a `summary` of the Education factor into the `barplot` function.
d. Fit a multiple regression model to predict the `Salary` using `Jobs` and two dummy variables representing education levels as predictor variables. Print a summary of the model.
e. Create a scatterplot of `Jobs` versus `Salary`. Overlay the scatterplot with the regression line for the `HS` level as given in the previous model. Also, overlay the scatterplot with the regression lines for the other two levels as given by the previous model.
f. How should the dummy variables be interpreted?

2. Job Changes Data with Interactions
Using the `job_changes.csv` file, answer the following questions:

a. Fit a multiple regression model on the dataset predicting `Salary` as a function of `Jobs` and `Education`. Include an interaction between the predictor variables. Print a summary of the model.
b. Interpret the interaction coefficients from the previous model.
c. Create a scatterplot of `Jobs` versus `Salary`. Overlay the scatterplot with the regression line for the `HS` level as given in the previous model. Also, overlay the scatterplot with the regression lines for the remaining levels as given by the previous model.

d. Using the multiple regression model with the interaction from part a, specify the simple linear regression equation that predicts `Salary` using `Jobs` as a predictor for high school graduates (`HS`). Also, specify the same simple linear regressions for college graduates (`bachelors`) and people with a graduate degree (`masters`).
e. Repeat the plot from part c and color the observations by `Education` level.
f. Create and add a variable called `JobsM` which mean-centers the `Jobs` variable.
g. Fit a multiple regression model that uses the following predictors:

- A dummy variable for observations at the `bachelors` level
- A dummy variable for observations at the `masters` level
- The mean-centered `Jobs`
- Interactions between each dummy variable and the mean-centered Jobs

h. Interpret the interactions from the previous part.

3. Automotive Tire Sales with Categorical Variables
Here, we will use the tires data to solve the questions below. Load in the `Tires.csv` file by running the code chunk below.

```
df = read.csv("tire_sales.csv")
```

a. Get a summary of the data using the `summary` function.
b. List the variables that are numeric and the variables that are categorical.
c. Generate a linear model to predict the response variable (`Sales`) from all of the other variables in the dataset.
d. Display a summary of the model given in the previous part.
e. Interpret the dummy variables from the summary in part d.

4. Facebook Analysis with Dummy Variables
Using the `facebook.csv` file, answer the following questions:

a. Fit a simple linear regression model to predict the number of Facebook friends (`Friends`) as a function of the number of friend requests sent (`Requests`). Print a summary of the model.
b. What percent of the variation in the number of friends can be explained by the model in the previous part?
c. Fit a multiple regression model to predict the number of friends (`Friends`) using the friend requests sent (`Requests`) and sex (`Sex`) as predictor variables. Print a summary of the model.
d. Create a scatterplot of `Requests` versus `Friends`. Overlay the scatterplot with the regression line for males as given in the previous model. Also, overlay the scatterplot with the regression line for females as given by the previous model.
e. How should the sex dummy variable be interpreted?

5. Facebook Analysis with Interactions
Using the `facebook.csv` file, answer the following questions:

a. Fit a multiple regression model on the dataset predicting `Friends` as a function of `Sex` and `Requests`. Include an interaction between the predictor variables. Print a summary of the model.
b. Interpret the interaction coefficient from the previous model.
c. Create a scatterplot of `Requests` versus `Friends`. Overlay the scatterplot with the regression line for males as given in the previous model. Also, overlay the scatterplot with the regression line for females as given by the previous model.
d. Using the multiple regression model with the interaction from part a, specify the simple linear regression equation which predicts the number of `Friends` using `Requests` for males. Also, specify the same simple linear regression equation for females.
e. Repeat the plot from the previous part and color the observations by `Sex`.
f. Fit a multiple regression model that uses the following predictors:

- A dummy variable for females
- The mean-centered friend requests
- An interaction

Chapter 8
Model Diagnostics

All generalizations are false, including this one.

—Mark Twain

8.1 Introduction

Knowing the correct mathematical model can be tremendously helpful when trying to make predictions. In many cases, however, a mathematical model that is relatively close to the true state may suffice. Yet this process brings its challenges. The quote from Mark Twain shown above alludes to the difficulty of attempting to identify an acceptable model.

In previous chapters, we discussed how to find and evaluate models. In this chapter, we explore methods for evaluating regression models—one important tool is known as diagnostic plots—and determine if they are acceptable for a given situation. We will discuss diagnostic plots after a review of the mathematical model for multiple regression. While diagnostic plots indicate when the regression model assumptions are violated, addressing these violations requires additional analysis. We introduce practical methods for fixing some common model assumption violations. This chapter further addresses unusual observations of the data that may occur. Lastly, a case study about an automotive sales conglomerate allows us to demonstrate how to run the analysis in detail with the relevant diagnostic plots and python source code.

8.2 Multiple Regression Model Revisited

In Chap. 5, Eq. (5.1) describes how the response variable Y relates to the predictor variables X_1, X_2, ..., X_p and an error term ε as specified below:

$$Y = \beta_0 + \beta_1 X_1 + \beta_2 X_2 + \cdots + \beta_p X_p + \varepsilon,$$

D. P. McGibney, *Applied Linear Regression for Business Analytics with Python*,
International Series in Operations Research & Management Science 378,
https://doi.org/10.1007/978-3-032-23806-1_8

where $\beta_0, \beta_1, \beta_2, ..., \beta_p$ are model parameters. As alluded to in the chapter introduction, generalizations must be met with skepticism since the actual mathematical model can be elusive or even impossible to find. In Chap. 7, we explored the general linear regression model, which provides some techniques for modeling nonlinear relationships by manipulating the data to conform to the model from Eq. (5.1). After making such manipulations, it is important to determine if the model assumptions are met or if another model should be considered.

8.3 Model Assumptions

In Chap. 4, the assumptions for the simple linear regression model were stated. Here, we extend the assumptions to the multiple predictor case. The four assumptions of the multiple regression model are listed below.

1. The relationship between the response and each predictor variable is linear (or $E[\varepsilon_i] = 0$ for each ε_i).
2. Each value of ε_i is independent.
3. The random variable, ε_i, is normally distributed.
4. The variance of ε_i, denoted as σ^2, is the same for all observations.

The assumptions are summarized in the statement below:

$$\varepsilon_i \overset{\text{i.i.d.}}{\sim} \text{N}(0, \sigma^2).$$

The notation can be summarized as "each error is independent and identically distributed following a normal distribution with mean of zero and variance of $sigma^2$."

In this chapter, we discuss these assumptions in detail, as they are frequently violated in practice. The consequences depend on which assumption fails to hold and to what extent. In certain situations, the coefficients might be biased, or some of the fundamental statistics from the regression may be invalid.

8.4 Violations of the Model Assumptions

If the relationship between a response and one or more predictor variables is nonlinear, then a linear model is not appropriate for the relationship. To resolve this issue, we can transform the predictor variable as discussed in Chap. 7, which allows some nonlinear relationships to fit into the framework of the General Linear Model. In some cases, it may be necessary to transform the response variable.

If the errors (ε) are dependent, then the relationship between the response and predictors is not correctly captured in the confidence and prediction intervals because the standard error estimates become unreliable.

Separately, if the variance of the errors changes as the response increases, then the estimate of the standard deviation is invalid. This violation, known as heteroscedasticity, also leads to invalid confidence and prediction intervals.

8.5 Residual Analysis

To assess whether the assumptions of linear regression are met, we can apply a variety of diagnostic tools. One popular method is referred to as residual analysis. To analyze the residuals of a regression model, we can look at a plot of the residuals on the y-axis and the fitted values ($\hat{Y}$) on the x-axis. In the simple linear regression case, the predictor variable can be plotted on the x-axis in place of the fitted values.

Figure 8.1a shows a residual plot in which the residuals have roughly the same variation across all fitted values, from left to right. In addition, the trend of the residuals and the fitted values is zero, indicating the linearity assumption is met.

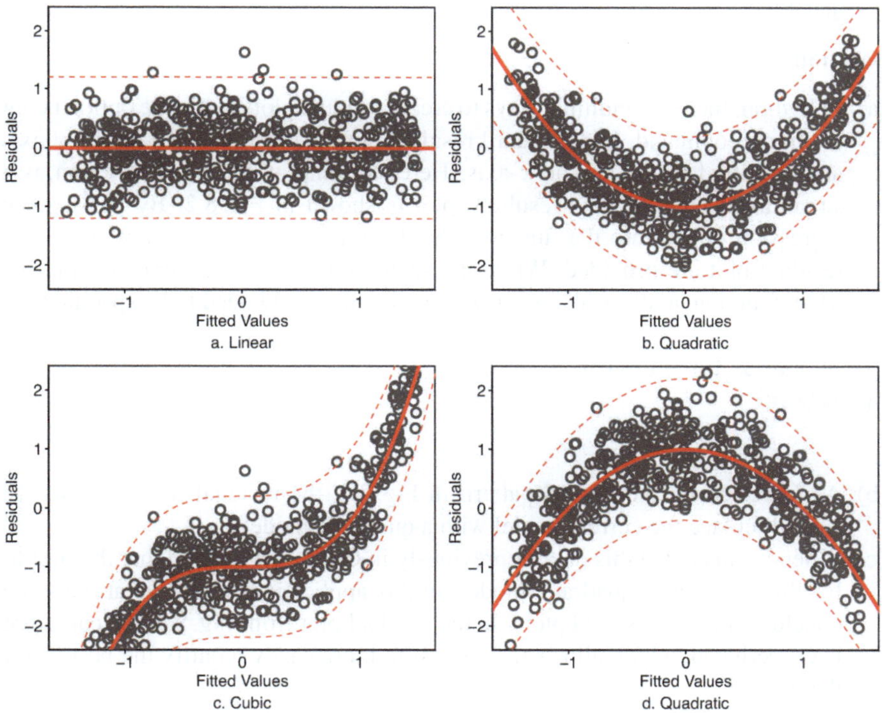

Fig. 8.1 Residual plots

Figure 8.1b and d, on the other hand, demonstrate a quadratic or parabolic relationship. The residuals in Fig. 8.1c form a cubic relationship with the fitted values. These patterns in the residuals indicate that the linearity assumption has been violated.

8.6 Sales Performance Application: Residual Analysis

As discussed in Chap. 7, a quadratic function provided the best fit for the sales data in Table 7.2. The linear model was fit using the following python commands.

```
from ravix import ols, get_data
df = get_data("employment.csv")
reg = ols("Sales ~ Employment", data = df)
```

From the simple linear regression model, do the following:

(a) Plot the residuals on the y-axis and the fitted values on the x-axis.
(b) Discuss the fit of the model based on the residual plot.
(c) What advantage does a residual plot have over a standard scatterplot between X and Y?

Solution:

(a) In Python, there are multiple ways to create residual plots, with one option being the `plot` command. In a standard residual plot, the residuals are on the y-axis and the fitted values on the x-axis. Here, we employ the `plot` function as a streamlined method. The resulting plot is shown in Fig. 8.2. By passing the regression object into the function, the default colors and labels relevant to a residual plot are provided. While the same plot could be generated using the `plot` function, additional customization options would need to be specified.

```
from ravix import plot
plot(reg)
```

(b) As indicated by the curved pattern in Fig. 8.2, the residual plot suggests we should replace the current model with a quadratic model.
(c) When we analyzed this dataset previously in Chap. 5, we found that the best fit for this data was a quadratic model. In this application, we arrive at the same conclusion using residual plots. In the residual plot from Fig. 8.2, the curvature is exacerbated, which allows the analyst to more easily identify the best fit for the data.

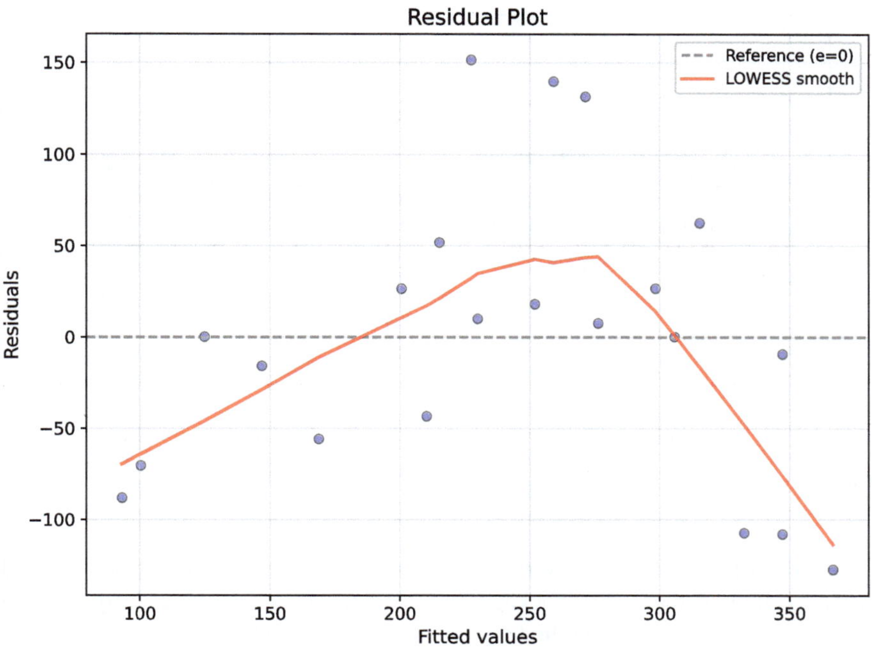

Fig. 8.2 Residual plot

8.7 Constant Variance

Model assumptions are commonly assessed by plotting residuals against the fitted values to detect patterns that may indicate nonlinearity or heteroscedasticity. The statistical term for constant variance is homoscedasticity, whereas heteroscedasticity refers to nonconstant variance.

In Fig. 8.3a, no discernible pattern exists between the residuals and the fitted values. The residuals have constant variance from left to right. The residuals from Fig. 8.3b increase in variation as the fitted values increase, resulting in a funnel-like shape as shown in Fig. 8.3b.

Identifying nonconstant variance from residual plots constitutes standard practice. For a more algorithmic procedure, a hypothesis test can be set up and concluded. This test, developed by Breusch and Pagan in 1979, is referred to as the Breusch-Pagan, or the nonconstant variance test. The null and alternative hypotheses would then be:

$$H_0 : \text{Constant variance}$$

$$H_1 : \text{Nonconstant variance.}$$

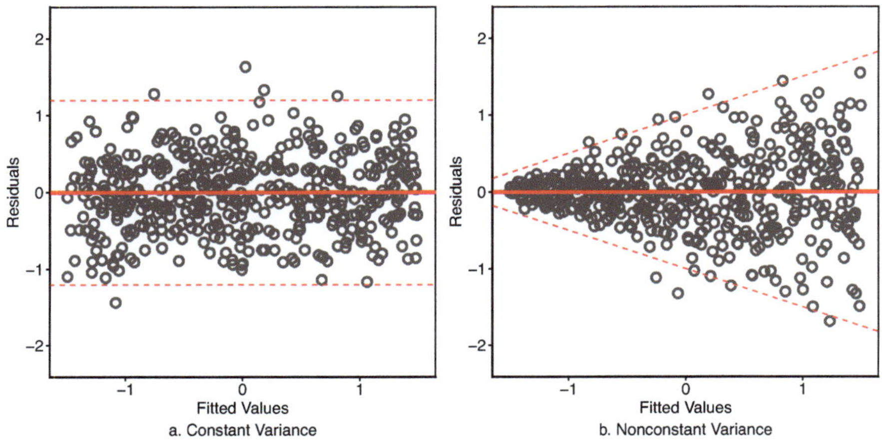

Fig. 8.3 Residual plots

In order to run a nonconstant variance test in Ravix, the `ncv` function can be used. After loading in the Ravix package and creating a model object `reg`, the `ncv` function can be used with the first argument being the model object, and there is an optional second argument to specify the level of significance (`alpha`). The syntax is given below.

```
from ravix import ncv
ncv(reg, alpha = 0.05)
```

```
# Nonconstant Variance Test
# ==========================================================
# Test Statistic       : 0.8408
# p-value              : 0.3592
# Result               : Constant variance (p >= 0.05)
# ==========================================================
```

This function calculates a test statistic and the corresponding p-value for the test. For the application in Sect. 8.6, the p-value generated for this test indicates that there the constant variance assumption appears to be satisfied. Specifically, there is not enough evidence to reject the H_0 of the test.

8.8 Tweet Engagement Application: Residual Variance

Lina Alvarez aspires to be a social media influencer. To achieve her goal, she investigates the performance of social media influencers to understand the relationship between the number of tweets that an influencer posts per year and the followers added in the same year. Understanding this relationship will help her manage her own social media account with greater efficiency. She collects the number of followers (`Followers`) and the corresponding number of tweets (`Tweets`) from influencers in similar domains and saves the data in the `tweet_engagement.csv` file.

Using the data in the `tweet_engagement.csv` file located on the companion site, do the following:

(a) Load in the data and plot `Followers` on the y-axis and the `Tweets` on the x-axis. Comment on the relationship between the plotted variables.
(b) Fit a linear regression model that predicts `Followers` using `Tweets`. Use the model to fit a residual plot. Comment on the variance of the residuals.
(c) Test the residuals for nonconstant variance by performing a nonconstant variance test. Comment on the findings.

Solution:

(a) A plot is easily generated by reading in the data and then using the `plot` function.

```
from ravix import get_data, plot
df = get_data("tweet_engagement.csv")
plot("Followers ~ Tweets", data = df)
```

While the `Followers` trends upward as `Tweets` increases, the variance of this trend increases as `Tweets` increases. This relationship can be seen by the upward-trending funnel shape of the scatterplot in Fig. 8.4.

(b) After fitting the linear model using `ols`, we again use the `plot` command with the model as the first argument.

```
reg = ols("Followers ~ Tweets", data = df)
plot(reg)
```

From the output in Fig. 8.5, the increase of the residual variance becomes obvious from the funnel shape of the residual plot.

(c) We use the `ncv` function with `reg` as the only argument.

```
ncv(reg)
```

```
# Nonconstant Variance Test
# ==========================================================
# Test Statistic        : 57.2324
# p-value               : 3.875e-14
# Result                : Nonconstant variance (p < 0.05)
# ==========================================================
```

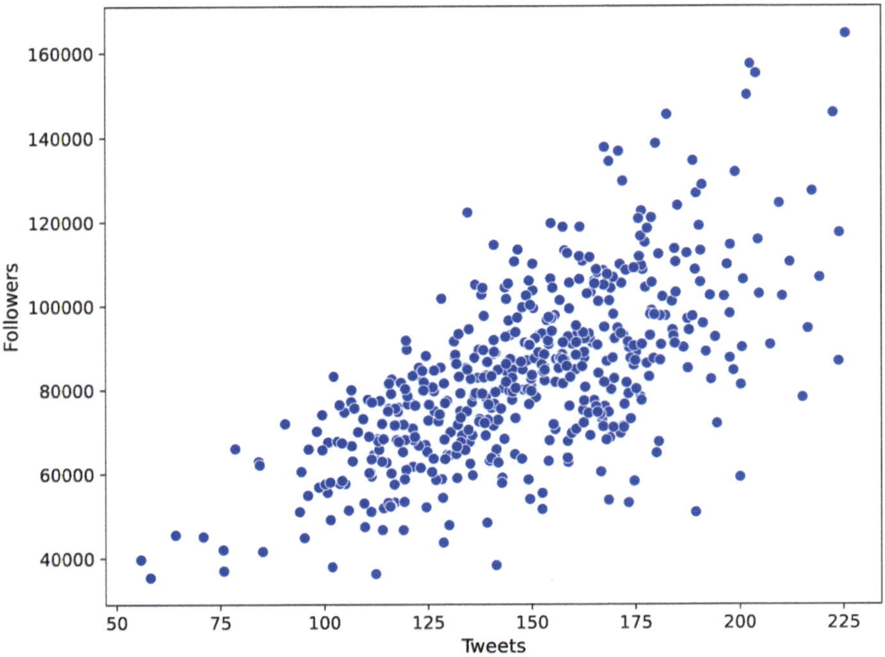

Fig. 8.4 Scatterplot of tweets vs. followers

Here the p-value (3.875×10^{-14}) is much less than $\alpha = 0.05$, which indicates that we reject the null hypothesis and conclude that the residuals have nonconstant variance. Further measures are required to address this violation and will be addressed further in Sect. 8.13.

8.9 Response Variable Transformations

In addition to predictor variable transformations, it may be helpful to transform the response variable. In fact, it is often the case that nonconstant variance can be corrected by transforming the response variable. The logarithmic transformation is a widely used transformation that we discuss here.

8.10 Logarithmic Transformations

In Fig. 8.6, the scatterplot on the left side indicates exponential growth as X increases. We recognize the exponential growth by the changing increase in Y as X increases. Specifically, the values of Y increase at a slow rate when X is negative,

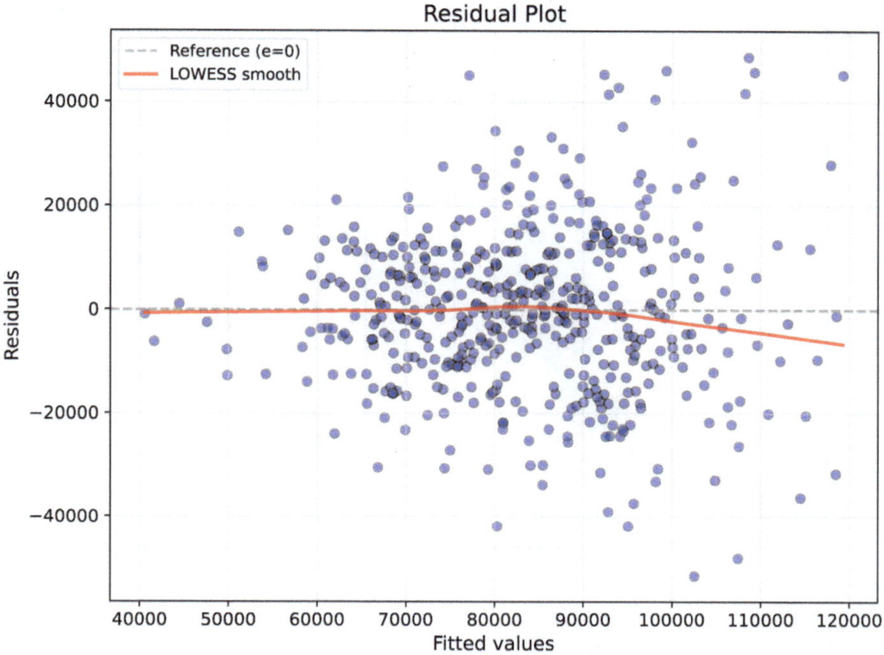

Fig. 8.5 Residual plot

but when X is positive, the points on the scatterplot increase more rapidly. Applying the natural logarithm of the response variable, Y, results in a linear relationship between $\log(Y)$ and X, as depicted on the right side of Fig. 8.6. This transformation is effective because taking the natural logarithm compresses large values of Y, reducing the curvature and revealing an underlying linear trend.

The left side of Fig. 8.7 exhibits a pattern of exponential decay in which the slope of the trend is initially large but gradually approaches zero. Noting the exponential decay, the logarithm of X is applied and plotted versus Y. The result is the linear relationship depicted on the right side of Fig. 8.7.

In some datasets, it may be helpful to take the logarithm of both the response and predictor variables to model a linear relationship. On the left side of Fig. 8.8, we note that many values appear bunched up between 0 and 2 for X and between 0 and 1 for Y. The logarithmic transform can "unbunch" these values. In the center of Fig. 8.8, we note that applying the logarithm of Y results in an exponential decay relationship. Therefore, the logarithm of X is applied, and the resulting relationship appears on the residual plotide of Fig. 8.8.

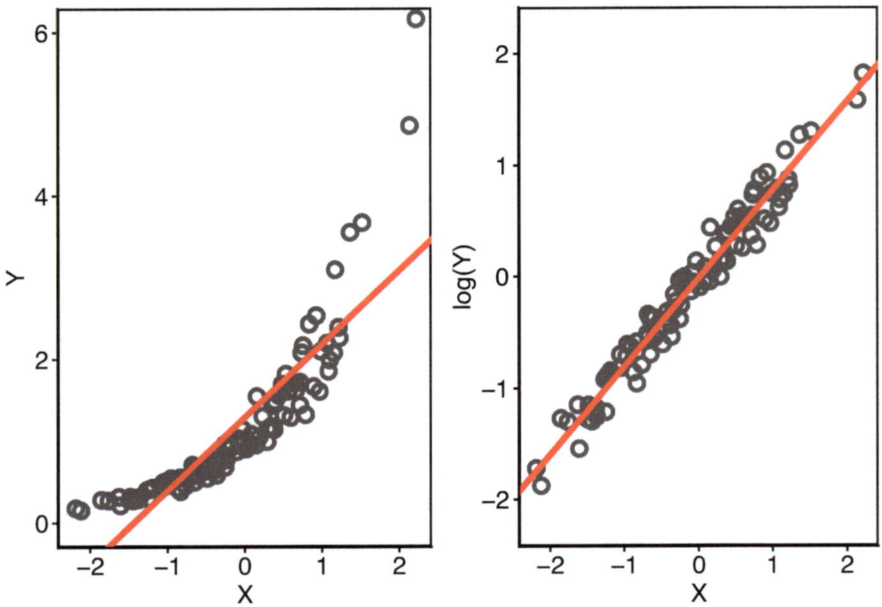

Fig. 8.6 Log(Y) transformation

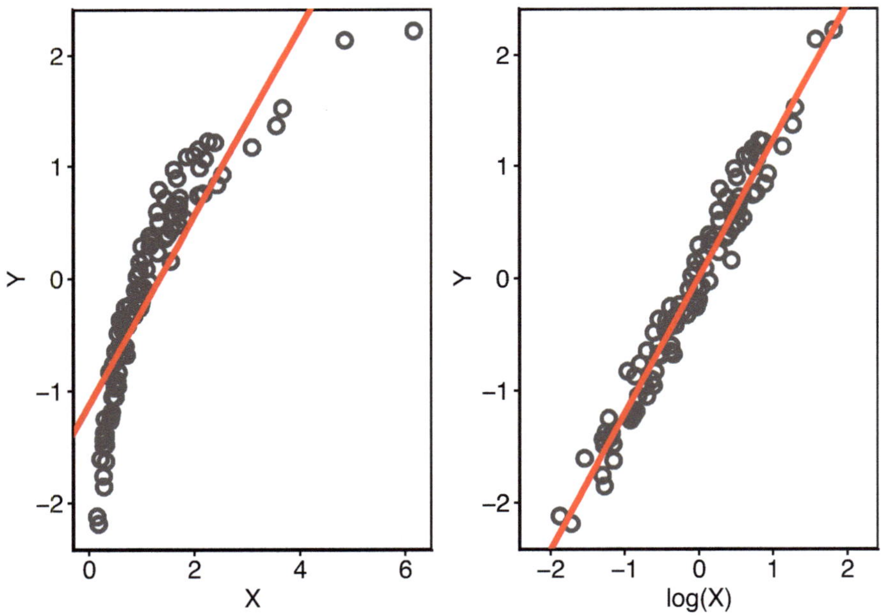

Fig. 8.7 Log(X) transformation

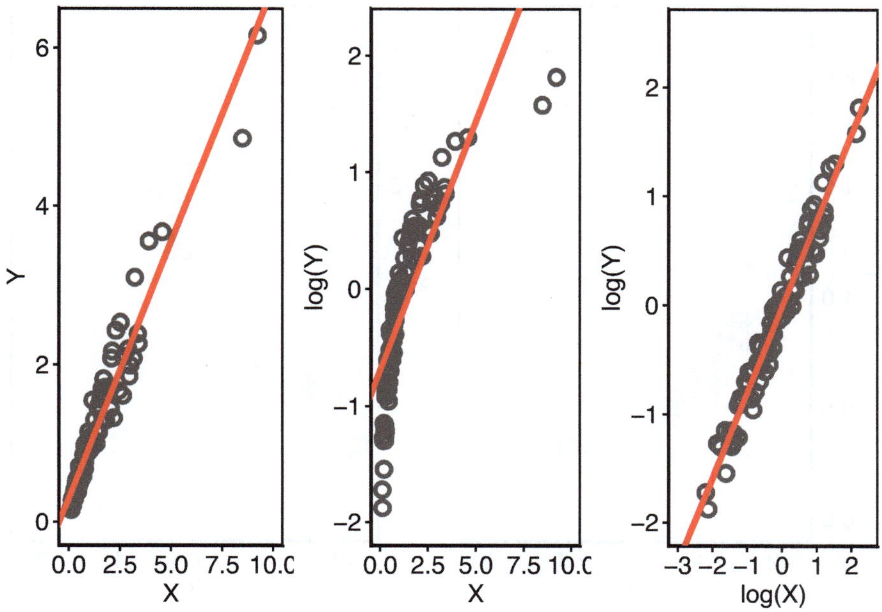

Fig. 8.8 Log-log transformation

8.11 Other Response Variable Transformations

Popular response variable transformations include:

- Logarithmic
- Reciprocal
- Square root

Of the three mentioned above, the logarithmic transformation is the most widely used. As an example of the reciprocal transformation, the left side of Fig. 8.9 depicts a relationship between X and Y for which applying a reciprocal transformation to Y is appropriate. The action results in the model fit depicted in the residual plotide plot of Fig. 8.9.

In Fig. 8.10, the observed X and Y values are plotted along with the untransformed regression fit on the left side of the figure. On the right side of Fig. 8.10, the plot depicts the square root transformation of the response variable. When variance violations cannot be resolved visually, formal transformation selection methods such as Box-Cox are useful.

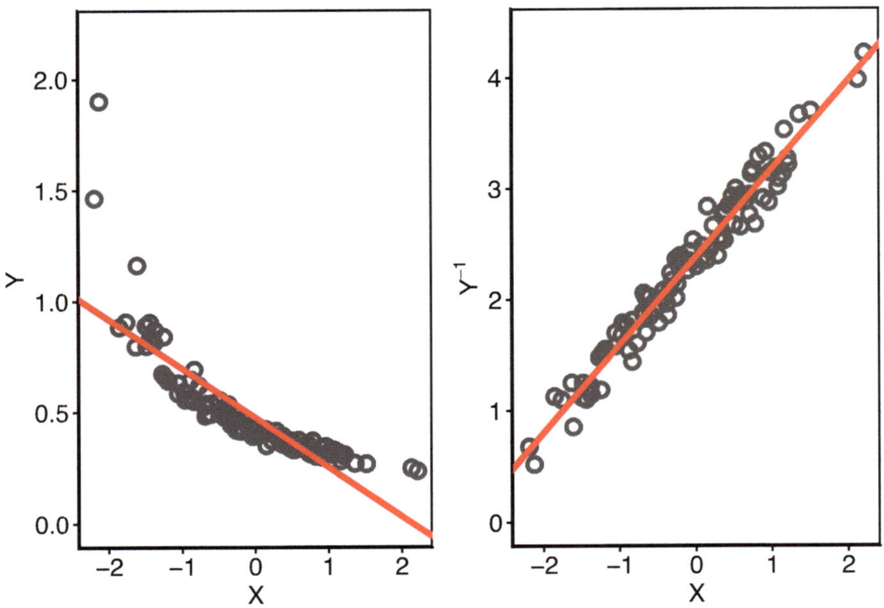

Fig. 8.9 Reciprocal transformation

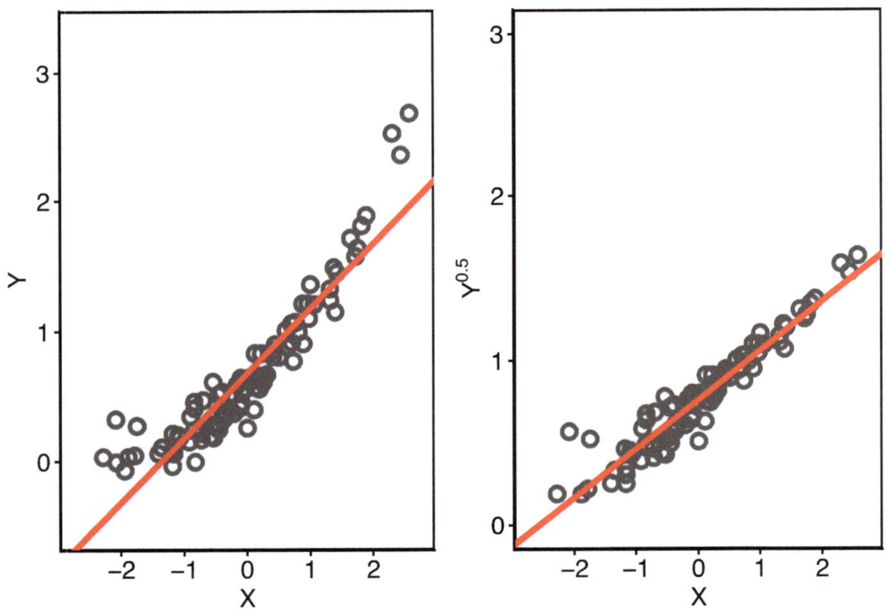

Fig. 8.10 Square root transformation

8.12 Box-Cox Transformation

Guidelines exist to help select the appropriate response variable transformation. One such set of guidelines is referred to as the Box-Cox method. Statisticians George E.P. Box and David R. Cox developed a procedure to identify the "optimal" power transformation of Y to stabilize the variance of the model residuals. In particular, the Box-Cox method finds the value of lambda (λ) that maximizes the log-likelihood function. Simply put, an analyst chooses the value of λ that is most likely to be the response variable exponent.

To use the Box-Cox method, we first find a model that predicts the untransformed Y, for example, the model `reg`. Then, the resulting model is passed along to the `box_cox` function.

```
from ravix import box_cox
box_cox(reg)
```

After passing the regression model into the `box_cox` function, the log-likelihood values are on the y-axis and values of λ are on the x-axis. Figure 8.11 shows an example of this plot.

While a more exact value of the estimate of λ can be found, an analyst commonly chooses the exponent of Y so that a common transformation is used. Therefore, we choose the exponent of Y to be one of the following:

- $\lambda = -1$ denotes the reciprocal transformation.
- $\lambda = 0$ denotes the logarithmic transformation.

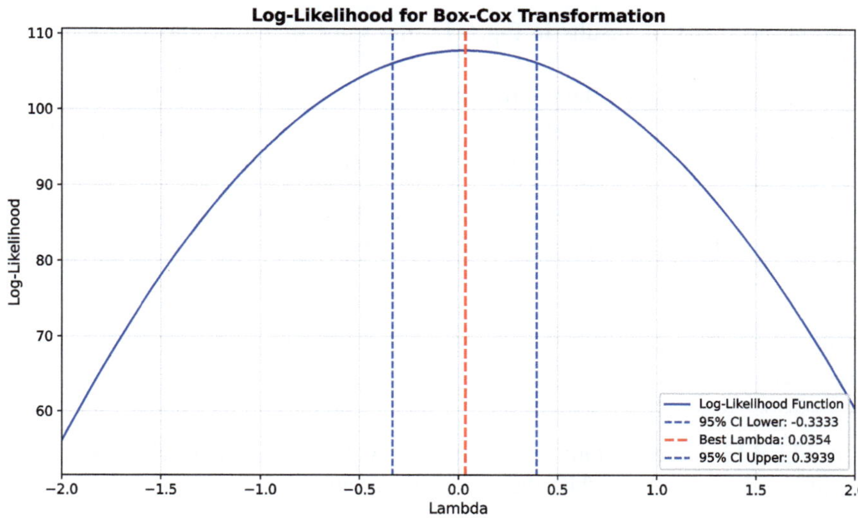

Fig. 8.11 Box-Cox results

- $\lambda = 0.5$ denotes the square root transformation.
- $\lambda = 1$ denotes that no transformation should be applied to Y.
- $\lambda = 2$ denotes a square transformation.

Figure 8.11 indicates that a logarithmic transformation would be most appropriate since $\lambda = 0$ is very close to the maximum log-likelihood.

8.13 Tweet Engagement Application: Box-Cox

Using the data from the previous application, use the techniques mentioned above to diagnose and rectify the nonconstant variance. Specifically, do the following:

(a) Apply the Box-Cox method to the linear model from the previous application and identify which transformation would be most appropriate.
(b) Fit a new model with the prescribed relationship and plot the residuals. Comment on the variance of the residuals.
(c) Perform a test for nonconstant variance on the revised model and compare the resulting p-value with the previous application's p-value.

Solution:

(a) The `reg` object from the previous application is used as the only argument of the `box_cox` function.

```
from ravix import box_cox
box_cox(reg)
```

From the Box-Cox results in Fig. 8.12, the value of λ that maximizes the log-likelihood function is close to 0.5. Therefore, we choose the response variable exponent of 0.5 to model the relationship between `Tweets` and `Followers`. Since taking a variable to the power of 0.5 is equivalent to taking the square root, this transformation is referred to as the square root transformation.

(b) The new model is fit by changing the response in the formula to `Followers ** .5`. After the transformed model is fit, we use the `plot` function to provide a residual plot.

```
Treg = ols("Followers ** .5 ~ Tweets", data = df)
plot(Treg)
```

In Fig. 8.13, there are only a few residuals with fitted values below 240. When the fitted values are above 240, the variance is relatively constant.

(c) Using the `ncv` function on the transformed regression model, we can conduct a test for nonconstant variance.

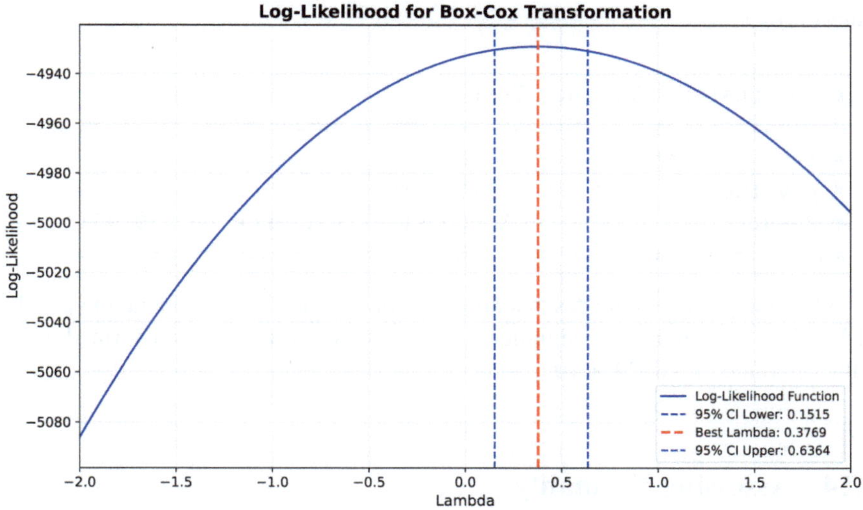

Fig. 8.12 Box-Cox results

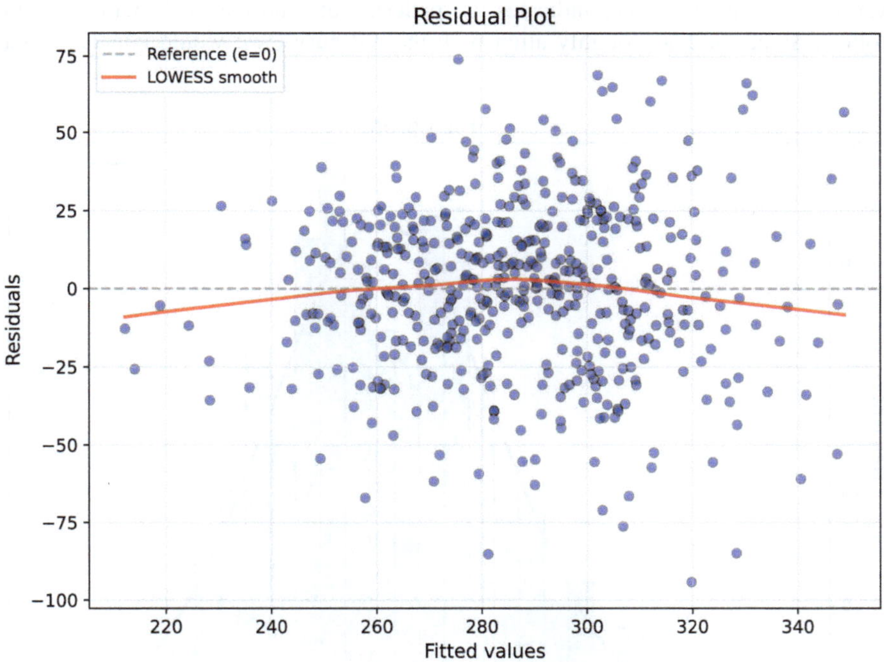

Fig. 8.13 Transformed response residual plot

```
ncv(Treg)
```

```
# Nonconstant Variance Test
# ========================================================
# Test Statistic        : 22.8458
# p-value               : 1.755e-06
# Result                : Nonconstant variance (p < 0.05)
# ========================================================
```

While the p-value indicates that nonconstant variance is still evident, the p-value of 1.755×10^{-6} indicates a significant improvement over the untransformed model with a p-value of 3.8724×10^{-14}.

8.14 Assessing Normality

Normality is often assessed by visually inspecting a histogram for a bell-shaped curve. Figure 8.14 shows a histogram with observed values of a continuous random variable X. The red curve indicates the pattern that a normal distribution would follow. Since the bars roughly align with the red curve, we would conclude that

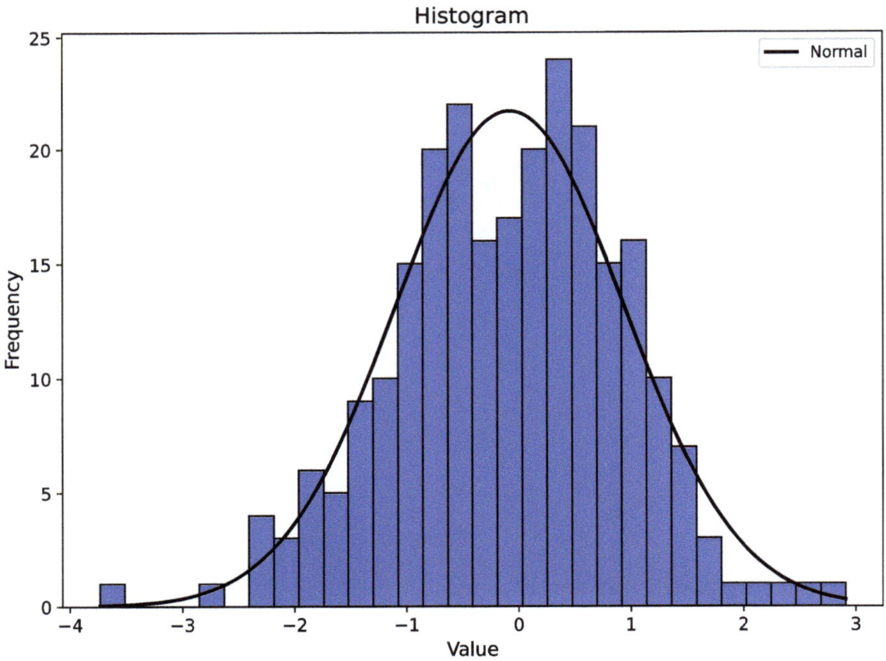

Fig. 8.14 Histogram of transformed residuals

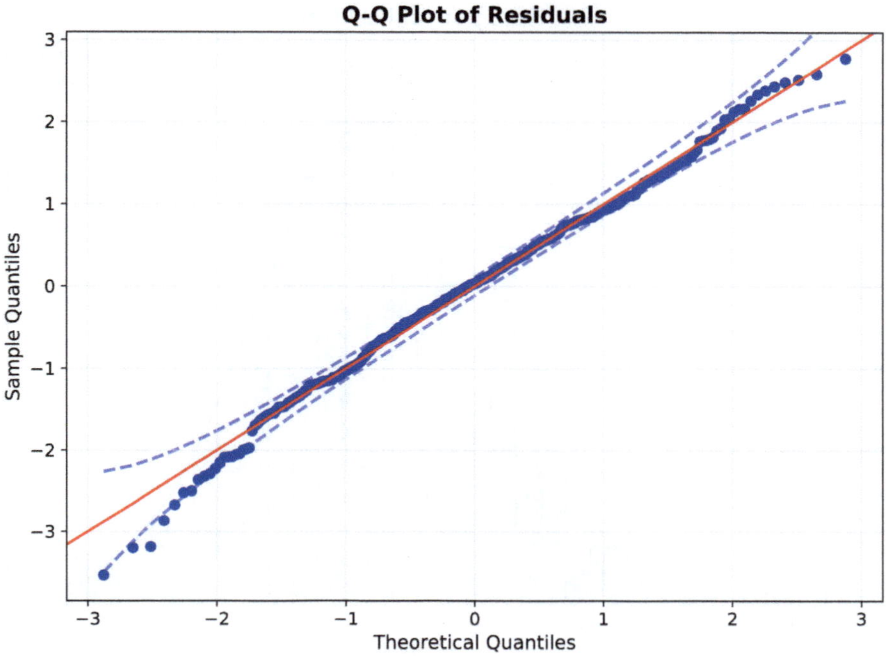

Fig. 8.15 Normal Q-Q plot

the distribution of X is approximately normal. This method of inspection is very intuitive since most people are familiar with the bell curve of a normal distribution.

Another popular method consists of plotting the ordered quantiles of the residuals on one axis versus the theoretical quantiles expected. This resulting plot is called a Quantile-Quantile plot (QQ plot). For normality to hold, the points should roughly fall on the line. The QQ plot in Fig. 8.15 shows the residuals on the x-axis, which lines up closely with the behavior of a normal distribution as seen by the straight line.

```
from ravix import qq
qq(Treg)
```

While plots are generally sufficient to test the residuals for normality, one might employ the Shapiro-Wilk hypothesis test. The null and alternative hypotheses would then be:

$$H_0 : \text{Normal distribution}$$

$$H_1 : \text{Non-normal distribution.}$$

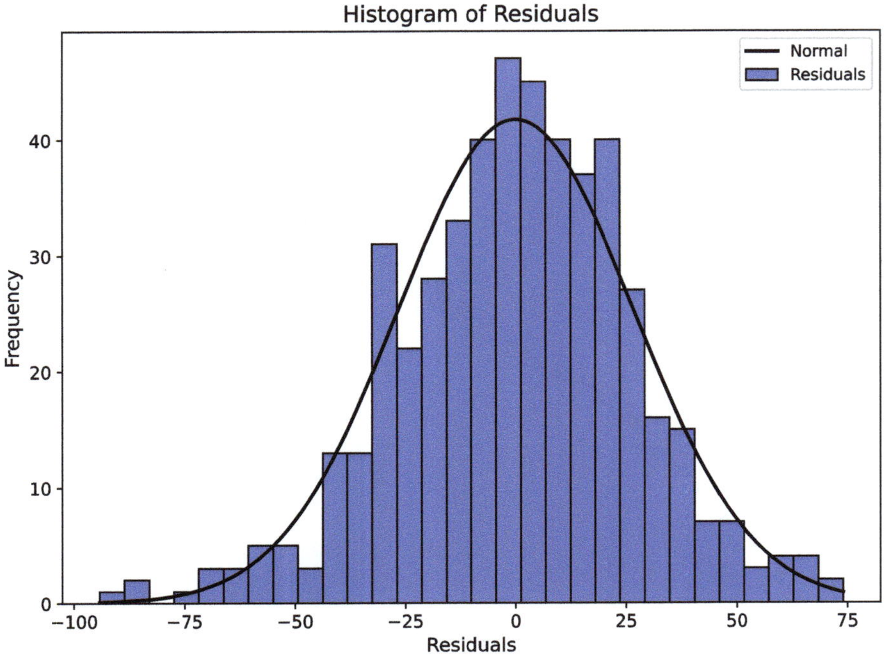

Fig. 8.16 Histogram of transformed residuals

The residuals are available in the model `Treg`. In Fig. 8.16, we visually inspect if the residuals are normal using the `hist` function.

```
from ravix import hist
hist(Treg)
```

The histogram of the transformed residuals indicates there is minimal visual evidence of non-normality. We investigate further using the Shapiro-Wilk test and get a p-value of 0.03293. This p-value indicates that the distribution of the transformed residuals is non-normal, or at least that the null hypothesis should be rejected at $\alpha = 0.05$.

```
from ravix import shapiro
shapiro(Treg)
```

```
# Shapiro-Wilk Test for Normality
# ========================================================
# Test Statistic      : 0.9936
# p-value             : 0.03293
# Result              : Non-normal (p < 0.05)
# ========================================================
```

Although the Shapiro-Wilk test detects mild deviation from normality, the histogram indicates the residuals are approximately bell-shaped, suggesting that the normality assumption is reasonably satisfied in practice.

8.15 Assessing Independence

Although statistical tests for independence exist, independence is usually established by understanding how the data were collected rather than by running a test. For example:

1. Time-ordered data
   Observations collected over time, such as stock prices, often show dependence because each value is related to previous values.
   Solution: Time series models (e.g., autoregressive integrated moving average (ARIMA), exponential smoothing) are specifically designed to model these time-based relationships.
2. Geographically clustered data
   Observations that are geographically close often display similar behavior due to shared economic, demographic, or market conditions. For example, housing prices in neighboring zip codes, retail sales across nearby stores, or insurance claim frequencies by region may be correlated because of local factors such as income levels, competition, or weather patterns.
   Solution: Spatial models (e.g., kriging, spatial autoregressive models) incorporate the structure of geographic correlations to improve predictions and inference.
3. Repeated measurements
   When multiple measurements are taken from the same subject (e.g., blood pressure readings over time), errors are often correlated due to individual-specific factors.
   Solution: This dependence can be handled by modeling within-subject patterns directly (e.g., mixed-effects models) or by transforming the repeated measurements, such as taking differences when appropriate.

Understanding whether your data are independent or dependent is critical for choosing the correct modeling approach. Mismatched models can lead to biased estimates, misleading predictions, and invalid inferences. Testing for dependence and exploring the structure of your data are essential steps before fitting a model.

8.16 Outliers and Influential Observations

Some observations are more extreme than others. An extreme observation is referred to as an outlier, which can be revealed by plotting the residuals against a predictor variable. Outliers will be easy to identify because the residual will lie far away from the rest of the plot.

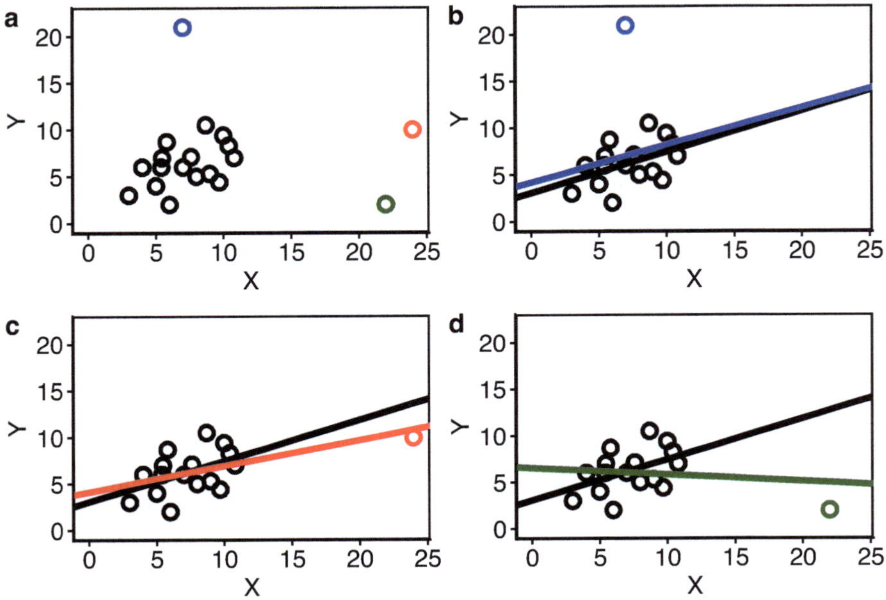

Fig. 8.17 Scatterplots illustrating the influence of outliers

If an outlier is influential, the results of the analysis may be affected. An influential observation refers to one that has a disproportionate effect on the value of the slope and intercept in the least-squares regression equation.

Figure 8.17a shows a scatterplot with three extreme observations colored blue, red, and green. Figure 8.17b specifically isolates the blue observation from Fig. 8.17a with the black observations and two lines. The black line is fit using only the black observations, while the blue line is fit using the black observations plus the isolated blue observation. Note that the blue line does not differ significantly from the black line. Because the blue dot lies near the black dots, the black dots indicate that the Y value should be lower, showing only a slight difference between the blue and black lines.

Notice the red and green dots in Fig. 8.17a, c, and d; both red and green dots have an extreme X value compared to the other observations. However, the red dot lies near the black regression line, whereas the green dot lies far from it. Since the green dot is an outlier and no other observations occur close to it, the regression line changes significantly when the green observation is included, as indicated by the green regression line. This drastic change to the regression line is in contrast to the red regression line, which remains somewhat similar to the black regression line. Therefore, because of this drastic change in slope and intercept, the green dot is considered an influential observation.

8.17 Residuals and Leverage

The leverage of an observation measures its ability to influence the fitted regression model. Observations with predictor values far from the center of the predictor data tend to have high leverage because they exert greater pull on the fitted regression line.

Leverage values always lie between 0 and 1, inclusive. A point with leverage near zero has minimal effect on the fitted model, whereas a point with leverage near 1 has the potential to strongly affect the regression line. Leverage alone does not imply influence; influence occurs when high leverage is paired with a large residual.

If we have a fitted model object called `model`, influence measures can be computed as follows.

```
from statsmodels.stats.outliers_influence import OLSInfluence
influence = OLSInfluence(model)
```

8.17.1 Leverage

Leverage can be attained for each predicted value in our dataset. Taking the ith predicted value, we compose a linear combination of weights and the response value in the dataset. For each predicted value, we have an equation of the form

$$\hat{Y}_i = h_{i1}Y_1 + h_{i2}Y_2 + ... + h_{in}Y_n, \tag{8.1}$$

where h_{ij} represents the weight assigned to the observed response Y_j, i indexes the predicted value $\hat{Y}_i$, and j indexes the observed response. From this relationship, we define the leverage of the ith value to be h_{ii}. If h_{ii} is large, then the y_i value has a large effect on $\hat{y}_i$, and we say that the ith value is highly leveraged.

Bringing together every value of h_{ij}, the hat matrix can be constructed, which is denoted as H.

```
hat_values = influence.hat_matrix_diag
```

In the Python code above, we extract the hat matrix from a Python `OLSInfluence` object.

8.17.2 Standardized Residuals

Prior to studying the residuals of the regression output, it is common practice to standardize them to compensate for differences in leverage. To standardize, one can simply divide by the standard deviation. The standardized residuals are given by:

$$r_i = \frac{e_i}{s\sqrt{1 - h_{ii}}} \tag{8.2}$$

```
standardized_resid = influence.resid_studentized_internal
```

where e_i is the residual, s is the RSE, and h_{ii} is the leverage for the ith observation. Internally, studentized residuals are more often referred to as standardized residuals.

8.17.3 Studentized Residuals

Standardized residuals do not follow a t-distribution since e_i and s are dependent. Therefore, we instead consider s_i, which is the RSE computed with every residual except the ith one. When we use s_i rather than s to standardize residuals, the resulting quantities are referred to as studentized residuals, which are defined as

$$r_i^* = \frac{e_i}{s_i\sqrt{1 - h_{ii}}}. \tag{8.3}$$

Studentized residuals can be computed in Python using the command here.

```
studentized_resid = influence.resid_studentized_external
```

Externally studentized residuals are more commonly referred to as studentized residuals. These residuals are computed by leaving out one observation at a time when calculating s_i.

8.17.4 Cook's Distance

An influential point is one that, if removed from the data, would significantly change the fit of the model. An influential point may either be an outlier, have large leverage, or both. Cook's distance refers to a commonly used measure of influence that combines these two properties. Points with $D_i > 1$ are considered to be influential.

$$D_i = \left(\frac{r_i^2}{p}\right)\left(\frac{h_{ii}}{1 - h_{ii}}\right) \tag{8.4}$$

The construct may look a little messy, but the main aspect to recognize is that Cook's distance depends on both the residual, e_i, and the leverage, h_{ii}. That is, both the x value and the y value of the data point play a role in the calculation of Cook's distance.

In short:

- D_i directly summarizes how much all of the fitted values change when the ith observation is deleted.
- A data point having a large D_i indicates that the data point strongly influences the fitted values.

We can compute Cook's distance using the following command:

```
cooks_d = influence.cooks_distance[0]
```

8.18 Case Study: Lead Generation

8.18.1 Problem Statement

Regression analysis is widely used across many business domains, including the evaluation of lead generation for sales revenue. Lead generation, or the methodology by which companies attract customers, results in sales and therefore generates revenue. For this reason, sales managers often work with analysts to find out which methods or tactics attract the most customers. Since data on customer acquisition is readily available, analysts serve a key role in understanding how to increase customer leads, thereby increasing sales revenue.

As a consulting analyst, you are tasked with helping Bergheger Automotive Group increase sales revenue. The group consists of 142 dealerships across the country. Management is interested in understanding their lead generation results from 2021. You request data on the various lead generation methods so that you can analyze and make recommendations. The sales team currently has several methods of customer acquisition available to them: radio ads, robocalls, email, and cold-calls. Each method carries its own strengths and weaknesses. In this analysis, you need to understand the relationship between sales revenue and the corresponding customer acquisition method. Specifically, your tasks consist of the following:

1. Identify revenue by lead generation method.
2. Describe the relationship between revenue and each lead generation method by fitting a simple linear regression line and interpreting the fitted model.
3. Make recommendations on how to increase revenue.

8.18.2 Data Description

You are given the `Sales.csv` file, which contains data from 142 dealerships with the following variables:

- `X1` is the number of radio ads.
- `Y1` is the sales revenue generated from the radio ads in dollars.
- `X2` is the number of robocalls in hundreds.
- `Y2` is the sales revenue generated from robocalls in dollars.
- `X3` is the number of emails in thousands.
- `Y3` is the sales revenue generated from emails in dollars.
- `X4` is the number of cold-calls in hundreds.
- `Y4` is the sales revenue generated from cold-calls in dollars.

As a first task, we will do some exploratory data analysis to understand the data. After loading in the data, we compute some descriptive statistics using the `describe` function.

```
from ravix import get_data
df = get_data("sales.csv")
df.describe()
```

```
#               X1           Y1          X2           Y2 \
# count 142.000000     142.000000 142.000000     142.000000
# mean  119.528169  121145.352113 124.315211  47823.063380
# std    47.027635   99474.818621 111.050273  18819.003897
# min    16.000000   15457.000000  14.110000   6294.000000
# 25%    83.250000   47529.750000  51.115000  33402.250000
# 50%   118.000000   84490.000000  88.770000  47140.000000
# 75%   158.000000  165425.000000 170.155000  63213.750000
# max   236.000000  740712.000000 880.470000  94410.000000
#
#               X3             Y3          X4             Y4
# count 142.000000     142.000000  142.000000     142.000000
# mean   54.224197  120832.823944   66.328732   75959.183099
# std    53.900692   98168.150760   18.029543    9709.366846
# min     4.498000   15674.000000   35.670000   57519.000000
# 25%    17.443250   47222.500000   50.112500   69128.000000
# 50%    34.699500   84123.000000   65.320000   74499.500000
# 75%    77.444250  166331.750000   80.750000   81188.750000
# max   368.485000  721275.000000   99.270000  111973.000000
```

The describe table indicates that the largest value among response variables is \$740,712, which can be observed by looking at the maximum value of `Y1` and the remaining variables. Here, the output is truncated, but can easily be observed. The `Y1` response variable, which corresponds to radio ad revenue, also shows the largest mean value. The mean and maximum values of `Y1` are followed closely by the `Y3` response variable, corresponding to email revenue. Revenue from cold-calls, represented by `Y4`, has the third lowest mean, followed by robocalls, represented by `Y2`.

In addition to viewing the `describe` function, an analyst may benefit from looking at some of the observations. As discussed previously, the `head` function will display the first five observations.

```
df.head()

#      X1      Y1      X2     Y2      X3      Y3     X4     Y4
# 0   126   72705   56.33  50375  40.757   75617  89.67  80251
# 1   116  110808  134.15  46315  33.269  107596  37.21  71193
# 2    56   35567   37.33  22573  10.151   35303  57.29  62825
# 3    95   74220   85.25  38017  21.971   72655  90.82  88127
# 4   140  138520  148.41  55891  53.701  137058  89.84  86198
```

Observing the head of the data provides some intuition about the data. For example, a reasonable value of `X1` is 100 radio ads, which would most likely generate revenue in the low one hundred thousand. Therefore, the revenue generated by an ad is roughly just above \$1,000. While this is a rough approximation, it can help understand the meaning of the `X1` variable and its relationship to `Y1`. Rough estimates of the other variable relationships can similarly aid in understanding the data.

8.18.3 Revenue by Lead Generation Method

Notice that the amount of revenue from the various sources is shown in the barplot in Fig. 8.18. The revenue variables (`Y1`,`Y2`,`Y3`,`Y4`) are chosen by the formula and aggregated using the `agg ="sum"` argument within the `barplot` function from Ravix. In addition to specifying the `ylab` and `title` arguments, the `color` argument is used to specify different colors for each revenue generation method. The order of the colors specified corresponds to the order of the sums. In particular, radio ads (`Y1`) will be dark red, robocalls (`Y2`) will be blue, emails (`Y3`) will be green, and cold-calls (`Y4`) will be orange. These colors were chosen since they are aesthetically pleasing and for easy viewing. We will adhere to this color coding in later plots throughout this case.

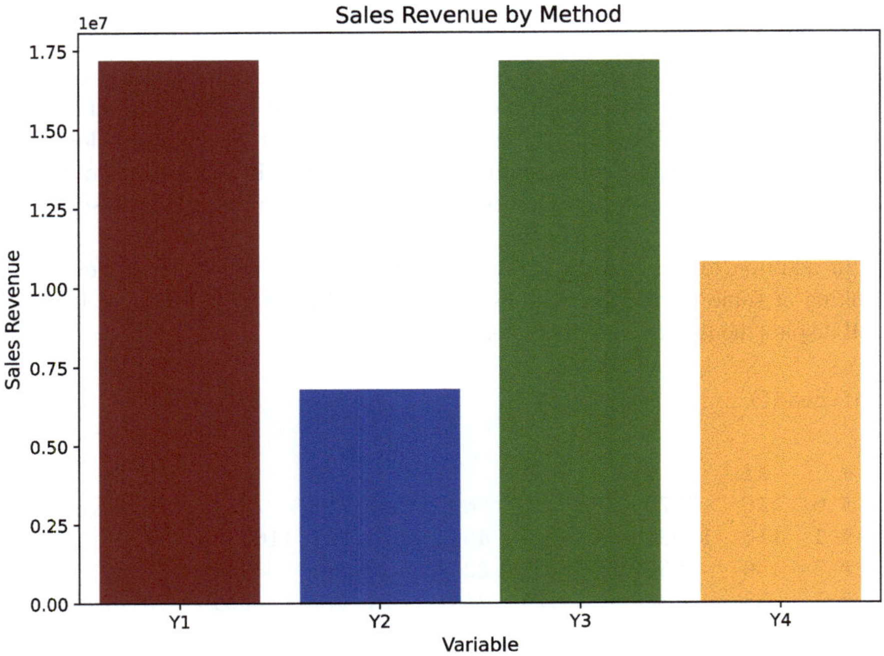

Fig. 8.18 Barplot of sales revenue by method

```
from ravix import barplot
barplot("~ Y1 + Y2 + Y3 + Y4", data = df, agg = "sum",
        color = ["darkred", "blue", "green", "orange"],
        ylab = "Sales Revenue",
        title = "Sales Revenue by Method")
```

Figure 8.18 supports our assessment of the response variables from observing the summary output. Specifically, the revenue from radio ads and emails is the highest, followed by the revenue from cold-calls and robocalls. Note that the largest total revenue for each response variable will have the same proportional value as the mean of said response variable. In fact, after dividing the sales revenue by the number of dealerships, each revenue variable will retain the same ratio as depicted in Fig. 8.18.

Boxplots of the different revenue amounts can be calculated using the `boxplot` function from Ravix. As a first attempt, we select the revenue variables using the `formula` argument and specify the dataframe. Rather than using the variable names `Y1`, `Y2`, `Y3`, and `Y4`, we could rename the colum-ns. We then pass the data subset to the `boxplot` function.

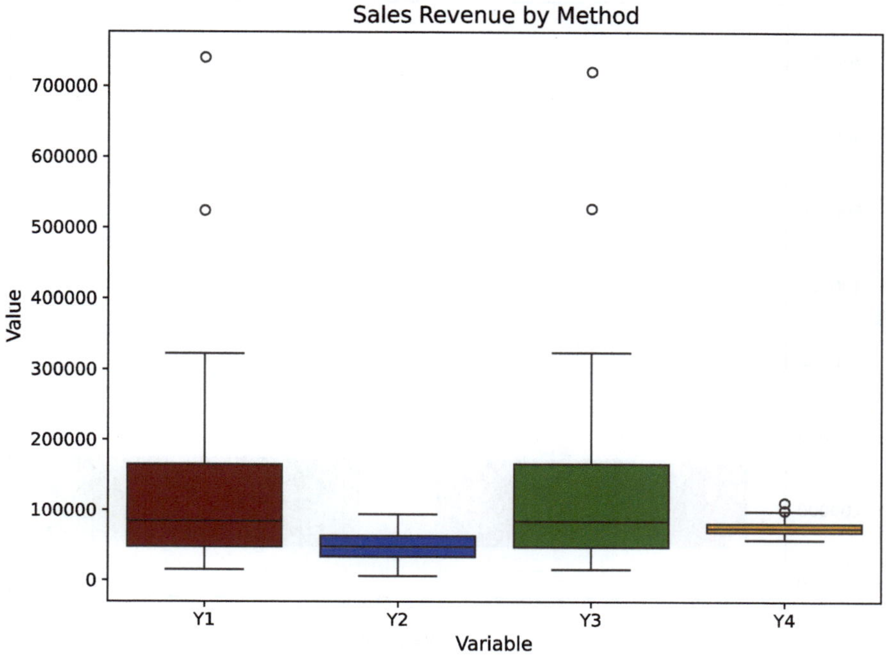

Fig. 8.19 Boxplot of sales revenue by method

```
from ravix import boxplot
boxplot("~ Y1 + Y2 + Y3 + Y4", data = df,
        color = ["darkred", "blue", "green", "orange"],
        title = "Sales Revenue by Method")
```

In Fig. 8.19, the variation of the radio ad and email variables can be observed. However, the variation of the robocalls and cold-calls is difficult to decipher since the y-axis is dictated by the revenue range from the radio ads and emails. As a simple remedy, we can plot the boxplot of radio ads and emails together using the same y-axis scale, and then plot the boxplot of robocalls and cold-calls on a separate y-axis scale. The result is shown in Figs. 8.20 and 8.21.

We use the formula argument to select the responses with similar values. Notice the different y-axis for the boxplots.

```
boxplot("Y1 ~ Y3", data = df, title = "",
        color = ["darkred", "green"])
```

```
boxplot("Y2 ~ Y4", data = df, title = "",
        color = ["blue", "orange"])
```

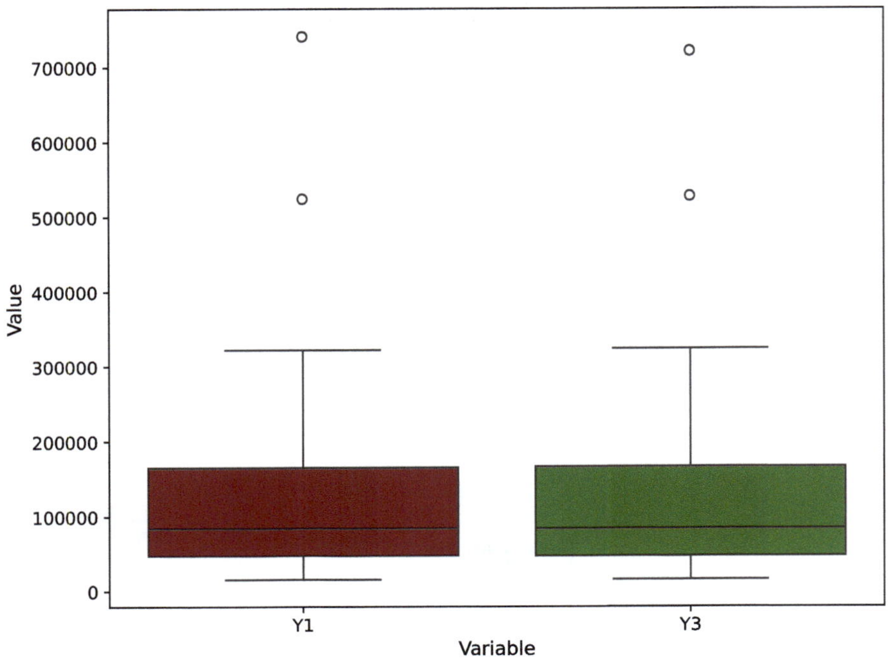

Fig. 8.20 Boxplots of sales revenue for Y1 and Y3

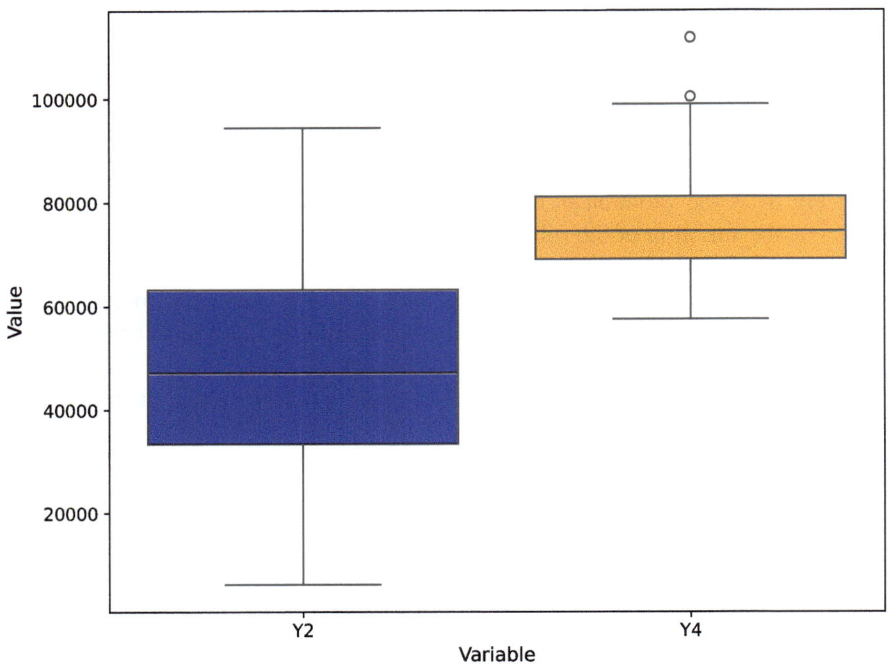

Fig. 8.21 Boxplots of sales revenue for Y2 and Y4

From Fig. 8.20, the revenue by radio ads and emails shows two similar boxplots, whereas Fig. 8.21 shows that revenue by robocalls is not very similar to that of cold-calls. For most dealerships, the revenue from cold-calls will be larger than the revenue from robocalls since the bottom whisker of cold-calls lands above the median line for robocalls.

8.18.4 Sales Versus Radio Ads

The relationship between sales and radio ads is clearly nonlinear as indicated by Fig. 8.22. We notice that observations are bunched up around the `Y1` axis. It can be argued that the trend is parabolic; however, a much better fit would be that of an exponential function.

```
from ravix import plot
plot("Y1 ~ X1", data = df, xlab = "Ads", ylab = "Sales",
     title = "Sales vs. Ads", color = "darkred")
```

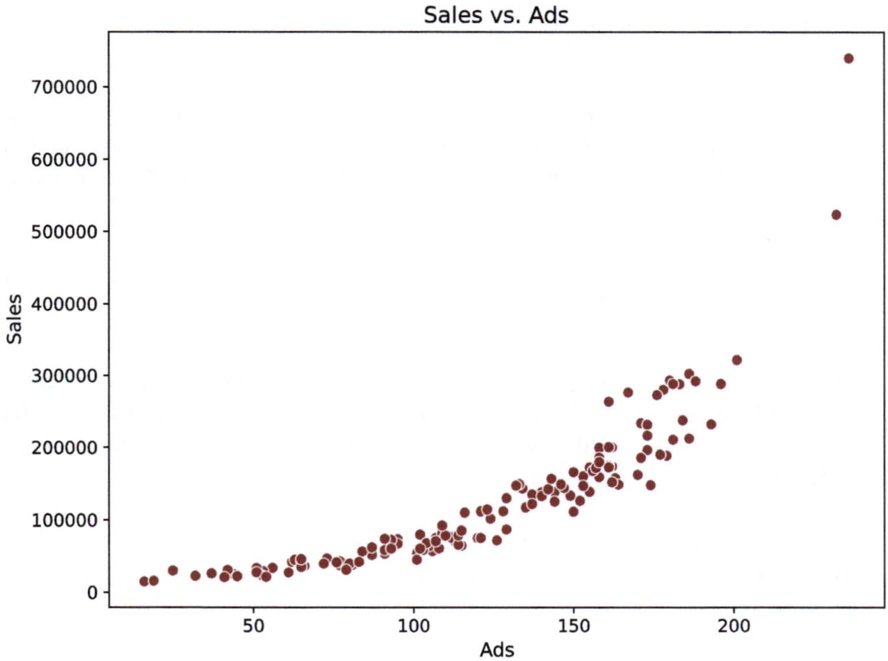

Fig. 8.22 Scatterplot of sales and ads

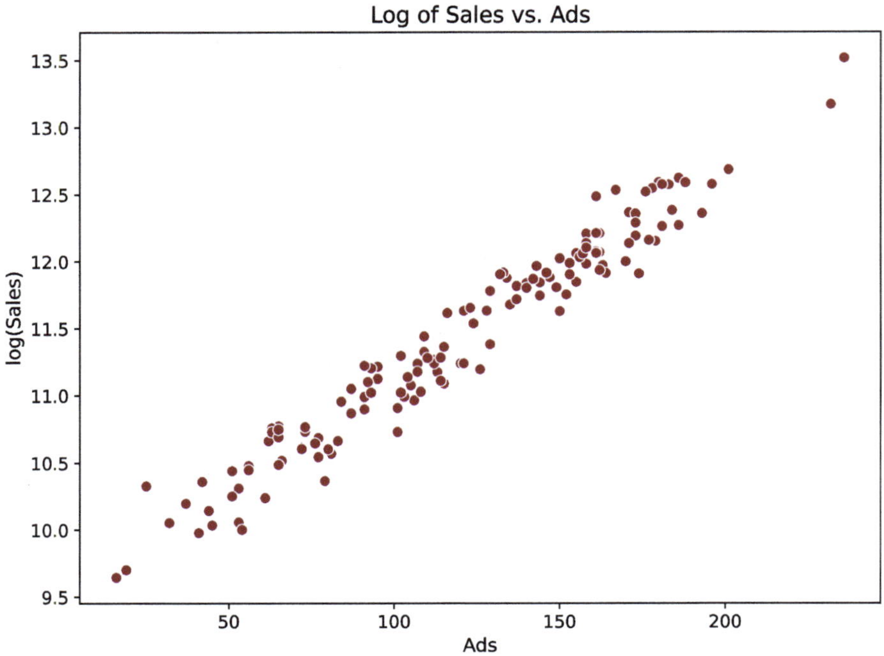

Fig. 8.23 Scatterplot of sales and ads

To account for this exponential, we can apply a logarithmic transform to the `Y1` variable, since applying a logarithmic transformation to `Y1` with `X1` as a predictor is equivalent to predicting `Y1` with $e = 2.718...$ raised to the power of `X1` for positive values of `Y1`.

```
plot("log(Y1) ~ X1", data = df, xlab = "Ads", ylab = "log(Sales)",
     title = "Log of Sales vs. Ads", color = "darkred")
```

In Fig. 8.23, the fit of the scatterplot follows a linear pattern, which indicates a fit of the logarithmic transformation is appropriate using the `ols` function. The summary of the transformed model is obtained by passing the `reg` object to the `summary` function.

```
from ravix import ols
reg = ols("log(Y1) ~ X1", data = df)
reg.summary()
```

```
# Summary of OLS Regression Analysis:
# ==========================================================
#
# Coefficients:
# ----------------------------------------------------------
#                Estimate  Std. Error  t-value    p-value
# Intercept        9.4878    0.038967  243.480    < 2e-16 ***
# X1             0.016134   0.0003035   53.157    < 2e-16 ***
#
# Model Statistics:
# ----------------------------------------------------------
# Residual Std. Error: 0.16949
# R-squared:        0.9528          AIC: -99.1245
# Adj. R-squared: 0.9525            BIC: -93.2129
# F-statistic: 2825.64 on 1 and 140 DF, p-value: < 2e-16
# ==========================================================
```

After fitting the transformed model, we inspect the residuals using a histogram and a residual plot. The histogram is generated using the `hist` function. Note that the residuals are within the regression object (`reg`), which allows us to create a histogram of the residuals using the `hist` function.

```
hist(reg, color = "darkred")
```

```
plot(reg, color = "darkred")
```

The diagnostic plots indicate the normality condition is met within reason, and the variance of the residuals does not violate any assumptions. More specifically, the histogram from Fig. 8.24 roughly exhibits a bell-shaped normal distribution, and the residual plot from Fig. 8.25 does not appear to show a pattern, hence meeting the linearity condition.

Since the conditions are met, the coefficients can be interpreted. Here, the coefficient is 0.0161339, which indicates that for every additional radio ad, the logarithm of revenue increases by 0.0161339. Alternatively, the revenue gets multiplied by $e^{0.0161339} = 1.016265$ for every additional radio ad. This amounts to a 1.6265% increase in the revenue per radio ad. Therefore, if the revenue is high, the increase will be more significant. This multiplicative increase may be due to the compounding effect of having multiple ads.

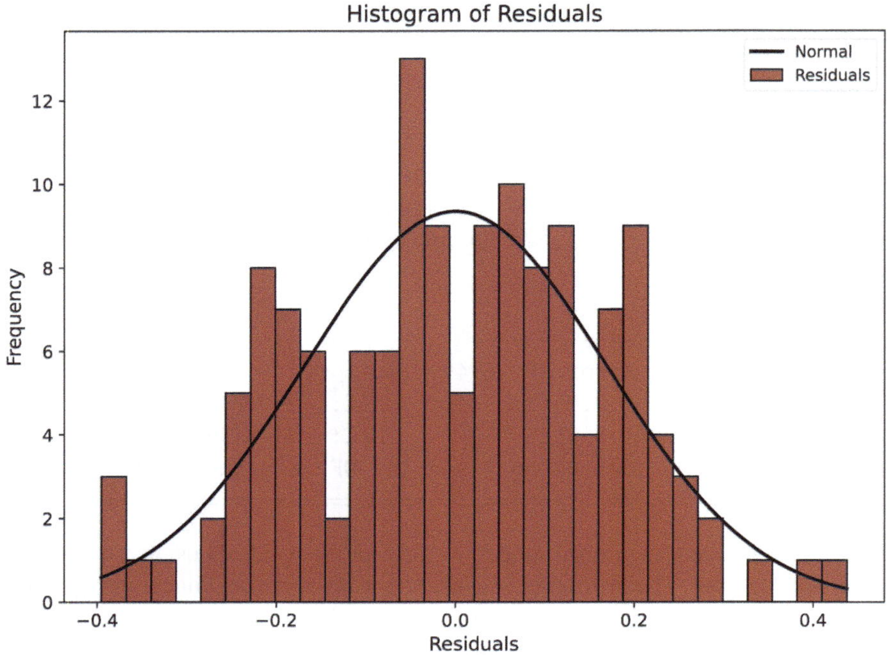

Fig. 8.24 Diagnostic plots of transformed ads model

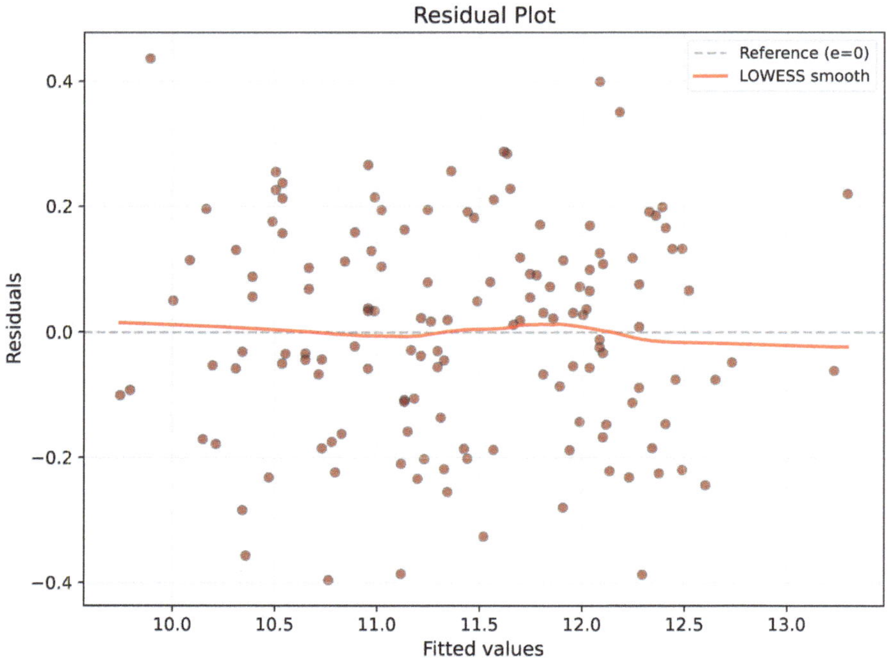

Fig. 8.25 Diagnostic plots of transformed ads model

8.18.5 Sales Versus Robocalls

From Fig. 8.26, we see that the relationship between robocalls and sales is also nonlinear. In this case, the relationship can be modified by transforming the X2 variable with a logarithmic transform since observations are bunched up around the X2 axis.

```
from ravix import plot
plot("Y2 ~ X2", data = df, xlab = "Robocalls (hundreds)",
     ylab = "Sales", title = "Sales vs. Robocalls",
     color = "blue")
```

After transforming the X2 variable, we observe the revised scatterplot in Fig. 8.27 with log(X2) on the x-axis.

```
plot("Y2 ~ log(X2)", data = df,
     xlab = "log(Robocalls (hundreds))",
     ylab = "Sales", title = "Sales vs. Log of Robocalls",
     color = "blue")
```

We fit the model with the revised variable transformation and print the model summary.

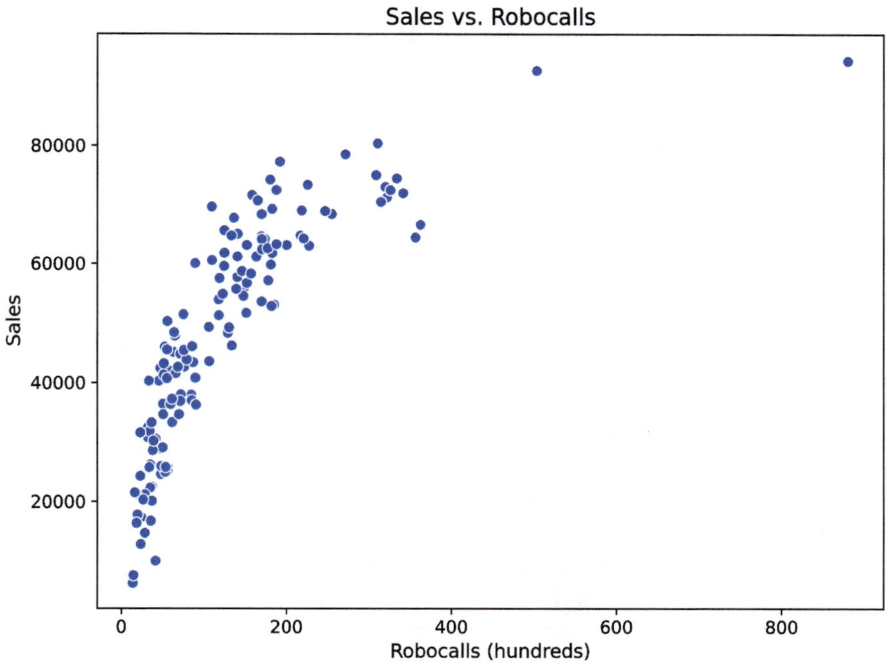

Fig. 8.26 Scatterplots of sales and robocalls

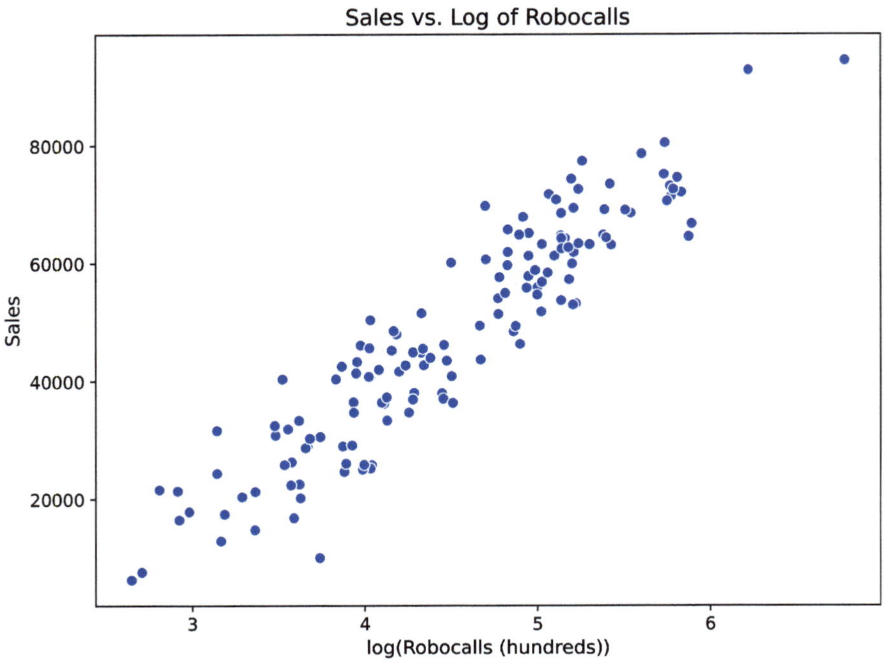

Fig. 8.27 Scatterplots of sales and transformed robocalls

```
reg = ols("Y2 ~ log(X2)", data = df)
reg.summary()
```

```
# Summary of OLS Regression Analysis:
# ======================================================
#
# Coefficients:
# ------------------------------------------------------
#                Estimate  Std. Error  t-value    p-value
# Intercept      -48407.2      3238.4  -14.948   < 2e-16 ***
# log(X2)         21379.3      707.92   30.200   < 2e-16 ***
#
# Model Statistics:
# ------------------------------------------------------
# Residual Std. Error: 6889.48
# R-squared:      0.8669          AIC: 2914.89
# Adj. R-squared: 0.8660          BIC: 2920.80
# F-statistic: 912.0575 on 1 and 140 DF, p-value: < 2e-16
# ======================================================
```

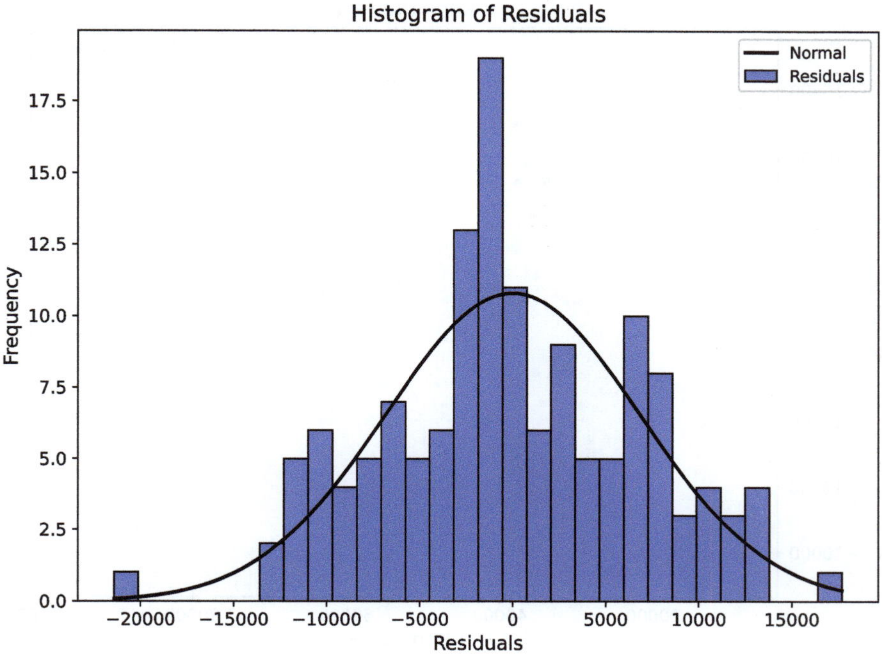

Fig. 8.28 Histogram of residuals for transformed robocall model

After fitting the revised model, we inspect the residuals using a histogram and a residual plot. The generated plots are shown in Figs. 8.28 and 8.29.

```
from ravix import hist
hist(reg, color = "blue")
```

```
plot(reg, color = "blue")
```

The diagnostic plots indicate that the residuals are normally distributed, and the model is linear with constant variance. Given that the diagnostic plots indicate the linear regression model assumptions are met, and the p-value of the coefficient is low, the coefficients can be interpreted. The coefficient of the log of robocalls is 21379.3, indicating that for every unit increase in the logarithm of robocalls, there will be an increase in revenue of \$21,379.30. The alternative interpretation holds that for every unit increase in the exponent of robocalls, the sales revenue increases by \$21,379.30. As demonstrated in the untransformed plot, the initial increase may be steep, but the returns diminish quickly. With robocalls, repeated calls are often ignored, so it's logical to use the robocalls sparingly, unless there is no penalty for making frequent calls.

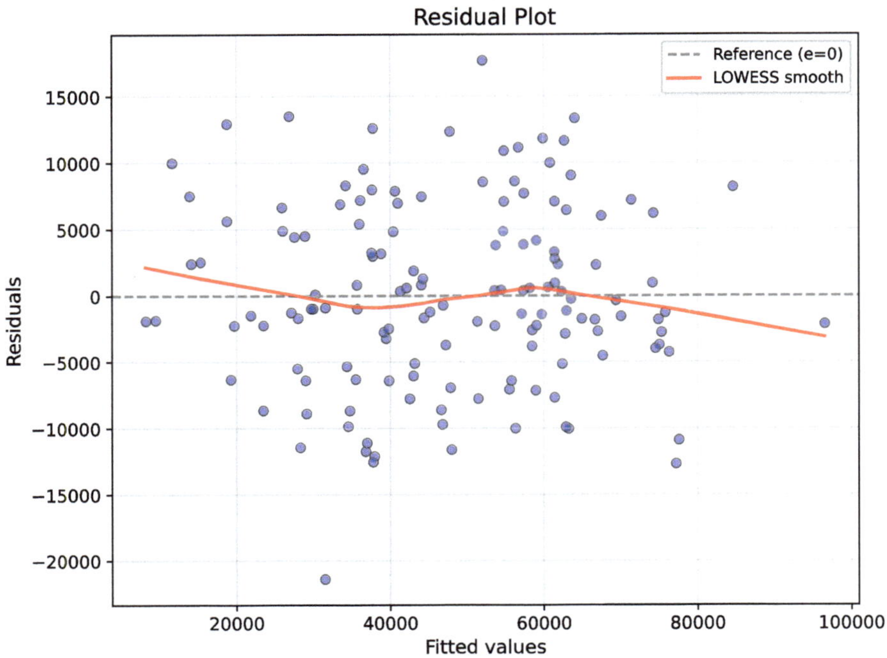

Fig. 8.29 Residual plot for transformed robocall model

8.18.6 *Sales versus Emails*

Figure 8.30 shows the untransformed scatterplot of sales and emails.

```
from ravix import plot
plot("Y3 ~ X3", data = df, xlab = "Emails",
     ylab = "Sales", title = "Sales vs. Emails",
     color = "green")


plot("log(Y3) ~ log(X3)", data = df, xlab = "log(Emails)",
     ylab = "log(Sales)", title = "log(Sales) vs. log(Emails)",
     color = "green")
```

Note that the observations are densely distributed around the lower values of the x-axis and become more dispersed as x increases. Similarly, the observations are densely distributed around the lower values of the y-axis and become more dispersed as y increases. This pattern indicates each axis can benefit from a logarithmic transformation. Applying logarithms to both x and y axes changes the pattern from Fig. 8.30 to a linear relationship, which can be observed in Fig. 8.31.

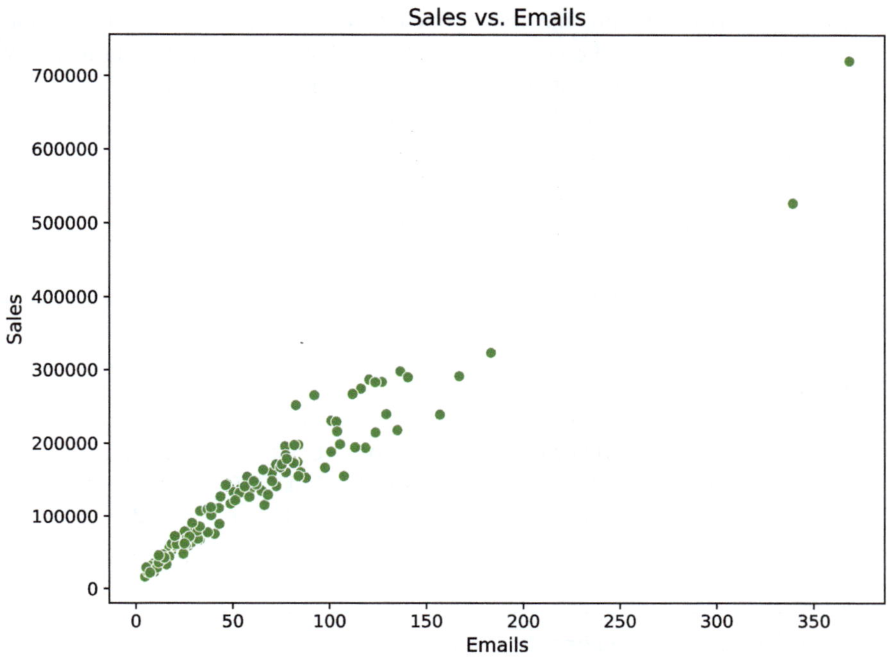

Fig. 8.30 Scatterplot of sales and emails

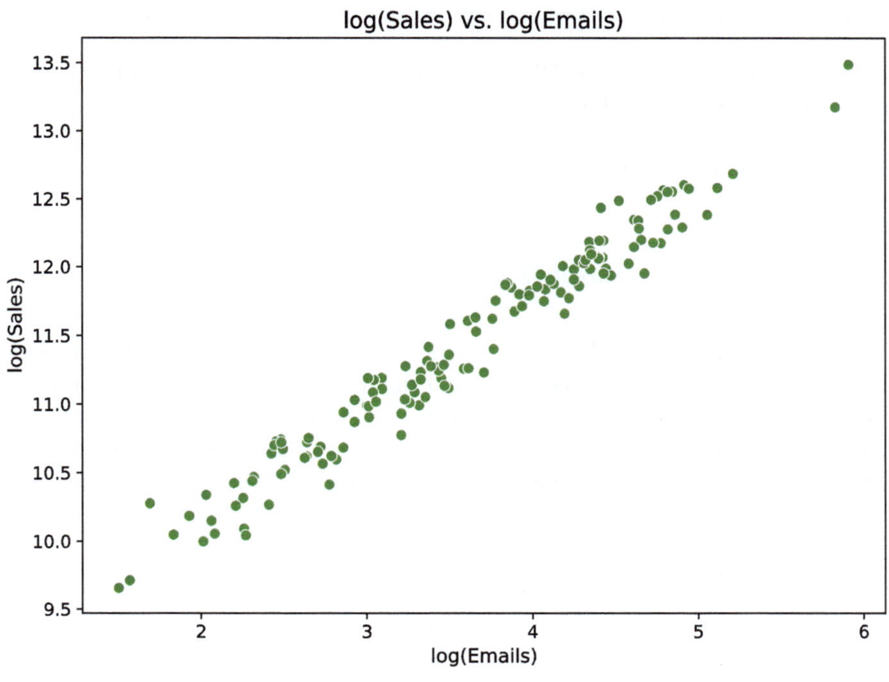

Fig. 8.31 Scatterplot of log(Sales) and log(Emails)

Using the `ols` function, the transformed model is fit. Since we named the regression object as `reg`, the previous regression object is replaced with this new fit. The summarized model is shown by using the `summary` function.

```
from ravix import ols
reg = ols("log(Y3) ~ log(X3)", data = df)
reg.summary()
```

```
# Summary of OLS Regression Analysis:
# ======================================================
#
# Coefficients:
# ------------------------------------------------------
#                Estimate  Std. Error  t-value    p-value
# Intercept        8.5336    0.049691  171.735    < 2e-16 ***
# log(X3)         0.80550    0.013427   59.991    < 2e-16 ***
#
# Model Statistics:
# ------------------------------------------------------
# Residual Std. Error: 0.150023
# R-squared:       0.9626           AIC: -133.7745
# Adj. R-squared: 0.9623            BIC: -127.8629
# F-statistic: 3598.87 on 1 and 140 DF, p-value: < 2e-16
# ======================================================
```

After fitting the revised model, we inspect the residuals using a histogram and a residual plot, as done previously.

```
from ravix import hist
hist(reg, color = "green")
```

```
plot(reg, color = "green")
```

The diagnostic plots in Figs. 8.32 and 8.33 indicate normally distributed residuals with constant variance and a linear pattern; therefore, the linear regression assumptions are met. Furthermore, the coefficient for the log of emails is significant, indicating that the coefficient can be interpreted. For a log-log model, the slope can be interpreted as a 1% increase in emails is associated with a 0.8055% increase in revenue.

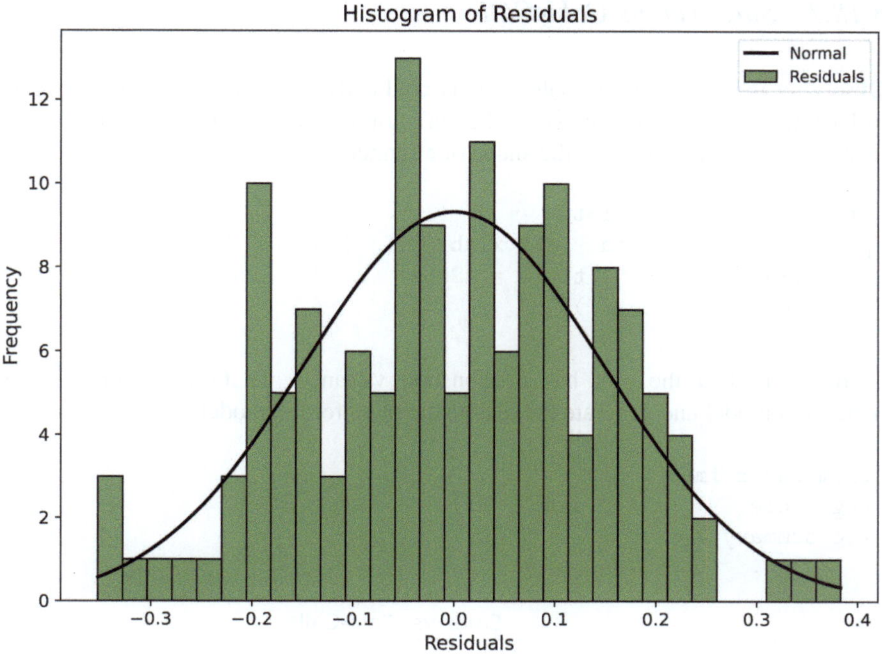

Fig. 8.32 Histogram of residuals for transformed email model

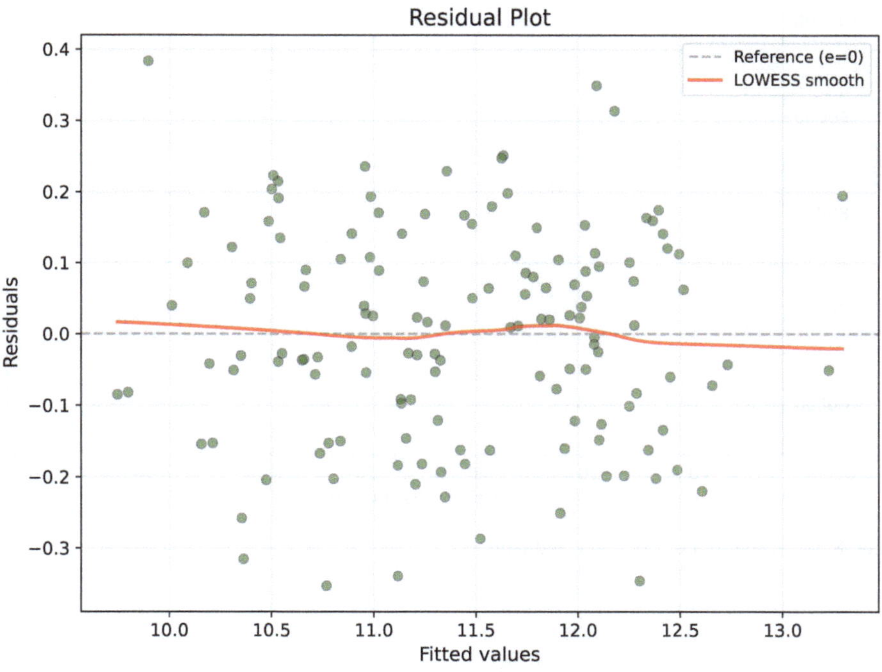

Fig. 8.33 Residual plot of transformed email model

8.18.7 *Sales versus Cold-Calls*

Observing the scatterplot of sales versus cold-calls in Fig. 8.34, the plot appears to have nonconstant variance since the variation increases from left to right. This increase in variation takes on the shape of a funnel.

```
from ravix import plot
plot("Y4 ~ X4", data = df, xlab = "Cold-Calls (hundreds)",
     ylab = "Sales", title = "Sales vs. Cold-Calls",
     color = "orange")
```

To verify that the plot has nonconstant variance, we fit the simple linear regression model and generate the diagnostic plots for the model.

```
from ravix import ols
reg = ols("Y4 ~ X4", data = df)
reg.summary()
```

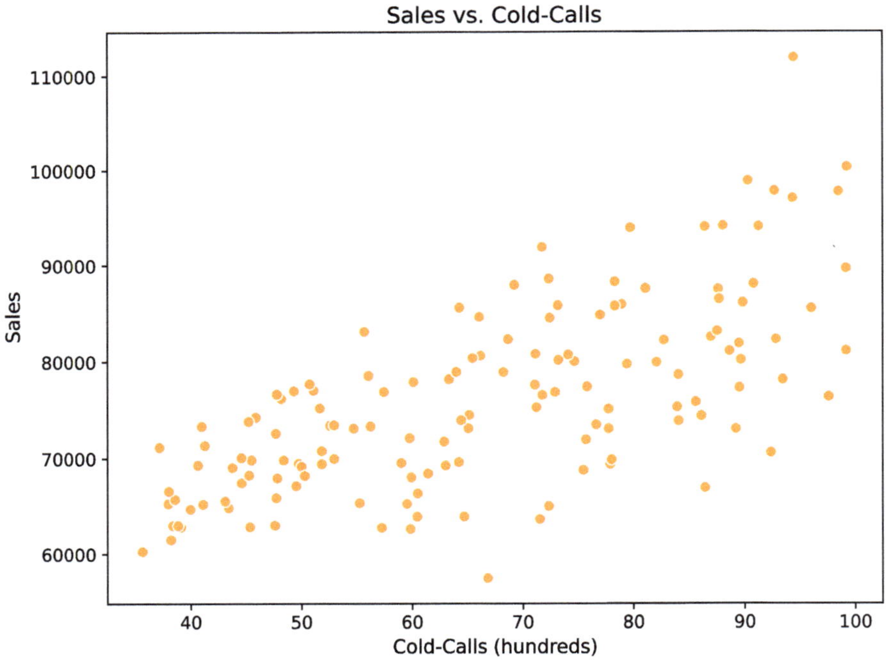

Fig. 8.34 Scatterplot of sales and cold-calls

```
# Summary of OLS Regression Analysis:
# ==========================================================
#
# Coefficients:
# ----------------------------------------------------------
#                 Estimate  Std. Error  t-value    p-value
# Intercept        51374.8      2269.0   22.642    < 2e-16 ***
# X4                370.64      33.019   11.225    < 2e-16 ***
#
# Model Statistics:
# ----------------------------------------------------------
# Residual Std. Error: 7068.93
# R-squared:        0.4737          AIC: 2922.19
# Adj. R-squared: 0.4699            BIC: 2928.10
# F-statistic: 126.0072 on 1 and 140 DF, p-value: < 2e-16
# ==========================================================
```

Inspecting the residuals using a histogram and a residual plot is done easily using the code here.

```
from ravix import hist
hist(reg, color = "orange")
```

```
plot(reg, color = "orange")
```

Note that the residual plots in Figs. 8.35 and 8.36 clearly show nonconstant variance since the spread in the residuals increases from left to right. The increasing variance pattern is that of a funnel. While it is clear that nonconstant variance is present, a simple hypothesis test can be performed in order to verify the results. Within the Ravix package, the ncv function tests for nonconstant variance using a Chi-square test.

```
from ravix import ncv
ncv(reg)
```

```
# Nonconstant Variance Test
# ===========================================================
# Test Statistic      : 21.6346
# p-value             : 3.299e-06
# Result              : Nonconstant variance (p < 0.05)
# ===========================================================
```

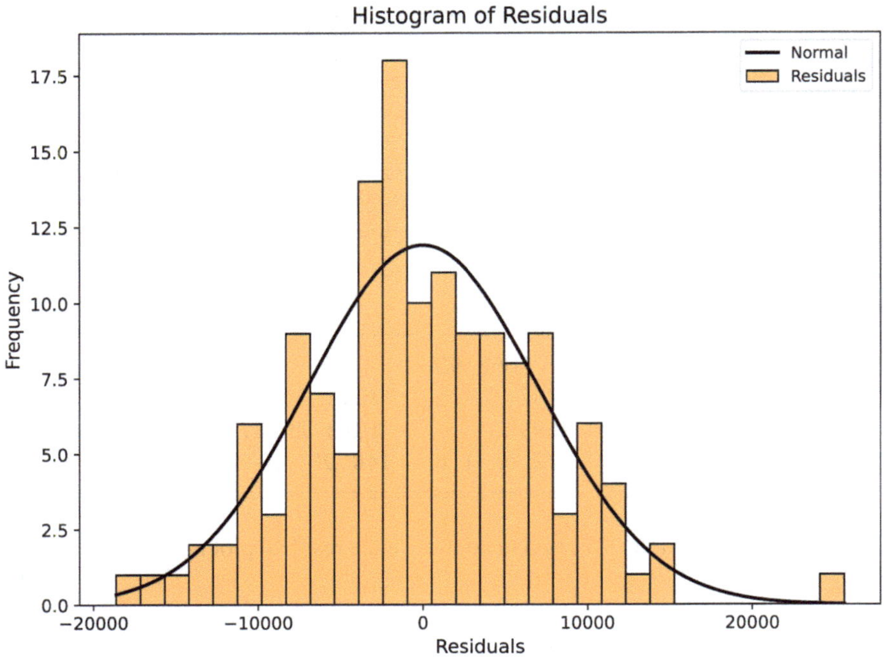

Fig. 8.35 Histogram of residuals for cold-calls

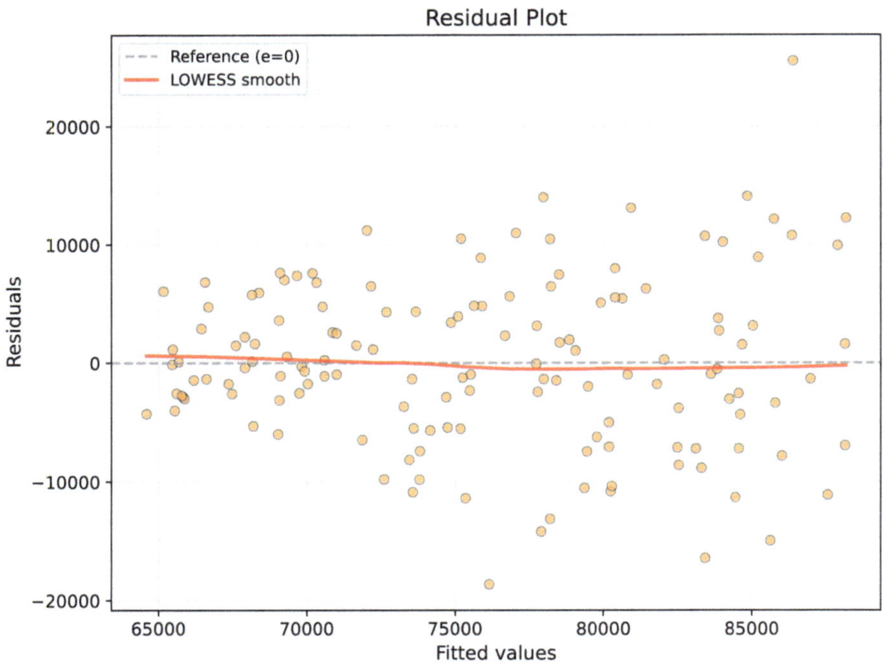

Fig. 8.36 Residual plot of cold-calls

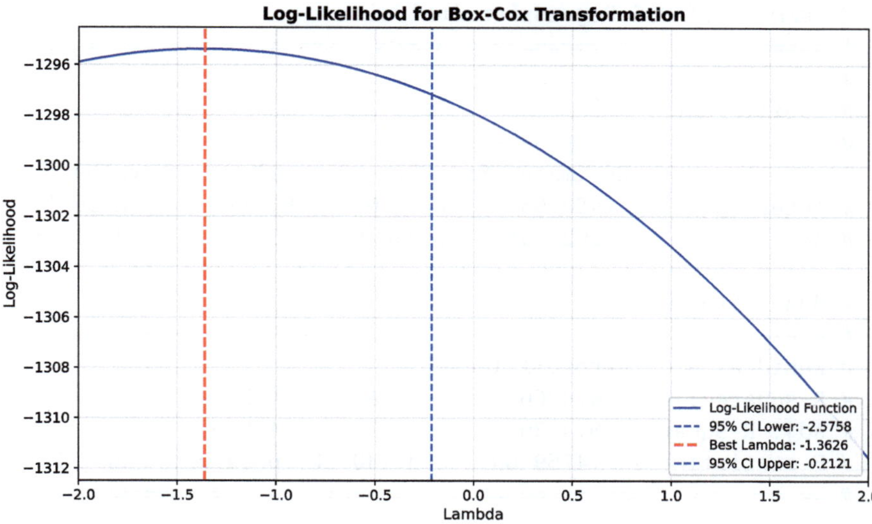

Fig. 8.37 Box-Cox output

Running the nonconstant variance test confirms there is nonconstant variance since the p-value from the test is extremely low. To remedy this condition, we will apply the Box-Cox transform using the `box_cox` function.

```
from ravix import box_cox
box_cox(reg)
```

From the Box-Cox output in Fig. 8.37, we observe that the exponent of the response variable (λ) maximizes the log-likelihood function when $\lambda = -1.36$. Therefore, we take revenue to the power of -1 as the response variable here. The transformed response becomes `1/Y4`, and we use `inv(Y4)` in the formula.

```
reg = ols("inv(Y4) ~ X4", data = df)
reg.summary()
```

```
# Summary of OLS Regression Analysis:
# =========================================================
#
# Coefficients:
# ---------------------------------------------------------
#                  Estimate  Std. Error  t-value    p-value
# Intercept       1.748e-05   3.775e-07   46.311    < 2e-16 ***
# X4             -6.202e-08   5.493e-09  -11.291    < 2e-16 ***
#
# Model Statistics:
# ---------------------------------------------------------
# Residual Std. Error: 1.18e-06
# R-squared:      0.4766          AIC: -3472.59
# Adj. R-squared: 0.4729          BIC: -3466.68
# F-statistic: 127.4769 on 1 and 140 DF, p-value: < 2e-16
# =========================================================
```

After fitting the revised model, we inspect the residuals using a histogram and a residual plot to check the model assumptions.

```
hist(reg, color = "orange")
```

```
plot(reg, color = "orange")
```

The diagnostic plots in Figs. 8.38 and 8.39 indicate that the linear regression assumptions are somewhat violated. Specifically, the histogram indicates non-normality given its distribution, and the residual plot indicates there is still evidence of nonconstant variance. While the residuals appear to be slightly normally distributed, the `shapiro` function can be used to assess normality with a hypothesis test.

```
from ravix import shapiro
shapiro(reg)
```

```
# Shapiro-Wilk Test for Normality
# ==========================================================
# Test Statistic      : 0.9846
# p-value             : 0.1129
# Result              : Normal (p >= 0.05)
# ==========================================================
```

Since the p-value of the Shapiro-Wilk test is not more extreme than $\alpha = 0.05$, the test indicates the residual distribution is normal. While the nonconstant variance

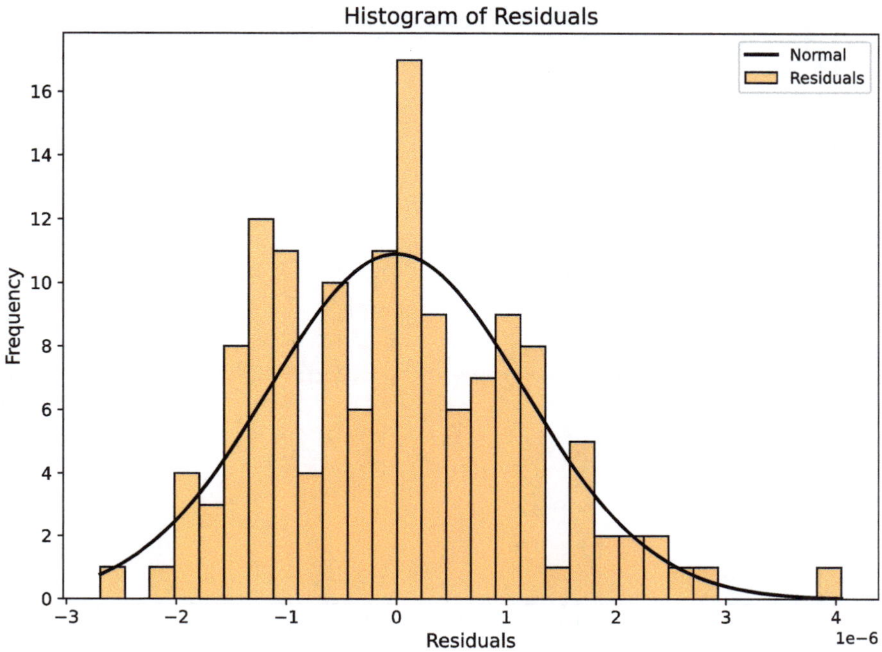

Fig. 8.38 Histogram of residuals for transformed cold-calls model

appears to have improved significantly from the revised residual plot, there appear to be traces of nonconstant variance. The ncv is used here on the revised model.

```
ncv(reg)

 # Nonconstant Variance Test
 # =======================================================
 # Test Statistic       : 4.0932
 # p-value              : 0.04306
 # Result               : Nonconstant variance (p < 0.05)
 # =======================================================
```

The p-value of the revised model drastically increased from the untransformed model. While the revised p-value is still below $\alpha = 0.05$, the nonconstant variance test may actually be better suited for lower values of α.

Here, we interpret the coefficient for the transformed model since the conditions of linear regression are nearly met, and the p-value of the coefficient indicates significance. The coefficient indicates that for every 100 calls, there is a 6.202×10^{-6} decrease in the inverse of revenue.

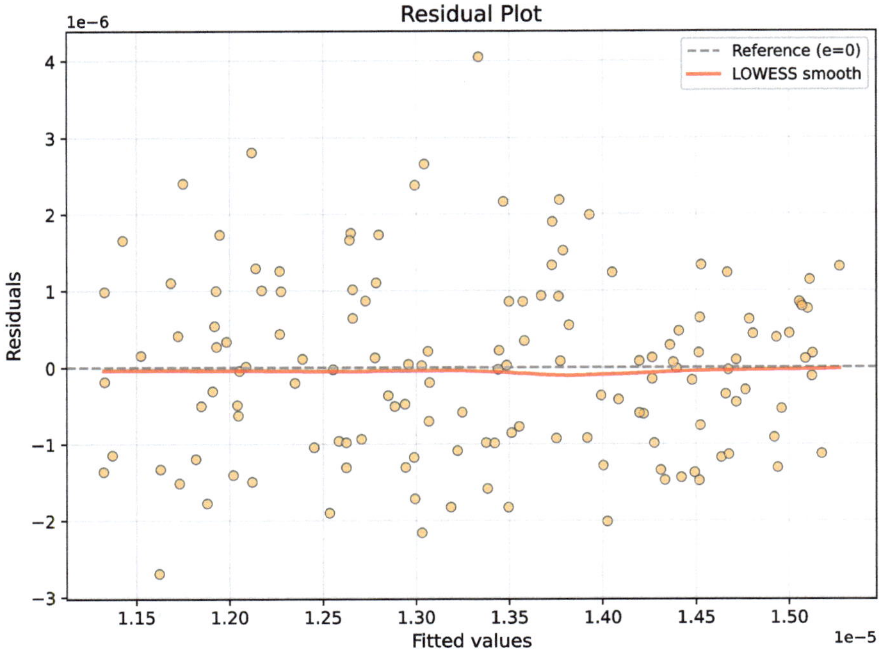

Fig. 8.39 Residual plot for transformed cold-calls model

8.18.8 Recommendations and Findings

From this analysis come multiple insights. In particular, regarding the tasks that were given, we find the following:

1. Identify revenue by lead generation method.

Task 1 was answered through the use of careful observation and plots, which made clear that ads and emails were the most effective in generating revenue, while cold-calls also provided some benefit, followed by robocalls.

2. Describe the relationship between revenue and each lead generation method by fitting a simple linear regression line and interpreting the fitted model.

Each regression model was fit and interpreted. Some interpretations, however, were more practical than others. The interpretations are as follows:

- The ad revenue increases by a multiple of 1.6265% for every additional ad.
- For every unit increase in the exponent of robocalls, the sales revenue increases by $21,379.30.
- For a 1% increase in emails, there is a 0.8055% increase in revenue.
- For every 100 call increase in the number of cold-calls, there is a 6.202e-08 decrease in the inverse of revenue from cold-calls.

3. Make recommendations on how to increase revenue.

- From observing the dealerships by revenue, dealership 88 can provide insights to other dealerships to increase revenue from ads, email, and robocalls.
- Also, dealership 6 has the most effective means by which to generate revenue from cold-calls, which could be helpful to share with other dealerships.
- Since ads increase revenue exponentially, dealerships should focus on increasing the number of ads, assuming that ads are not cost-prohibitive.
- While robocalls may be effective in the short run, if there is a cost associated with having extra calls, then the number of robocalls should be limited. Also, the negative publicity of robocalls should be considered when determining the number of calls to make.
- Both emails and cold-calls are beneficial methods of lead generation and should be continued if they are not cost-prohibitive.

8.18.9 Case Conclusion

This analysis demonstrates how studying marketing data can lead to increasing revenue streams for a car dealership conglomerate. Analyzing marketing methods can help optimize profits and other performance metrics. Analytics is becoming more relevant in the field of marketing as data becomes more abundant.

Real-world data, such as marketing data, can be messy and require variable transformations and diagnostic measures to better understand the relationship between responses and predictors. In this case, the responses required transformations in order to adhere to the assumptions of linear regression. We used many Ravix commands to model and analyze the data, such as `plot`, `boxplot`, `ols`, `hist`, `qq`, `shapiro`, `box_cox`, and `ncv`.

Problems

1. Sales from Web Ads Diagnostics
Using the `sales_ads.csv` file, answer the following questions:

a. Fit a simple linear regression model to predict the sales in thousands of USD (`Sales`) as a function of the number of website ads (`Ads`) without using transformations. Print a summary of the model.
b. Generate a scatterplot of `Sales` and `Ads`. Include the linear regression line in the plot.
c. Construct a scatterplot and histogram of the residuals for the linear model from part a.

d. From the scatterplot and histogram in part c, which model assumptions appear to be violated?
e. Determine an appropriate transformation for correcting the problems found in part d and fit the corresponding regression model.
f. Plot the transformed data and the regression line.
g. Construct a scatterplot and histogram of the residuals for the linear model from part f.

2. Healthcare Customers Diagnostics
Using the `insurance.csv` file, answer the following questions:

a. Fit a simple linear regression model to predict the number of customers per day (`Customers`) as a function of the number of people with Aetna insurance plans (`Plans`) within a 5 mile radius of each clinic without using transformations. Print a summary of the model.
b. Generate a scatterplot of `Customers` and `Plans`. Include the linear regression line in the plot.
c. Construct a scatterplot and histogram of the residuals for the linear model from part a.
d. From the scatterplot and histogram in part c, which model assumptions appear to be violated?
e. Since the observations appear to be bunched up around the x-axis, apply a logarithmic transformation to the x-axis and generate a model summary.
f. Plot the transformed data and the regression line from part e.
g. Construct a scatterplot and histogram of the residuals for the linear model from part e.

3. House Price Prediction Diagnostics
Using the `price_size.csv` file, answer the following questions:

a. Fit a simple linear regression model to predict the price in thousands of USD (`Price`) as a function of the size of the house (`Size`) in square feet without using transformations. Print a summary of the model.
b. Generate a scatterplot of `Price` and `Size`. Include the linear regression line in the plot.
c. Construct a scatterplot and histogram of the residuals for the linear model from part a.
d. From the scatterplot and histogram in part c, which model assumptions appear to be violated?
e. Since the observations appear to be bunched up around the x and y axes, apply logarithmic transformations to both X and Y. Generate a model summary.
f. Plot the transformed data and the regression line from part e.
g. Construct a scatterplot and histogram of the residuals for the linear model from part e.

Chapter 9
Variable Selection

All models are wrong, but some are useful.

—George E. P. Box

9.1 Introduction

While it may be a bit strict to say that *all* models are wrong, it is often the case that a model is imperfect. However, an imperfect model may still provide a great amount of value. When attempting to find the best model from the data given, being able to select the predictor variables is of utmost importance in the model-building process. In fact, one of the most important aspects of model creation is knowing which predictor variables to use, a process sometimes called feature selection or variable selection. Variable selection can be tremendously helpful when an analyst is attempting to find a mathematical model that is relatively close to the true state.

In this chapter, a methodology for finding a subset of predictor variables to create an acceptable model is explored. Particularly, iterative approaches for finding a good model are explored along with the best subset selection method. We also discuss model evaluation criteria to assist the analyst in selecting a model. Commonly used metrics include adjusted R^2, Akaike's Information Criterion (AIC), and Bayesian Information Criterion (BIC), each of which balances model fit against model complexity. To further motivate the concepts discussed in this chapter, an Airbnb dataset is used. Finally, the chapter case study consists of finding an appropriate model to predict healthcare costs using an insurance dataset.

9.2 Parsimonious Models

The primary goals of regression analysis are to produce accurate predictions while maintaining interpretability. A model that achieves both is referred to as

D. P. McGibney, *Applied Linear Regression for Business Analytics with Python*,
International Series in Operations Research & Management Science 378,
https://doi.org/10.1007/978-3-032-23806-1_9

parsimonious. In practice, parsimony is achieved by selecting the smallest set of predictor variables that still provides an adequate model fit.

9.3 Airbnb Pricing Application

Victoria would like to list her property on Airbnb in Edinburgh, the capital of Scotland. In order to price her property competitively, she collects data on listings to analyze the contributing factors of price. The `Airbnb.csv` dataset consists of 10,370 rental listings from Airbnb in Edinburgh for a period from June 25, 2019, to June 24, 2020.

The variables for each listing are:

- **Bathrooms**—Number of bathrooms
- **Bedrooms**—Number of bedrooms
- **Beds**—Number of beds
- **Accommodates**—Number of guests the listing can accommodate
- **Guests**—Number of guests included without an additional fee
- **MinNights**—Minimum number of nights required for booking
- **MaxNights**—Maximum number of nights the listing can be rented
- **ExtraPeople**—Average fee for each additional person in British pounds
- **HostListings**—Number of listings the host manages
- **ResponseRate**—Average host response rate
- **Deposit**—Average security deposit required for booking in British pounds
- **CleaningFee**—Average cleaning fee charged in British pounds
- **FeeMissing**—A dummy variable that is 1 if the cleaning fee is missing and 0 otherwise
- **Price**—Average price of the listing in British pounds

Using the Edinburgh Airbnb data file, do the following:

(a) Load in the data and print the first five observations to the console.
(b) Fit a regression model predicting `Price` using all other variables and print the model summary.
(c) Repeat the previous part, but exclude the predictor variable with the lowest individual significance.
(d) Comment on the model fit for the models in parts b and c. Which model is more parsimonious?

Solution:

(a) The `Airbnb.csv` data is available within Ravix. After importing the package, the `get_data` function is used to create the dataframe `df` is created. The first five observations are printed using the `head` function.

```
from ravix import get_data
df = get_data("airbnb.csv")
df.head()
```

```
#    Bathrooms  Bedrooms  Beds  Accommodates  Guests  \
# 0        1.0         1     1             2       1
# 1        1.5         2     2             4       4
# 2        1.0         0     2             2       2
# 3        1.0         1     1             2       2
# 4        1.0         1     1             2       2
#
#    MinNights  MaxNights  ExtraPeople  HostListings  \
# 0        2.9       30.0            0             1
# 1        2.0      365.0           25             1
# 2        4.2       60.0           10             2
# 3        2.0       21.0            0             1
# 4        1.0       31.0            0             1
#
#    ResponseRate  Deposit  CleaningFee  FeeMissing  \
# 0           1.0    200.0         40.0           0
# 1           1.0    250.0         30.0           0
# 2           1.0    100.0         15.4           1
# 3           1.0      0.0         10.0           0
# 4           1.0     75.0         10.0           0
#
#         Price
# 0  101.054794
# 1  111.816438
# 2   49.983562
# 3   33.827397
# 4   80.241096
```

(b) The `reg1` model is fit using the `ols` function. We specify that `Price` is predicted using all other variables in the dataframe using the formula as the first argument. The model summary is printed using the `summary` function.

```
from ravix import ols
reg1 = ols("Price ~ . ", data = df)
reg1.summary()
```

```
# Summary of OLS Regression Analysis:
# ======================================================
#
# Coefficients:
# ------------------------------------------------------
#                 Estimate   Std. Error  t-value    p-value
# Intercept         15.755       3.8052    4.140   3.49e-05 ***
# Bathrooms         12.656       1.0325   12.258    < 2e-16 ***
# Bedrooms          8.5513      0.94134    9.084    < 2e-16 ***
# Beds              3.2999      0.67044    4.922   8.70e-07 ***
# Accommodates      12.882      0.52156   24.698    < 2e-16 ***
# Guests         -0.36181      0.37421   -0.967     0.3336
# MinNights         2.0150      0.35169    5.730   1.03e-08 ***
# MaxNights    -0.0014954    0.0007393   -2.023     0.0431 *
# ExtraPeople    -0.16668     0.035897   -4.643   3.47e-06 ***
# HostListings    0.39766     0.024771   16.054    < 2e-16 ***
# ResponseRate    -12.952       3.5793   -3.619     0.0003 ***
# Deposit        0.060526    0.0052226   11.589    < 2e-16 ***
# CleaningFee    0.080951     0.023535    3.440     0.0006 ***
# FeeMissing       2.1638      0.93926    2.304     0.0213 *
#
# Model Statistics:
# ------------------------------------------------------
# Residual Std. Error: 40.322
# R-squared:        0.4764          AIC: 106116.44
# Adj. R-squared: 0.4758            BIC: 106217.90
# F-statistic: 724.9262 on 13 and 10356 DF, p-value: < 2e-16
# ======================================================
```

(c) From the model summary in the previous part, we note that all predictor variables are significant with the exception of `Guests`, which has an individual significance p-value of 0.333641. Here, we fit `reg2` by specifying a formula that includes all of the predictor variables except the `Guests` variable.

```
reg2 = ols("Price ~ . - Guests", data = df)
reg2.summary()
```

```
# Summary of OLS Regression Analysis:
# =======================================================
#
# Coefficients:
# -------------------------------------------------------
#                    Estimate  Std. Error  t-value    p-value
# Intercept            15.568      3.8002    4.097   4.23e-05 ***
# Bathrooms            12.657      1.0325   12.259    < 2e-16 ***
# Bedrooms             8.5485     0.94133    9.081    < 2e-16 ***
# Beds                 3.2672     0.66959    4.879   1.08e-06 ***
# Accommodates         12.798     0.51440   24.880    < 2e-16 ***
# MinNights            2.0375     0.35092    5.806   6.58e-09 ***
# MaxNights        -0.0014906   0.0007393   -2.016     0.0438 *
# ExtraPeople        -0.17785    0.033987   -5.233   1.70e-07 ***
# HostListings        0.39391    0.024465   16.101    < 2e-16 ***
# ResponseRate        -12.962      3.5793   -3.621     0.0003 ***
# Deposit            0.060696   0.0052196   11.629    < 2e-16 ***
# CleaningFee        0.078094    0.023349    3.345     0.0008 ***
# FeeMissing           2.2486     0.93516    2.404     0.0162 *
#
# Model Statistics:
# -------------------------------------------------------
# Residual Std. Error: 40.3219
# R-squared:         0.4764          AIC: 106115.38
# Adj. R-squared: 0.4758             BIC: 106209.59
# F-statistic: 785.2638 on 12 and 10357 DF, p-value: < 2e-16
# =======================================================
```

(d) While the multiple R^2 and adjusted R^2 values are relatively unchanged, the `reg2` model is more parsimonious since it has fewer predictors.

9.4 Assessing Model Performance

We take into consideration several metrics to assess our models. Since each metric evaluates a different aspect of model performance, we examine several of the more popular metrics in greater detail here.

Multiple R-Squared and Adjusted R-Squared

In Chap. 3, the R^2 value was introduced for simple linear regression and later, in Chap. 5, the multiple R^2 and adjusted R^2 were introduced and built off of the knowledge of R^2. Naturally, the multiple R^2 represents the amount of explained variation over the total variation,

$$R^2 = SSR/SST. \tag{9.1}$$

However, multiple R^2 never decreases when additional predictors are added to the model, even if those predictors have no real relationship with the response. As a result, irrelevant variables can artificially inflate R^2.

To counteract this, the adjusted R^2 penalizes unnecessary variables by accounting for the number of predictors (p) relative to the number of observations (n):

$$R_a^2 = 1 - (1 - R^2)\frac{n-1}{n-p-1}, \tag{9.2}$$

which gives the adjusted R^2 value. Both the multiple R^2 and the adjusted R^2 are given from the model summary.

9.4.1 Akaike Information Criterion

A widely used metric for model selection is the Akaike Information Criterion (AIC). This criteria takes into account the SSE, the number of observations (n), and the number of predictors (p),

$$AIC = n \log \frac{SSE}{n} + 2p. \tag{9.3}$$

To use the AIC criterion, we find AIC values for each model we would like to compare and then choose the model with the lower AIC value. The AIC value is calculated and is an attribute within the model object (`aic`).

9.4.2 Bayesian Information Criterion

Another commonly used metric for model selection is the Bayesian Information Criterion (BIC). The equation for BIC is similar to Eq. (9.3),

$$BIC = n \log \frac{SSE}{n} + p \log n. \tag{9.4}$$

Note that while AIC uses $2p$ as the last component of the equation, BIC uses $p \log n$. This subtle difference will increase BIC more rapidly for an increase in the number of predictors (p). Thus, BIC is penalized more heavily for higher values of p, resulting in fewer predictor variables in the chosen model. The BIC value is also stored in the regression object, accessible through the attribute `bic`.

9.4.3 Mallows's C_p

One additional measure that is popular for model selection is Mallows's C_p. This measure is calculated by

$$C_p = \frac{SSE}{MSE_{full}} - n + 2p, \tag{9.5}$$

where SSE is calculated from the model in question and MSE_{full} is the mean squared error calculated from the model with all considered predictors. Like the previous metrics, Mallows's C_p can be used to find a good model with only a few variables. While we seek to find the smallest value of C_p, we consider a model to be good if $C_p \leq p$.

In practice, Mallows's C_p is most useful if the full model is not a serious candidate for the final model. For the full model, it can be proven that $C_p = p$ so Mallows's statistic will always deem the full model good.

Mallows's C_p isn't used here because the full model is already under consideration, making C_p less informative. AIC and BIC are more suitable for comparing models in this context.

9.5 Airbnb Pricing Application: Model Comparison

The models from the previous application were fit using the following code:

```
reg1 = ols("Price ~ .", data = df)
reg2 = ols("Price ~ . - Guests", data = df)
```

Using the models from the previous application, answer the following:

(a) How do we account for the difference between the multiple R^2 and the adjusted R^2 in the first model? Is this difference relatively large or small?
(b) Calculate the AIC and BIC for the first model.
(c) Calculate the AIC and BIC for the second model.
(d) Which model has the lower AIC? Which model has the lower BIC?

Solution:

(a) The adjusted R^2 takes into account the number of predictors and the number of observations. There is a very small difference between the multiple R^2 and the adjusted R^2 in the first model.
(b) The AIC and BIC values are attributes of the regression model (`reg1`). While they can be found in the model summary, they can also be output directly from the model.

```
reg1.aic
```

```
# 106116.44414135069
```

```
reg1.bic
```

```
# 106217.89755356782
```

(c) Here, we repeat the calculations from the previous part using `reg2`.

```
reg2.aic
```

```
# 106115.3801631882
```

```
reg2.bic
```

```
# 106209.58690310411
```

(d) The `reg2` model has a lower AIC and a lower BIC, indicating that `reg2` is a better model than `reg1`.

9.6 Backward Elimination

In the first application of this chapter, we fit a model using all of the predictor variables in the dataset, and then we used a model without the least significant predictor variable. In essence, we were performing backward elimination with the individual significance as the selection criteria. However, most backward elimination algorithms use AIC, BIC, or adjusted R^2 as the selection criteria.

Backward elimination begins with a model that includes all predictor variables the analyst would like to consider. After fitting this full model, the analyst evaluates whether removing any one variable improves the model selection metric; a variable is only excluded if doing so results in a better model than the current one. The process of excluding variables one at a time is repeated until no improvement can be made in the model selection metric. From this process, it should be noted that a predictor variable cannot reenter at a subsequent step or iteration. A structural outline of this algorithm is depicted in Fig. 9.1.

The `step` function in Ravix implements backward elimination. The first argument of `step` is a formula that includes all predictor variables under consideration. The `data` source is also a required argument. The `direction` argument's default is `"backward"`, indicating backward elimination.

```
from ravix import stepwise
stepwise(formula, data)
```

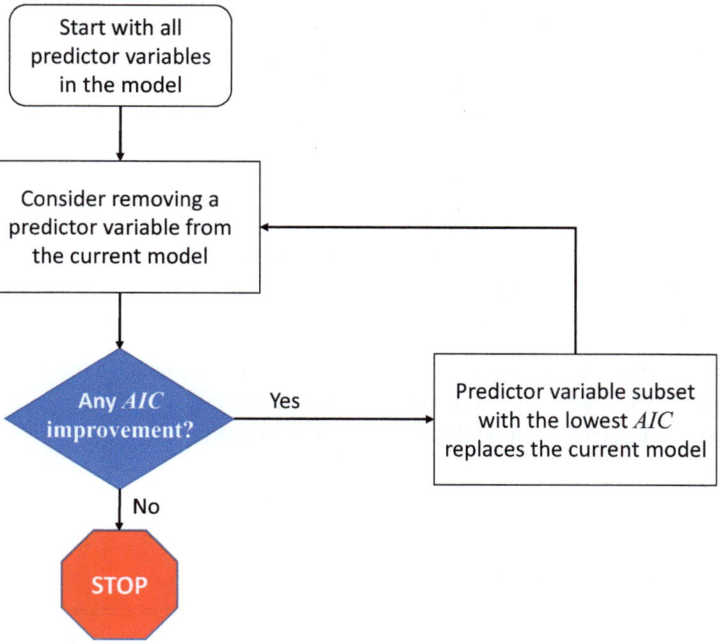

Fig. 9.1 Backward elimination

The `step` function uses `metric = "aic"` as the default. The procedure begins with a full model and considers removing one predictor variable at a time. The AIC is calculated for each model with one variable removed, and the model chosen for the next iteration is the one with the lowest AIC. The algorithm is repeated on this reduced model and continues in this fashion. The process stops when removing a variable fails to improve the AIC.

While `aic` is used by default, other selection criteria can be specified in the `metric` argument. The options for `metric` are `"aic"`, `"bic"`, `"pvalue"`, and `"adjr2"`.

9.7 Airbnb Pricing Application: Backward Elimination

Using the Edinburgh Airbnb data file, do the following:

(a) Run backward elimination beginning from the full model in the previous part.
(b) Specify which variable is eliminated in the first iteration of backward elimination. How many variables are eliminated in total using this process?
(c) Print a model summary of the backward elimination model.

Solution:

(a) Backward elimination is run using the `step` function. The formula and the data are the only necessary arguments to run backward elimination using this function, but including the `verbose = True` argument allows the steps to be printed.

```
from ravix import stepwise
BE = stepwise("Price ~ .", data = df, verbose = True)
```

```
# Initial AIC = 106116.4441
# Step 1: remove Guests (AIC=106115.3802)
# No improvement found. Stopping at step 2.
# Final AIC = 106115.3802
```

(b) As seen from the output, the `Guests` variable is eliminated in the first step. The `Guests` variable is the only variable eliminated.

(c) Here we pass the `BE` model to the `summary` function.

```
BE.summary()
```

```
# Summary of OLS Regression Analysis:
# ========================================================
#
# Coefficients:
# --------------------------------------------------------
#                  Estimate  Std. Error  t-value    p-value
# Intercept          15.568      3.8002    4.097   4.23e-05 ***
# Bathrooms          12.657      1.0325   12.259    < 2e-16 ***
# Bedrooms           8.5485     0.94133    9.081    < 2e-16 ***
# Beds               3.2672     0.66959    4.879   1.08e-06 ***
# Accommodates       12.798     0.51440   24.880    < 2e-16 ***
# MinNights          2.0375     0.35092    5.806   6.58e-09 ***
# MaxNights      -0.0014906   0.0007393   -2.016     0.0438 *
# ExtraPeople      -0.17785    0.033987   -5.233   1.70e-07 ***
# HostListings      0.39391    0.024465   16.101    < 2e-16 ***
# ResponseRate      -12.962      3.5793   -3.621     0.0003 ***
# Deposit          0.060696   0.0052196   11.629    < 2e-16 ***
# CleaningFee      0.078094    0.023349    3.345     0.0008 ***
# FeeMissing         2.2486     0.93516    2.404     0.0162 *
#
# Model Statistics:
# --------------------------------------------------------
# Residual Std. Error: 40.3219
# R-squared:        0.4764          AIC: 106115.38
# Adj. R-squared: 0.4758            BIC: 106209.59
# F-statistic: 785.2638 on 12 and 10357 DF, p-value: < 2e-16
# ========================================================
```

9.8 Forward Selection

Forward selection begins without any predictor variables in the model. This model without predictor variables is referred to as the null model and only contains the intercept. Beginning with this null model, we consider all possible models with one predictor variable. Once we choose the best model, we continue adding in predictor variables, one at a time, until there is no longer an improvement in the selection criterion. In this case, we are bringing variables into the model, and we don't consider removing them in the forward selection algorithm. A structural outline of this algorithm is depicted in Fig. 9.2.

To implement forward selection in Ravix, we use the `step` function and specify the `formula`, `data`, and `direction` arguments. Note that the `formula` includes all the predictor variables under consideration.

```
from ravix import stepwise
stepwise(formula, data, direction = "forward")
```

Since `direction` defaults to `"backward"`, it is necessary to specify the `direction` argument to execute forward selection.

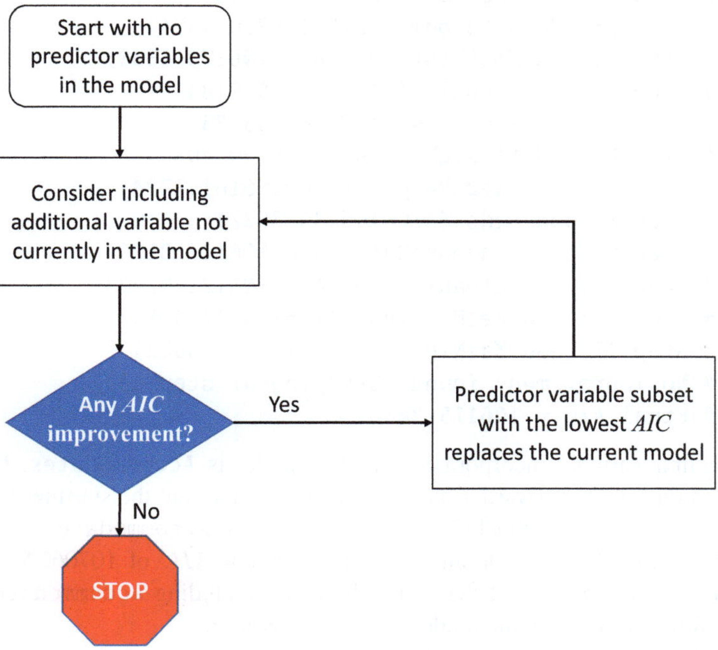

Fig. 9.2 Forward selection

9.9 Airbnb Pricing Application: Forward Selection

Using the Edinburgh Airbnb data file, do the following:

(a) Run forward selection and print the steps using the `verbose` argument.
(b) Specify which variable is incorporated in the first iteration of forward selection.
(c) Print a model summary of the forward selection model.

Solution:

(a) The `step` function is used to perform forward selection by beginning without any predictor variables. Then, the process evaluates which variable to include in the model, considering all the predictor variables. Evaluating variables to include is repeated until there is no longer an improvement in AIC. The `verbose` argument is set to ensure that the output is printed. The complete syntax is specified here to create the `FS` object.

```
from ravix import stepwise
FS = stepwise("Price ~ .", data = df, direction = "forward",
              verbose = True)
```

```
# Initial AIC = 112800.9472
# Step 1: add Accommodates (AIC=107062.8332)
# Step 2: add Bathrooms (AIC=106800.4324)
# Step 3: add HostListings (AIC=106578.1330)
# Step 4: add Deposit (AIC=106375.0134)
# Step 5: add Bedrooms (AIC=106233.8317)
# Step 6: add MinNights (AIC=106190.4080)
# Step 7: add ExtraPeople (AIC=106161.7234)
# Step 8: add Beds (AIC=106139.8622)
# Step 9: add ResponseRate (AIC=106127.8120)
# Step 10: add CleaningFee (AIC=106121.0762)
# Step 11: add FeeMissing (AIC=106117.4503)
# Step 12: add MaxNights (AIC=106115.3802)
# No improvement found. Stopping at step 13.
# Final AIC = 106115.3802
```

(b) The first variable incorporated into the model is `Accommodates`. From the beginning of the forward selection output, we note that the starting AIC takes on a value of 112800.9472. Note that including `Accommodates` denotes the inclusion of this variable and the corresponding AIC of 107062.8332 is the lowest AIC on the first iteration. Therefore, including `Accommodates` is the best improvement on the model at the first iteration.
(c) The `FS` model is easily printed using the `summary` function.

```
FS.summary()
```

```
# Summary of OLS Regression Analysis:
# ======================================================
#
# Coefficients:
# ------------------------------------------------------
#                    Estimate  Std. Error  t-value    p-value
# Intercept            15.568      3.8002    4.097   4.23e-05 ***
# Accommodates         12.798     0.51440   24.880    < 2e-16 ***
# Bathrooms            12.657      1.0325   12.259    < 2e-16 ***
# HostListings       0.39391    0.024465   16.101    < 2e-16 ***
# Deposit           0.060696   0.0052196   11.629    < 2e-16 ***
# Bedrooms           8.5485     0.94133    9.081    < 2e-16 ***
# MinNights          2.0375     0.35092    5.806   6.58e-09 ***
# ExtraPeople      -0.17785    0.033987   -5.233   1.70e-07 ***
# Beds               3.2672     0.66959    4.879   1.08e-06 ***
# ResponseRate     -12.962      3.5793   -3.621     0.0003 ***
# CleaningFee     0.078094    0.023349    3.345     0.0008 ***
# FeeMissing        2.2486     0.93516    2.404     0.0162 *
# MaxNights     -0.0014906   0.0007393   -2.016     0.0438 *
#
# Model Statistics:
# ------------------------------------------------------
# Residual Std. Error: 40.3219
# R-squared:         0.4764          AIC: 106115.38
# Adj. R-squared: 0.4758             BIC: 106209.59
# F-statistic: 785.2638 on 12 and 10357 DF, p-value: < 2e-16
# ======================================================
```

9.10 Stepwise Regression

While stepwise regression could possibly refer to backward elimination or forward selection, we will use it here to refer to the more general case of stepwise regression where predictor variables could be added or removed from the model. The stepwise algorithm starts with a null model and then adds in variables, one at a time, but also considers removing variables from the model.

The stepwise method differs from the forward selection method because we consider removing variables from the model as well. In datasets with many highly correlated predictors, a variable may enter the model at an early stage but later be removed as additional predictors are added. This occurs because the new variables

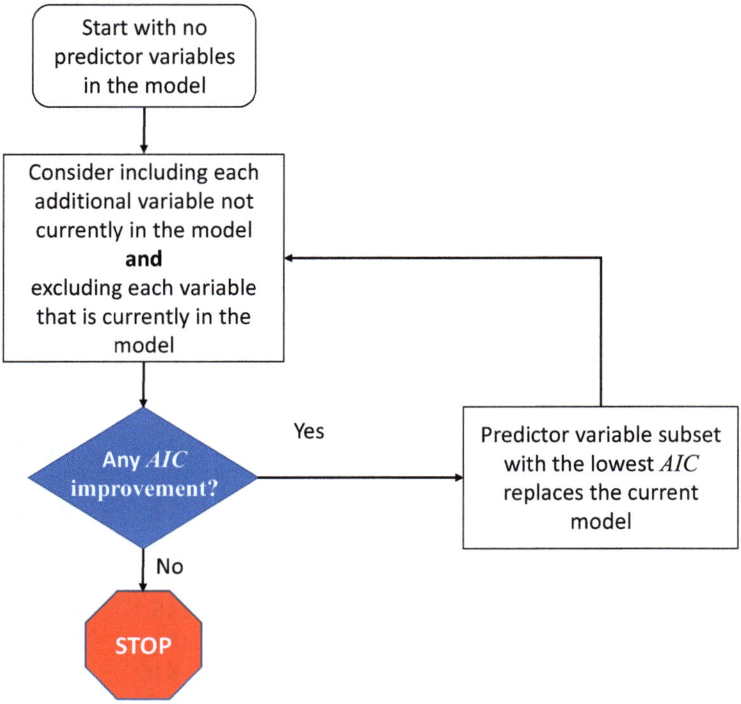

Fig. 9.3 Stepwise regression

explain much of the same variation, making the earlier predictor redundant once the model updates. A structural outline of this algorithm is depicted in Fig. 9.3.

Since the stepwise algorithm is similar to forward selection, the code to implement stepwise is similar to that of forward selection. The `step` function can be used to invoke the stepwise algorithm:

```
from ravix import stepwise
SW = stepwise(formula, data, direction = "both")
```

Notice the `direction` argument is specified as `"both"` to allow variables to be added or removed in each step.

9.11 Airbnb Pricing Application: Stepwise Regression

Using the Edinburgh Airbnb data file, do the following:

(a) Run stepwise regression and print the steps using the `verbose` argument.

(b) Specify which variable is incorporated in the first iteration of the stepwise regression.
(c) Print a model summary of the stepwise regression model.
(d) Clearly state the difference (if any) between the backwards elimination, forward selection, and the stepwise regression models.

Solution:

(a) Stepwise regression is performed using the `step` function. Note that the syntax is identical to that of the other selection methods with the exception of the `direction` argument, which is specified as "`both`".

```
from ravix import stepwise
SW = stepwise("Price ~ .", data = df, direction = "both",
              verbose = True)
```

```
    # Initial AIC = 112800.9472
    # Step 1: add Accommodates (AIC=107062.8332)
    # Step 2: add Bathrooms (AIC=106800.4324)
    # Step 3: add HostListings (AIC=106578.1330)
    # Step 4: add Deposit (AIC=106375.0134)
    # Step 5: add Bedrooms (AIC=106233.8317)
    # Step 6: add MinNights (AIC=106190.4080)
    # Step 7: add ExtraPeople (AIC=106161.7234)
    # Step 8: add Beds (AIC=106139.8622)
    # Step 9: add ResponseRate (AIC=106127.8120)
    # Step 10: add CleaningFee (AIC=106121.0762)
    # Step 11: add FeeMissing (AIC=106117.4503)
    # Step 12: add MaxNights (AIC=106115.3802)
    # No improvement found. Stopping at step 13.
    # Final AIC = 106115.3802
```

(b) The `Accommodates` variable is incorporated in the first iteration, which is identical to the process for forward selection. In fact, the variables are included in the same order as forward selection. Even though stepwise had the option of eliminating variables, the AIC would not have improved by eliminating variables from the model.
(c) A summary of the `SW` object is done here.

```
SW.summary()
```

```
# Summary of OLS Regression Analysis:
# ======================================================
#
# Coefficients:
# ------------------------------------------------------
#                 Estimate  Std. Error  t-value   p-value
# Intercept         15.568      3.8002    4.097  4.23e-05 ***
# Accommodates      12.798     0.51440   24.880   < 2e-16 ***
# Bathrooms         12.657      1.0325   12.259   < 2e-16 ***
# HostListings     0.39391    0.024465   16.101   < 2e-16 ***
# Deposit         0.060696   0.0052196   11.629   < 2e-16 ***
# Bedrooms          8.5485     0.94133    9.081   < 2e-16 ***
# MinNights         2.0375     0.35092    5.806  6.58e-09 ***
# ExtraPeople     -0.17785    0.033987   -5.233  1.70e-07 ***
# Beds              3.2672     0.66959    4.879  1.08e-06 ***
# ResponseRate     -12.962      3.5793   -3.621    0.0003 ***
# CleaningFee   0.078094    0.023349    3.345    0.0008 ***
# FeeMissing        2.2486     0.93516    2.404    0.0162 *
# MaxNights   -0.0014906   0.0007393   -2.016    0.0438 *
#
# Model Statistics:
# ------------------------------------------------------
# Residual Std. Error: 40.3219
# R-squared:      0.4764           AIC: 106115.38
# Adj. R-squared: 0.4758           BIC: 106209.59
# F-statistic: 785.2638 on 12 and 10357 DF, p-value: < 2e-16
# ======================================================
```

(d) All three methods come up with the same solution. While it may seem that the algorithms should come to the same solution, it is not always the case. For convenience of syntax and computation, consider beginning with backward elimination.

9.12 Best Subsets Regression

The best subsets regression algorithm consists of evaluating all possible predictor variable subsets. Therefore, best subsets regression guarantees that the chosen model will be the best model according to the criteria that you choose. For instance, the best subsets will find the model with the best adjusted R^2 for all variable combinations if that is our criteria.

The implementation of best subsets regression in Ravix is `bsr`, which requires at minimum a formula and a dataset. An optional argument of `max_var` specifies the number of variables to be considered. By default, `bsr` considers variable combinations up to 8. In addition, a `metric` can be specified of either `"adjr2"`, `"aic"`, `"bic"`, or `"rmse"`. The default `metric` is `"aic"`.

```
from ravix import bsr
bsr(formula, data, max_var = 8, metric = "aic")
```

9.13 Airbnb Pricing Application: Best Subsets Regression 1

Using the Edinburgh Airbnb data file, do the following:

(a) Run the best subsets method on the data using all predictor variables using the default value of `max_var` and `metric`.
(b) Plot the results of the best subsets method from part a using the `plot_bsr` function.
(c) Interpret the plot from part b by specifying the best number of variables.
(d) Plot the results of the best subsets method from part a using the `plot_bsr` function using `type = "predictors"`.
(e) Interpret the plot from part d by specifying the best model in the plot.
(f) Print the model summary resulting from the best subsets.
(g) Find the corresponding AIC, BIC, and adjusted R^2 of this best model.

Solution:

(a) The best subsets object, BSR1, is created using the `bsr` function. The primary input arguments for this function are the same as those of the `ols` function. By only setting the primary input arguments, the default argument of `max_var` of 8 is used. The default `metric` is `"aic"`.

```
from ravix import bsr
BSR1 = bsr("Price ~ .", df)
```

(b) We can simply pass BSR1 to the `plot_bsr` function with `type = "line"` to get a line plot of the best subsets with AIC as the metric. To use adjusted R^2 as the metric, set the `metric` argument to `"adjr2"` in the `bsr` function and rerun the `plot_bsr`.

```
from ravix import plot_bsr
plot_bsr(BSR1, type = "line")
```

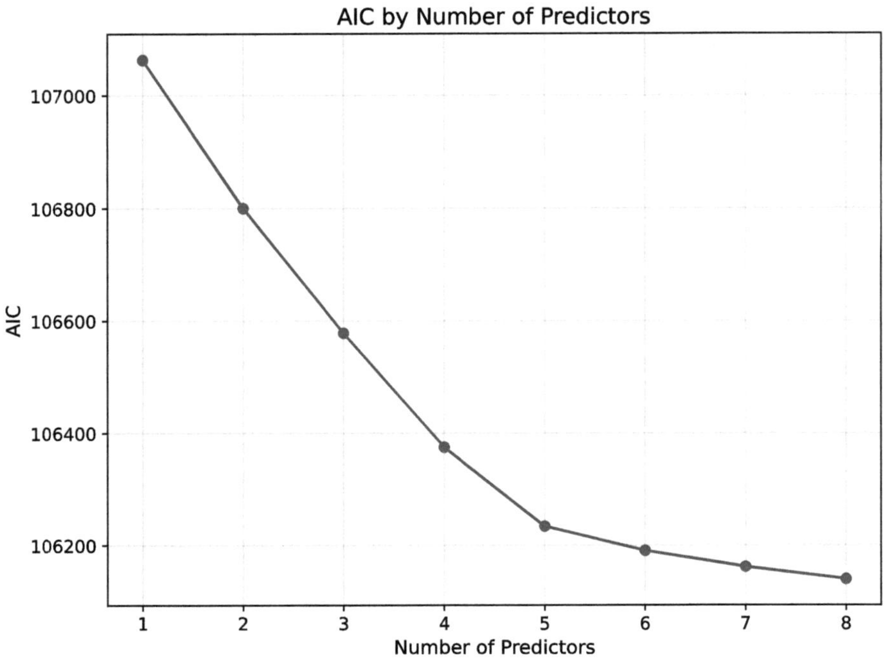

Fig. 9.4 Best subsets regression results 1

(c) From this best subsets analysis, the best model contains 8 predictors since we see the lowest AIC value of 106139.86 corresponds to 8. This information can be seen from Fig. 9.4.
(d) The top models from best subsets can be plotted in a variable inclusion plot by using the default `type = "predictors"` within the `plot_bsr` function.

```
plot_bsr(BSR1)
```

(e) From the plot above, we see that the lowest AIC is 106,139.862, which is the top row of Fig. 9.5. Note that the top row has 8 rectangles filled with green, denoting their inclusion in the model. This best model consists of the predictor variables: `Accommodates`, `Bathrooms`, `Bedrooms`, `Beds`, `Deposit`, `ExtraPeople`, `HostListings`, and `MinNights`.
(f) The `bsr` function returns the optimal model under the chosen metric along with comparative information across candidate models. The `summary` function from Ravix can print a model summary.

```
BSR1.summary()
```

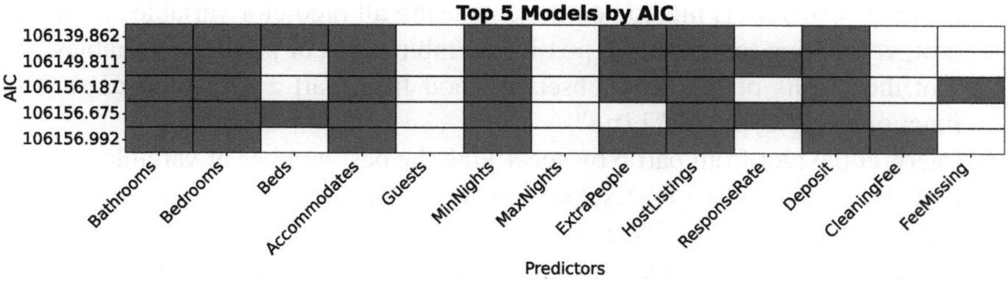

Fig. 9.5 Best subsets regression results 2

```
# Summary of OLS Regression Analysis:
# ==========================================================
#
# Coefficients:
# ----------------------------------------------------------
#                   Estimate  Std. Error  t-value     p-value
# Intercept           3.1627      1.3624    2.321      0.0203 *
# Bathrooms           12.732      1.0334   12.321    < 2e-16 ***
# Bedrooms            8.9855     0.93560    9.604    < 2e-16 ***
# Beds                3.2631     0.66792    4.885   1.05e-06 ***
# Accommodates        12.877     0.49785   25.865    < 2e-16 ***
# MinNights           2.2471     0.34864    6.446   1.20e-10 ***
# ExtraPeople       -0.18647    0.033345   -5.592   2.30e-08 ***
# HostListings       0.38642    0.024382   15.848    < 2e-16 ***
# Deposit           0.064506   0.0051560   12.511    < 2e-16 ***
#
# Model Statistics:
# ----------------------------------------------------------
# Residual Std. Error: 40.3773
# R-squared:          0.4748          AIC: 106139.86
# Adj. R-squared: 0.4743              BIC: 106205.08
# F-statistic: 1170.62 on 8 and 10361 DF, p-value: < 2e-16
# ==========================================================
```

By inspection of the summary output, we see that the AIC is 106139.8622, the BIC is 106205.0822, and the adjusted R^2 value is 0.4743.

9.14 Airbnb Pricing Application: Best Subsets Regression 2

Using the Edinburgh Airbnb data file, do the following:

(a) Run the best subsets method on the data using all predictor variables using the `max_var` option to include all possible combinations of predictor variables.
(b) Plot the results of the best subsets method from part a using the `plot_bsr` function using `type = "line"`.
(c) Interpret the plot from part b by specifying the best number of variables.
(d) Plot the results of the best subsets method from part a using the `plot_bsr` function using `type = "predictors"`.
(e) Interpret the plot from part d by specifying the best model in the plot.
(f) Print the model summary resulting from the best subsets.

Solution:

(a) Here, we again use `bsr` in a similar fashion to part a in the previous application, but specify `max_var` to include all 13 variables.

```
from ravix import bsr
BSR2 = bsr("Price~.", df, max_var = 13)
```

(b) Here, we plot a best subsets line plot using AIC as the metric.

```
from ravix import plot_bsr
plot_bsr(BSR2, type = "line")
```

(c) Figure 9.6 indicates that the best number of predictors is above 9, but it is difficult to tell which one is the best from the graph. The lowest AIC value of 106115.38 is shown in the plot, indicating the best number of predictors is 12.
(d) The top models from best subsets can be plotted in a variable inclusion plot by specifying the default `type = "predictors"` within the `plot_bsr` function as done in the previous application (Fig. 9.7).

```
plot_bsr(BSR2)
```

(e) The best model according to AIC corresponds to the top row, which contains all 13 predictor variables except the `Guests` variable. The resulting AIC is 106,115.38.
(f) The model from the `BSR2` object chooses the best AIC because we used the default metric when creating the object in part a. The model summary shows the best AIC model.

```
BSR2.summary()
```

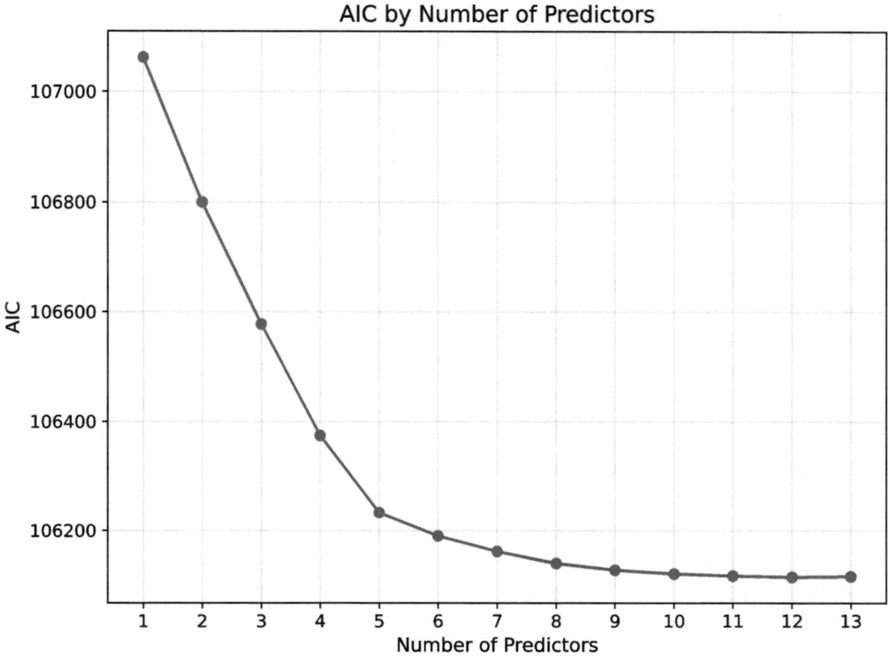

Fig. 9.6 Best subsets regression results 1

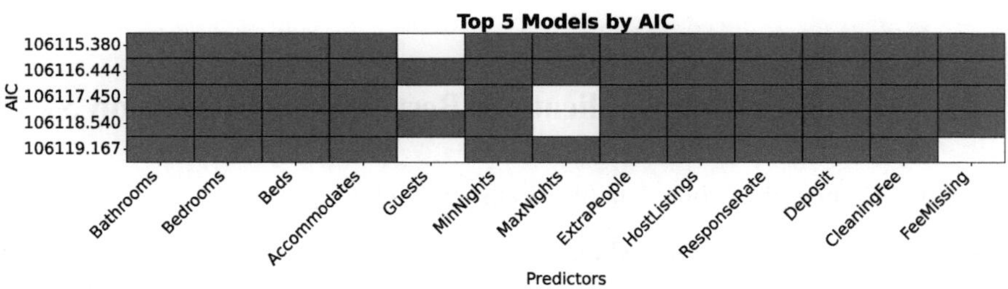

Fig. 9.7 Best subsets regression results 2

```
# Summary of OLS Regression Analysis:
# ======================================================
#
# Coefficients:
# ------------------------------------------------------
#                 Estimate  Std. Error  t-value    p-value
# Intercept         15.568      3.8002    4.097   4.23e-05 ***
# Bathrooms         12.657      1.0325   12.259    < 2e-16 ***
# Bedrooms          8.5485     0.94133    9.081    < 2e-16 ***
# Beds              3.2672     0.66959    4.879   1.08e-06 ***
# Accommodates      12.798     0.51440   24.880    < 2e-16 ***
# MinNights         2.0375     0.35092    5.806   6.58e-09 ***
# MaxNights    -0.0014906   0.0007393   -2.016     0.0438 *
# ExtraPeople     -0.17785    0.033987   -5.233   1.70e-07 ***
# HostListings     0.39391    0.024465   16.101    < 2e-16 ***
# ResponseRate     -12.962      3.5793   -3.621     0.0003 ***
# Deposit         0.060696   0.0052196   11.629    < 2e-16 ***
# CleaningFee     0.078094    0.023349    3.345     0.0008 ***
# FeeMissing        2.2486     0.93516    2.404     0.0162 *
#
# Model Statistics:
# ------------------------------------------------------
# Residual Std. Error: 40.3219
# R-squared:        0.4764          AIC: 106115.38
# Adj. R-squared: 0.4758            BIC: 106209.59
# F-statistic: 785.2638 on 12 and 10357 DF, p-value: < 2e-16
# ======================================================
```

9.15 Airbnb Pricing Application: Best Subsets Regression 3

Using the Edinburgh Airbnb data file, do the following:

(a) Run the best subsets method on the data using all predictor variables using the `max_var` option to include all possible combinations of predictor variables with `metric="bic"`.
(b) Plot the results of the best subsets method from part a using the `plot_bsr` function using `type = "predictors"` with `top_n = 10`.
(c) Print the model summary resulting from the best subsets.
(d) How do different selection criteria influence the choice of a best subsets model, and what are the trade-offs between them?

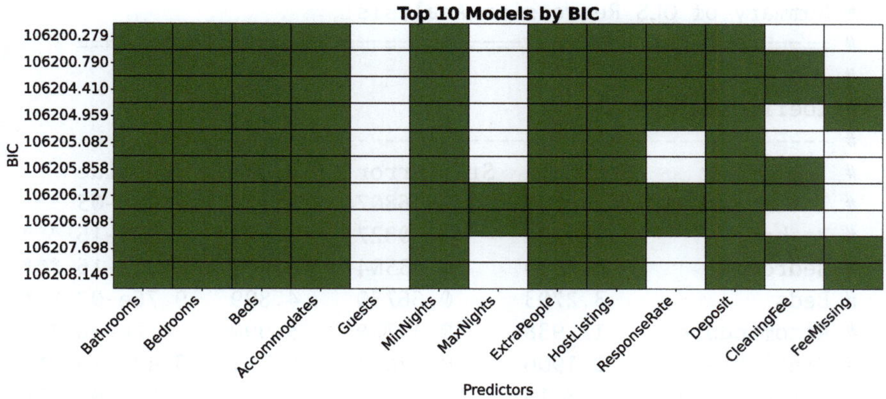

Fig. 9.8 Best subsets regression results 3

Solution:

(a) Including the `metric = "bic"` argument within the `bsr` function allows for the algorithm to find the best model according to the BIC metric.

```
from ravix import bsr
BSR3 = bsr("Price ~ .", data = df, max_var = 13, metric = "bic")
```

(b) We modify the variable inclusion plot from part d by specifying `metric = "bic"`. Note that `type` is an unnecessary argument to specify since we are using the default `type`. Specifying `top_n = 10` will list the top 10 models as ranked by the BIC metric (Fig. 9.8).

```
from ravix import plot_bsr
plot_bsr(BSR3, top_n = 10)
```

(c) The model summary shows the best BIC model, which consists of 9 predictors as shown in the variable inclusion plot.

```
BSR3.summary()
```

```
# Summary of OLS Regression Analysis:
# ======================================================
#
# Coefficients:
# ------------------------------------------------------
#                  Estimate  Std. Error  t-value    p-value
# Intercept          15.979      3.6807    4.341   1.43e-05 ***
# Bathrooms          12.724      1.0327   12.321    < 2e-16 ***
# Bedrooms           8.9579     0.93504    9.580    < 2e-16 ***
# Beds               3.2703     0.66750    4.899   9.76e-07 ***
# Accommodates       12.938     0.49780   25.990    < 2e-16 ***
# MinNights          2.1906     0.34874    6.281   3.49e-10 ***
# ExtraPeople      -0.18518    0.033325   -5.557   2.81e-08 ***
# HostListings      0.38690    0.024367   15.878    < 2e-16 ***
# ResponseRate      -13.350      3.5621   -3.748     0.0002 ***
# Deposit          0.064040   0.0051543   12.425    < 2e-16 ***
#
# Model Statistics:
# ------------------------------------------------------
# Residual Std. Error: 40.3519
# R-squared:        0.4755          AIC: 106127.81
# Adj. R-squared: 0.4750            BIC: 106200.28
# F-statistic: 1043.42 on 9 and 10360 DF, p-value: < 2e-16
# ======================================================
```

(d) In best subsets regression, the model that includes all predictor variables will often appear to have the best fit, simply because more variables can always improve the apparent performance—regardless of whether those variables are meaningful. This is similar to how multiple R^2 always increases with additional predictors, even if they aren't truly useful. Adjusted R^2 corrects for this by accounting for model complexity.

Both adjusted R^2 and AIC select the model with 12 predictors as the best. In the earlier example, we removed a low-significance variable, and while the adjusted R^2 did not visibly change due to rounding, best subsets confirms a slight improvement. Eliminating unnecessary variables results in a more parsimonious model.

BIC, on the other hand, favors a model with just 9 predictors. Its stronger penalty for model complexity often selects simpler models that may be easier to interpret while still maintaining good predictive performance.

9.16 Stepwise and Best Subsets Regression

The backward elimination, forward selection, and stepwise algorithms selected variables in an iterative manner but did not consider all of the possible predictor variable subsets. While these iterative methods are sufficient in most cases, they may be inadequate when many variables are involved because there is no guarantee that the resulting model will be the best one. Best subsets regression, on the other hand, will guarantee that the best possible metric is found across all predictor variable combinations. We note that the best subsets regression can be limited in the `bsr` function using the `max_var` argument. This argument can be adjusted as discussed in the previous application; however, the number of subset models needed to be fit grows at a rapid rate as the number of variables gets large. Since the number of variable subsets becomes very large when the number of variables is large, the best subsets regression algorithm can be computationally expensive. More specifically, each variable subset results in a regression model fit, which can take a long time to compute. In the Airbnb applications, all of the iterative approaches had the same resulting model, which would also yield the same model for best subsets regression using the AIC metric (this exercise is left to the reader).

Note that the selection methods from this chapter are only a starting place for what variables should be included. After determining a good subset of variables, the analyst should continue their analysis to see if interaction or higher-order terms should be included in the final model.

9.17 Case Study: Cancer Treatment Cost Analysis

9.17.1 Problem Statement

An insurance company heavily relies on various factors to determine what an individual's premium should be. A profitable insurance company charges more in premiums, on average, than the amount of health care expenses incurred by beneficiaries. Hence, insurance company analysts must study the profiles of individual beneficiaries to properly estimate the future expenses that each may incur, and then properly calculate the appropriate premium for that individual.

In this case study, we focus on the factors of healthcare costs that can be used by an insurance company to determine premiums. Healthcare costs can vary for a variety of reasons. Among these, cancer treatment represents a particularly costly medical expense that insurance pays for. This case examines the data on charges billed by health insurance for the treatment of different types of cancer.

Carla is a junior analyst at Quality Insurance (QI). She was recently tasked with examining insurance charges made by policyholders who were diagnosed with cancer. She fit a regression model from the data, but some issues regarding the model need to be resolved. After Carla finished her analysis, you were hired as an independent contractor to check the quality of her analysis and answer a few additional questions. Specifically, your tasks consist of the following:

1. Assess the regression model fit from the preliminary analysis.
2. Find a parsimonious model that predicts insurance charges and interpret the coefficients.
3. Test for differences in charges across the different BMI categories for smokers and nonsmokers.
4. Test for income differences across the different insurance plans.

9.17.2 Data Description

QI provides data concerning cancer patients from 2021. Several variables included within the data can be used to predict the amount of charges billed to QI.

The dataset consists of the variables listed below:

- **ID**—The row or observation number, where each row represents an individual
- **Age**—The age of the individual receiving treatment
- **BMI**—Body Mass Index (BMI), a measure of an individual's weight in reference to their height
- **Income**—The income of the individual
- **Plan**—A categorical variable (values 1, 2, or 3) indicating the type of medical insurance plan selected by the individual
- **Smoke**—A categorical variable representing whether an individual smokes or not
- **Sex**—A categorical variable representing whether an individual is male or female
- **Charges**—Medical costs billed to health insurance

The data is contained within the `Charges.csv` file.

9.17.3 Preliminary Analysis

Carla performed a preliminary analysis. She loaded in the data, and fit a regression model using all of the columns, other than `Charges`, as predictors. Her code and summary are shown here.

```
from ravix import get_data, ols
df = get_data("charges.csv")
reg = ols("Charges ~ .", data = df)
reg.summary()
```

```
# Summary of OLS Regression Analysis:
# =====================================================
#
# Coefficients:
# -----------------------------------------------------
#                    Estimate  Std. Error  t-value    p-value
# Intercept         -74855.0     17533.6   -4.269   2.15e-05 ***
# ID                 -6.9671      4.8058   -1.450     0.1475
# Age                 940.32      189.27    4.968   7.96e-07 ***
# BMI                 3979.4      307.23   12.953    < 2e-16 ***
# Plan                2190.4      2315.9    0.946     0.3445
# Sex_male            4913.5      2796.7    1.757     0.0793 .
# Smoke_Y            52844.5      3385.9   15.607    < 2e-16 ***
# Income             0.12896    0.072220    1.786     0.0745 .
#
# Model Statistics:
# -----------------------------------------------------
# Residual Std. Error: 43821.41
# R-squared:       0.3006          AIC: 24221.60
# Adj. R-squared: 0.2957           BIC: 24260.86
# F-statistic: 60.9047 on 7 and 992 DF, p-value: < 2e-16
# =====================================================
```

Unfortunately, there are several errors in the analysis. The residual plots indicate that the linear regression assumptions have been violated.

```
from ravix import plot
plot(reg)
```

In Fig. 9.9, the residuals clearly show a pattern that changes as the fitted values increase. The pattern of the residuals violates the assumption of linearity. Section 8.3 lists the model assumptions.

```
from ravix import hist
hist(reg)
```

In Fig. 9.10, the histogram deviates from the normal curve, in part due to right skewness. As a result, the normality assumption of the regression model appears to be violated. Here, we use the `shapiro` function from Ravix to invoke the Shapiro-Wilk test.

```
from ravix import shapiro
shapiro(reg)
```

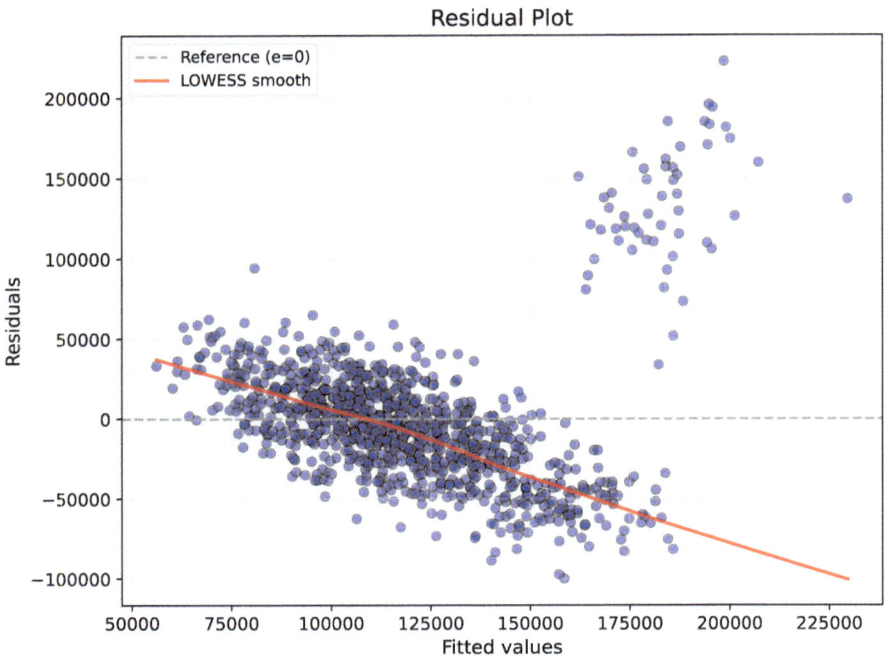

Fig. 9.9 Residual plot for preliminary model

```
# Shapiro-Wilk Test for Normality
# ==========================================================
# Test Statistic        : 0.8845
# p-value               : 1.201e-26
# Result                : Non-normal (p < 0.05)
# ==========================================================
```

The results of the Shapiro-Wilk test further verify the claim that the residuals follow a non-normal distribution. In particular, the Shapiro-Wilk p-value is well below 5%, indicating that the distribution of the residuals proves to be non-normal. Although the estimated coefficients remain the best linear unbiased estimates under the model, the violation of the normality assumption suggests that further investigation is warranted.

9.17.4 Preprocessing

After the preliminary analysis, some of Carla's senior employees advise you that a more careful inspection of the variables is required before fitting a regression

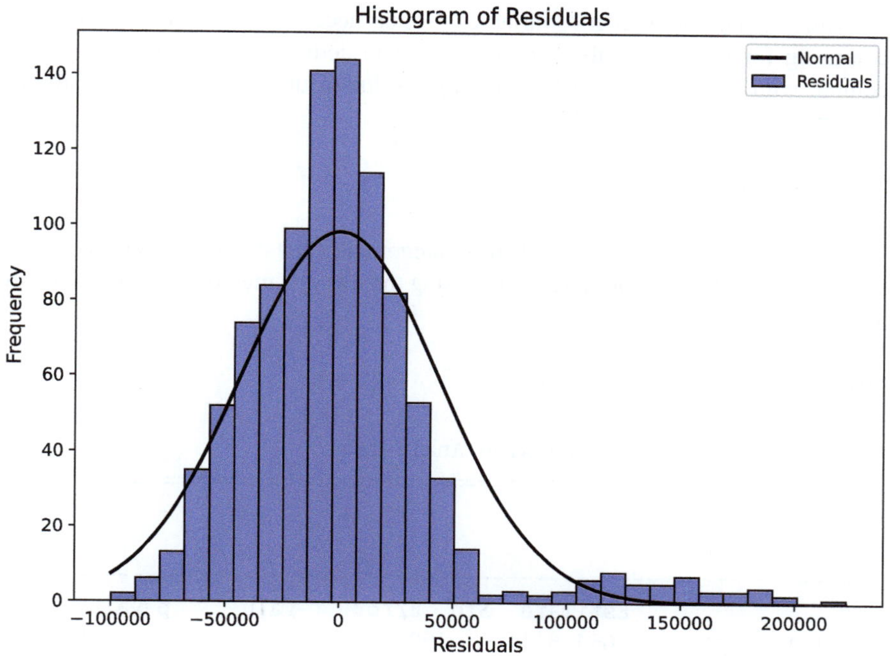

Fig. 9.10 Histogram of residuals for preliminary model

model. Several functions exist to observe the data. Here, we will make use of the head function.

```
df.head()
```

```
#     ID  Age    BMI  Plan     Sex Smoke        Income  \
# 0    1   59  26.97     2    male     N    77965.0000
# 1    2   68  26.48     1  female     N    61048.0000
# 2    3   73  25.59     2  female     N    42460.0000
# 3    4   61  33.69     1    male     Y    84107.0000
# 4    5   79  25.15     3    male     Y   136426.4714
#
#        Charges
# 0  126966.7743
# 1  118915.9842
# 2  108496.4245
# 3  302738.5223
# 4  129749.9500
```

Notice that `Smoke` and `Sex` are categorical variables with character string values. `ID` and `Plan` are also categorical variables but they have numeric values.

While taking on numeric variables is not necessarily incorrect, it can be confusing for the data analyst and can result in coding errors. Dropping the ID column with the drop function is a simple solution that removes the possibility of using ID as a predictor.

```
df = df.drop(columns = ["ID"])
```

Since it was determined that Plan is categorical, care should be taken so that it is not mistaken for a numeric variable. Using the astype method, we specify Plan as a category.

```
df['Plan'] = df['Plan'].astype('category')
```

```
# Summary of OLS Regression Analysis:
# =====================================================
#
# Coefficients:
# -----------------------------------------------------
#                Estimate  Std. Error  t-value    p-value
# Intercept      -66436.1     19916.2   -3.336     0.0009 ***
# Age              867.51      198.97    4.360   1.44e-05 ***
# BMI              3970.5      307.37   12.918    < 2e-16 ***
# Plan_2          -414.30      3495.3   -0.119     0.9057
# Plan_3           5771.8      4799.1    1.203     0.2294
# Sex_male         4928.1      2798.7    1.761     0.0786 .
# Smoke_Y         52781.2      3388.2   15.578    < 2e-16 ***
# Income         0.082722    0.083169    0.995     0.3202
#
# Model Statistics:
# -----------------------------------------------------
# Residual Std. Error: 43844.92
# R-squared:         0.2998          AIC: 24222.67
# Adj. R-squared: 0.2949             BIC: 24261.94
# F-statistic: 60.6875 on 7 and 992 DF, p-value: < 2e-16
# =====================================================
```

9.17.5 Best Subsets Model

Note that the full model has an adjusted R^2 very similar to the model fit in the preliminary analysis. In fact, by handling the Plan and ID columns correctly,

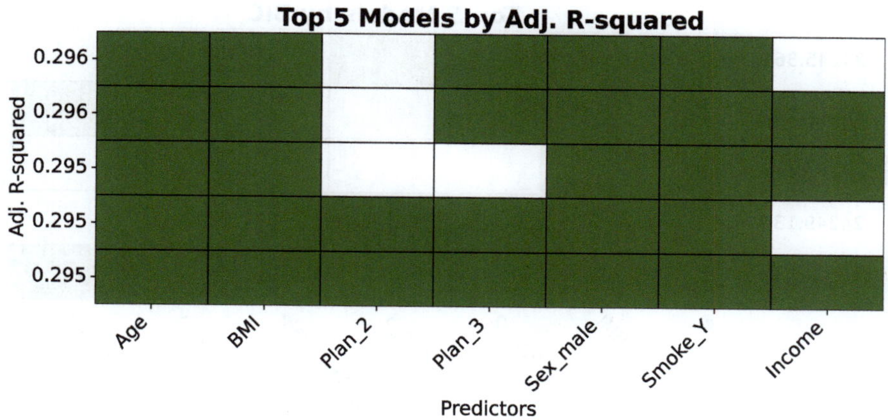

Fig. 9.11 Best subsets variable plot

there appeared to be no benefit even though the preliminary model includes many predictors that should not be used. To find a more parsimonious model, we can utilize best subsets. In particular, the Ravix package allows us to use the `bsr` function.

```
full = ols("Charges ~ .", data = df)
full.summary()
```

The results from the best subsets object, BSR, can be observed by using the `plot_bsr` function. Here, we plot the models ranked according to their adjusted R^2 value in Fig. 9.11.

```
from ravix import bsr, plot_bsr
BSR = bsr("Charges ~ .", data = df, metric = "adjr2")
plot_bsr(BSR)
```

In Fig. 9.12, we plot another best subsets object (BSR2), except this time ranked according to its BIC values.

9.17.6 *Stepwise Methods*

To run backward elimination on the initial dataset, we employ the `step` function specifying a `formula` and the `data`. The default `metric` is AIC.

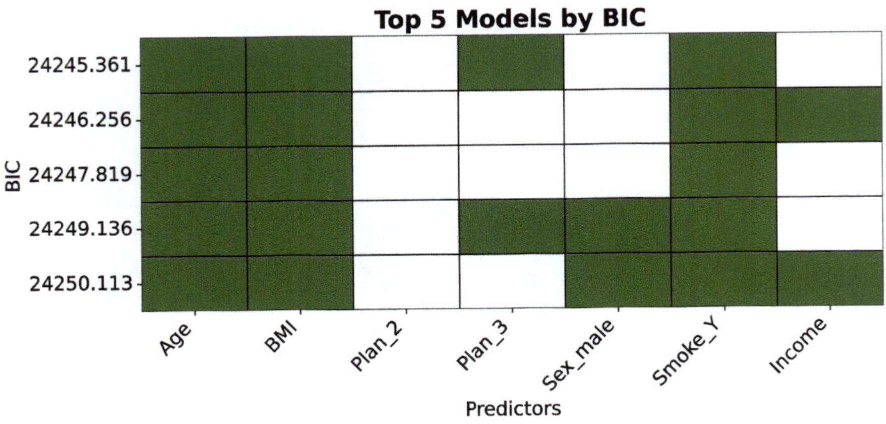

Fig. 9.12 Best subsets variable plot

```
BSR2 = bsr("Charges ~ .", data = df, metric = "bic")
plot_bsr(BSR2)
```

Forward selection can also use the `step` function, but we need to specify the `direction` as `"forward"`.

```
from ravix import stepwise
BE = stepwise("Charges ~ .", data = df)
```

Stepwise regression can go in both directions, either eliminating or adding variables. Therefore, we specify `direction = "both"` to invoke stepwise.

```
FS = stepwise("Charges ~ .", data = df, direction = "forward")
```

For this dataset, the resulting models (`BE`, `FS`, and `SW`) result in the same model. The summary of these can be seen by passing the model to Ravix's `summary` function.

```
SW = stepwise("Charges ~ .", data = df, direction = "both")
```

```
# Summary of OLS Regression Analysis:
# ========================================================
#
# Coefficients:
# --------------------------------------------------------
#                 Estimate  Std. Error  t-value    p-value
# Intercept       -52260.5     13401.5   -3.900     0.0001 ***
# Smoke_Y          52721.4      3385.4   15.573    < 2e-16 ***
# BMI               3966.5      307.17   12.913    < 2e-16 ***
# Age               743.88      155.44    4.786   1.96e-06 ***
# Plan_3            9129.8      2962.1    3.082     0.0021 **
# Sex_male          4938.7      2796.8    1.766     0.0777 .
#
# Model Statistics:
# --------------------------------------------------------
# Residual Std. Error: 43823.06
# R-squared:       0.2991           AIC: 24219.69
# Adj. R-squared: 0.2956            BIC: 24249.14
# F-statistic: 84.8453 on 5 and 994 DF, p-value: < 2e-16
# ========================================================
```

While the stepwise procedures (forward, backward, and both) return the same model in this instance, this outcome is not guaranteed in more complex scenarios. As data structures become more intricate—such as through interaction terms or multicollinearity—each method may follow a different path and arrive at a different final model. This reinforces the importance of understanding how selection criteria and model complexity can influence the behavior and results of automated procedures like stepwise regression.

```
SW.summary()
```

```
from ravix import plot
plot(SW)
```

The residual plot and histogram from the stepwise model given in Figs. 9.13 and 9.14 still reveal violations of regression assumptions, including non-linearity and heteroscedasticity. While the model improves statistical metrics such as adjusted R^2 and AIC compared to the preliminary model, the residuals still exhibit increasing variance with higher fitted values, indicating heteroscedasticity remains an issue.

With a slight decrease in adjusted R^2 and persistent violations in the residual diagnostics, a deeper exploration of the data is warranted to identify nonlinear patterns and inform potential transformations.

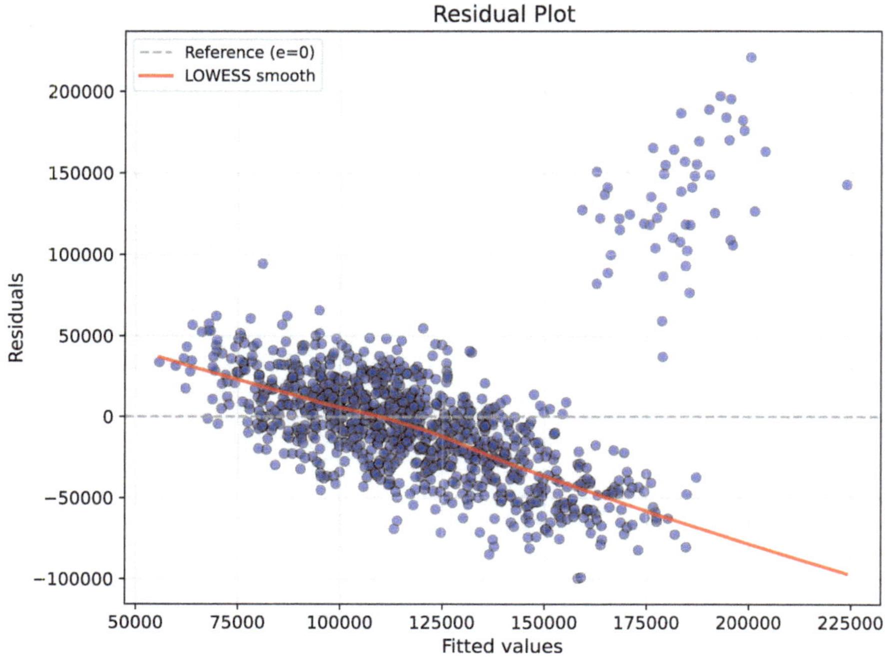

Fig. 9.13 Residual plot for stepwise model

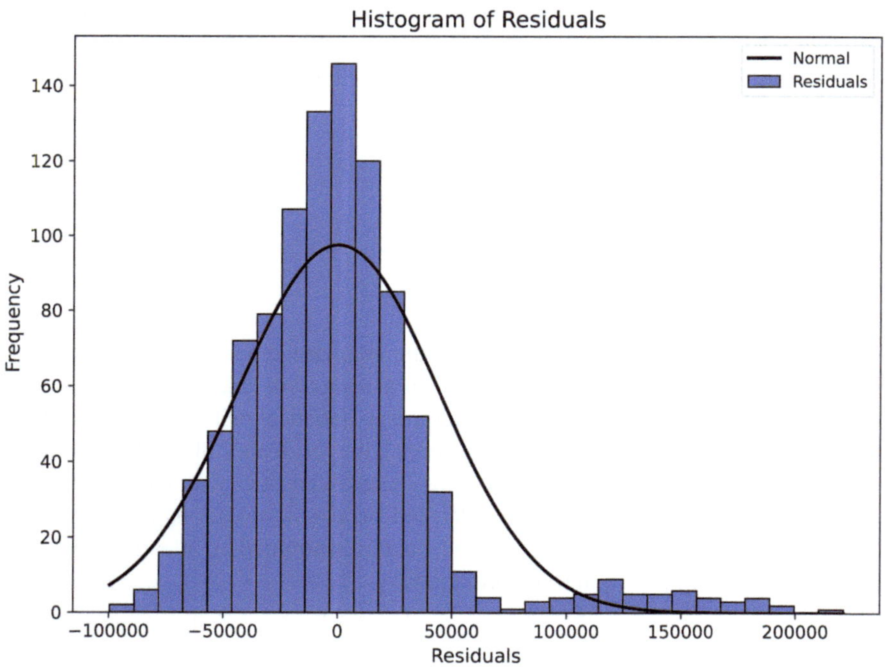

Fig. 9.14 Histogram of residuals for stepwise model

9.17.7 Exploratory Data Analysis

Before refining the regression model further, it is important to explore the individual variables and their relationships. This step helps identify potential transformations, nonlinear patterns, or variable interactions that may improve model accuracy. In this section, we examine the details of the data, including correlations and scatterplots, to better understand the structure of the data and guide subsequent modeling decisions.

A preliminary plot of the correlations is generated using `plot_cor` and shown in Fig. 9.15. Notice that we simply input the dataframe without the categorical variables. The plot shows that the strongest correlation is a negative value between `Age` and `Income`, while the second largest correlation is between `BMI` and `Charges`.

```
from ravix import hist
hist(SW)
```

While the correlation plot shows some valuable information, the correlation coefficient values determine the strength of a linear relationship and do not fully capture nonlinear relationships. Therefore, it is beneficial to look at the scatterplots of the variables in the data. Ravix allows for the easy generation of scatterplots in a matrix using the `plot` function. Again, we drop the categorical variables either using the formula or simply dropping them when specifying the dataframe.

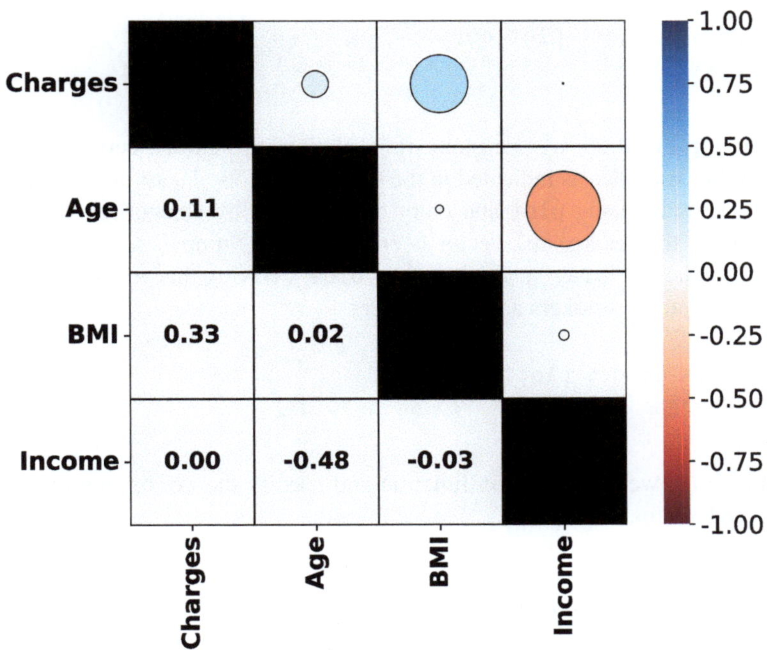

Fig. 9.15 Scatterplot matrix colored by smokers

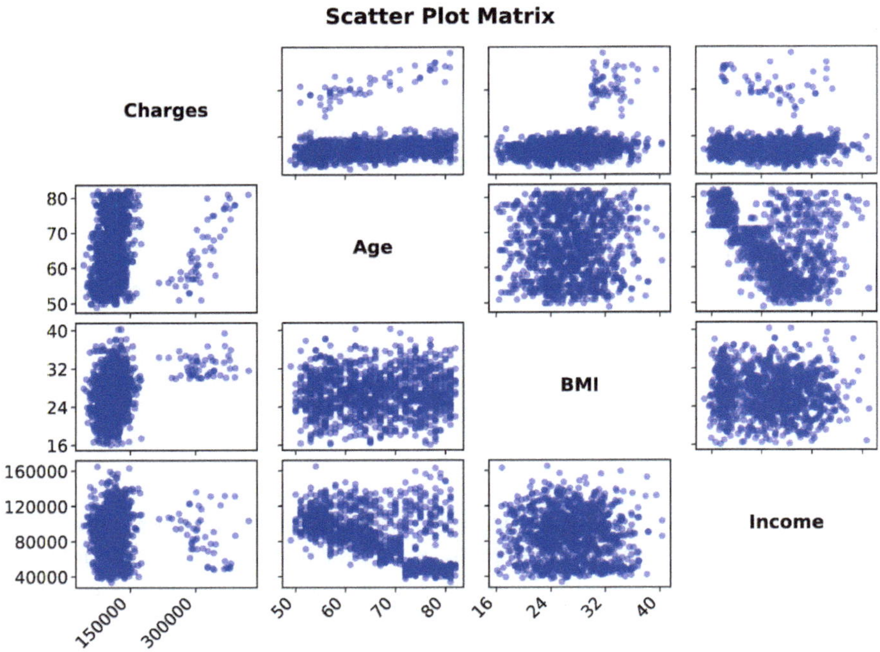

Fig. 9.16 Scatterplot matrix

```
from ravix import plot_cor
plot_cor("Charges ~ Age + BMI + Income", data = df)
```

Observing Fig. 9.16, we see interesting relationships shown between `Charges` and the other variables as indicated in the top row of plots. Therefore, we plot those scatterplots again using `plot` and color the variables by categorical variables. In this instance, the `Smoke_col` vector is created using Numpy's `where` function to denote a smoker as `"gray"` and otherwise `"blue"`. Having this vector will allow us to color the data by smokers and nonsmokers.

```
from ravix import plot
plot("Charges ~ Age + BMI + Income", df)
```

In Fig. 9.17, we use the `plot` function and specify the `color` argument as the `Smoke_col` vector.

```
Smoke_col = np.where(df.Smoke == "Y", "gray", "blue")
```

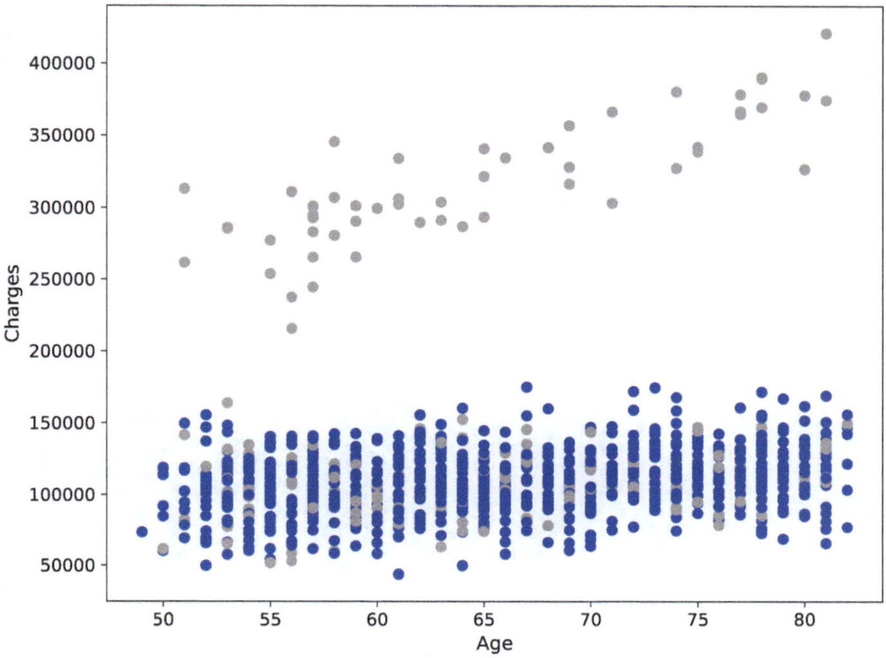

Fig. 9.17 Scatterplot of charges by age

Investigating Fig. 9.17, we see that there is a linear relationship between `Age` and `Charges` for nonsmokers as indicated by the blue points. Above the blue points, there is a linear trend consisting of gray points. Thus, it becomes clear that `Charges` and `Age` may have a different relationship for smokers and nonsmokers.

In Fig. 9.17, gray represents smokers and blue represents nonsmokers. Because smoking is a serious health risk, the charges incurred increase more rapidly with age for smokers as shown.

The scatterplots from Fig. 9.16 also reveal a relationship between `BMI` and `Charges`, which also warrants further investigation, and, therefore, is shown in Fig. 9.18.

```
from ravix import plot
plot("Charges ~ Age", data = df, color = Smoke_col)
```

Investigating the right side of Fig. 9.18, we find the divide between the `Charges` that smokers have when their `BMI` is lower or higher than 30. On the BMI scale, a value of 30 represents the threshold between overweight and obese, which is a health risk that could cause medical charges to be higher. The weight status categories from

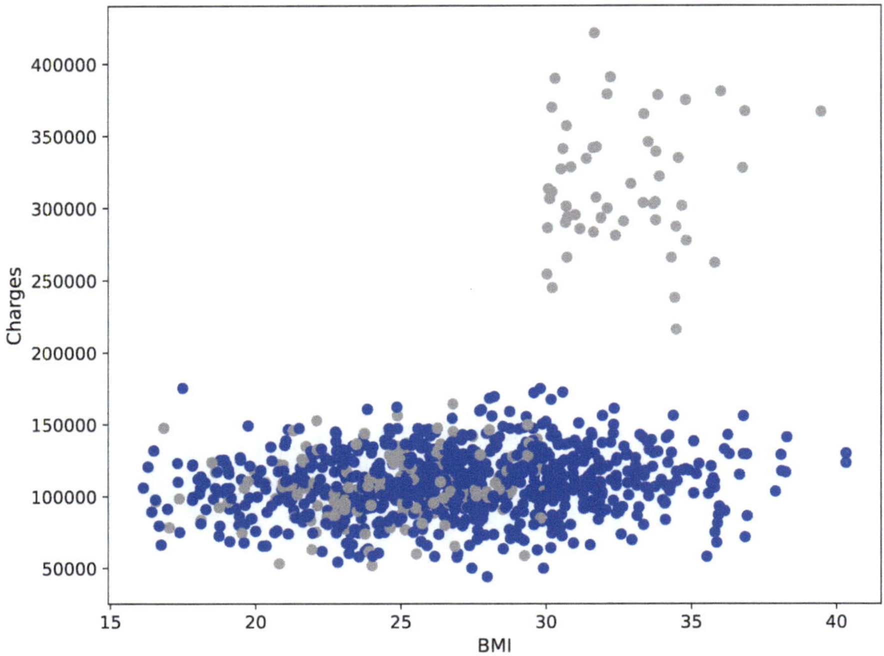

Fig. 9.18 Scatterplot of charges by BMI

Table 9.1 Weight status by body mass index

Body Mass Index (BMI)	Weight Status
BMI < 18.5	Underweight
18.5 ≤ BMI < 25	Normal
25 ≤ BMI < 30	Overweight
BMI ≥ 30	Obese

BMI levels are given in Table 9.1. From Fig. 9.18, we see clearly that a dummy variable for obese individuals could be included to more accurately model `Charges`.

Note that a similar set of scatterplot matrices can be created by using the numeric variables and coloring them by `Sex` and `Plan`. The code to create the plots is similar to that of the previous `plot` command.

These scatterplots are not shown since they do not reveal meaningful relationships in the data. However, the analyst should make that determination by plotting and observing the scatterplots.

The following line of code creates a dummy variable based on the numeric BMI column. The result of the inequality test is either 1 or 0 for each observation.

```
plot("Charges ~ BMI", data = df, color = Smoke_col)
```

While the BMI scale is relatively well-known, it may come as a surprise to some that the previous scatterplots show different trends when BMI is above 30. Our analysis therefore prompts the insurance company to identify obese individuals, who are more at-risk and will most likely have more insurance charges.

9.17.8 Modeling with Interactions

From the analysis in the previous sections, we note that smokers should be treated differently in modeling their medical charges to account for the effects of smoking. We achieve this goal by using interactions. Here, we would like to check the models consisting of all predictor variables, as well as the variable interactions with smokers. Thus, the interaction terms are used:

- Age:Smoke
- BMI:Smoke
- Income:Smoke
- Obese:Smoke

In addition, we would like to consider the effects of obese smokers since they are at a higher risk of incurring charges. The three-way interactions are included:

- Age:Obese:Smoke
- BMI:Obese:Smoke
- Income:Obese:Smoke

For convenience, the formula consisting of Charges as predicted by all other variables and the interactions is defined and named fmla.

```
df["Obese"] = np.where(df.BMI > 30, 1, 0)
```

Using the formula given by fmla, a full model is fit and summarized.

```
fmla = "Charges ~ . + Age:Smoke + BMI:Smoke + Income:Smoke \
        + Obese:Smoke + Age:Obese:Smoke + BMI:Obese:Smoke \
        + Income:Obese:Smoke"
```

```
# Summary of OLS Regression Analysis:
# ============================================================
#
# Coefficients:
# ------------------------------------------------------------
#                      Estimate Std. Error  t-value   p-value
# Intercept             45327.7    10983.5    4.127  3.99e-05 ***
# Age                    611.04     102.16    5.981  3.09e-09 ***
# BMI                    798.52     243.52    3.279    0.0011 **
# Plan_2                -1587.7     1644.3   -0.966    0.3345
# Plan_3                10895.1     2249.1    4.844  1.48e-06 ***
# Sex_male               3142.2     1315.7    2.388    0.0171 *
# Smoke_Y               -7190.3    23286.6   -0.309    0.7576
# Income            -0.026048   0.041322   -0.630    0.5286
# Obese                 -624.25     2498.4   -0.250    0.8027
# Age:Smoke_Y           -34.691     228.47   -0.152    0.8793
# BMI:Smoke_Y            245.43     580.85    0.423    0.6727
# Income:Smoke_Y     0.049313   0.074992    0.658    0.5110
# Obese:Smoke_Y         -1522.3    57992.7   -0.026    0.9791
# Age:Obese:Smo...       3784.1     449.64    8.416   < 2e-16 ***
# BMI:Obese:Smo...      -1910.1     1461.6   -1.307    0.1916
# Income:Obese:...    0.23979    0.15381    1.559    0.1193
#
# Model Statistics:
# ------------------------------------------------------------
# Residual Std. Error: 20513.92
# R-squared:        0.8480           AIC: 22711.47
# Adj. R-squared: 0.8456           BIC: 22789.99
# F-statistic: 365.8817 on 15 and 984 DF, p-value: < 2e-16
# ============================================================
```

The `full` model above returns a great fit, especially compared with the low adjusted R^2 from the initial model. However, the model is not easily interpreted since there are too many predictor variables. In the following section, other models will be considered.

```
from ravix import ols
full = ols(fmla, data = df)
full.summary()
```

```
from ravix import plot
plot(full)
```

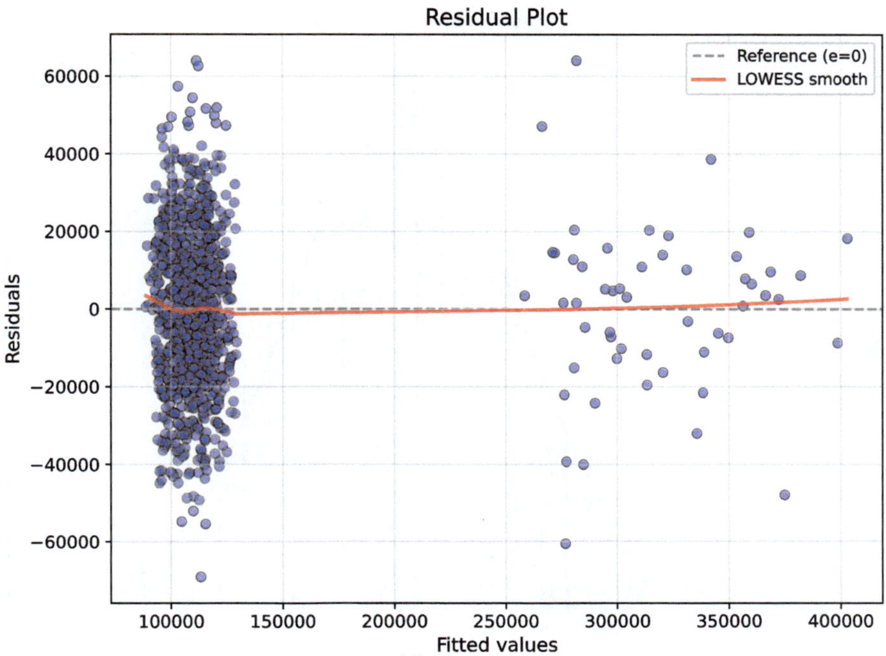

Fig. 9.19 Residual plot for full model

The diagnostic plots for the full model, shown in Figs. 9.19 and 9.20, indicate a noticeable improvement over the earlier models. The residual plot (Fig. 9.19) shows a more symmetric spread around zero, with reduced curvature compared to both the preliminary and stepwise models. Although a gap remains in the fitted values between 125,000 and 250,000, the residual pattern overall appears more stable, with fewer signs of heteroscedasticity. The histogram of residuals (Fig. 9.20) also shows a distribution that is closer to normality, exhibiting less skew and fewer extreme outliers. These diagnostics suggest that the inclusion of interaction terms has helped account for important structural variation in the data.

```
from ravix import hist
hist(full)
```

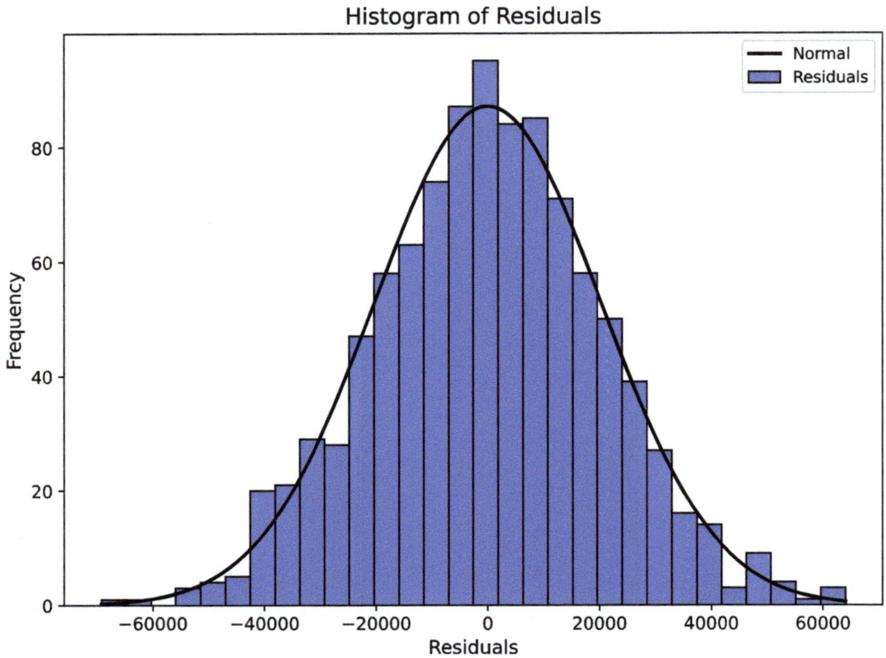

Fig. 9.20 Histogram of residuals for full model

```
# Summary of OLS Regression Analysis:
# ==========================================================
#
# Coefficients:
# ----------------------------------------------------------
#                       Estimate Std. Error  t-value   p-value
# Intercept              44879.0     6231.9    7.202  1.17e-12 ***
# Age                     614.80     72.472    8.483   < 2e-16 ***
# BMI                     740.36     150.90    4.906  1.08e-06 ***
# Plan_3                 11650.7     1389.6    8.384   < 2e-16 ***
# Age:Obese:Smo...        3164.0     46.449   68.118   < 2e-16 ***
#
# Model Statistics:
# ----------------------------------------------------------
# Residual Std. Error: 20549.99
# R-squared:       0.8457          AIC: 22704.10
# Adj. R-squared: 0.8451           BIC: 22728.63
# F-statistic: 1363.63 on 4 and 995 DF, p-value: < 2e-16
# ==========================================================
```

Note that in this case, if we were to use the same formula, data, and metric, the `step` function would give the same outcome.

```
from ravix import bsr
BSR = bsr(fmla, data = df, metric = "bic")
BSR.summary()
```

In the summary above, the best subsets model was recreated using the `bsr` function, which can be confirmed since the coefficients match the best subsets coefficients displayed using the `coef` function. Because the best model was selected using the BIC criterion, a more parsimonious model was returned, even though some alternative models achieved slightly higher adjusted R^2 values. This model is considered more parsimonious because it uses fewer predictors while still maintaining a reasonably high R^2. While the summary indicates the model is parsimonious, it is important to also check the diagnostic plots.

```
SW = stepwise(fmla, data = df, metric = "bic")
```

```
from ravix import plot
plot(BSR)
```

In Fig. 9.21, the residuals have the same spread from left to right even though the obese-smokers have larger predicted values. Figure 9.22 indicates that the residuals are distributed normally. By observing the residual plots of Figs. 9.21 and 9.22, we can conclude that the best subsets model does not violate the linear regression assumptions.

9.17.9 Coefficient Interpretation

Here, we interpret the coefficients of the best subsets model.

- Intercept—The intercept represents the `Charges` when all the predictor variables are 0. While the p-value is significant, having `Age` and `BMI` set to 0 does not make sense. If `Age` and `BMI` are mean-centered, then the intercept has more intuitive meaning.
- Age—Since the p-value indicates significance, the interpretation is:

  The `Charges` *is expected to increase by $614.80 for every year increase in* `Age`*, holding all other variables constant.*

- BMI—Since the p-value indicates significance, the interpretation is:

  The `Charges` *is expected to increase by $740.36 for every unit increase in* `BMI`*, holding all other variables constant.*

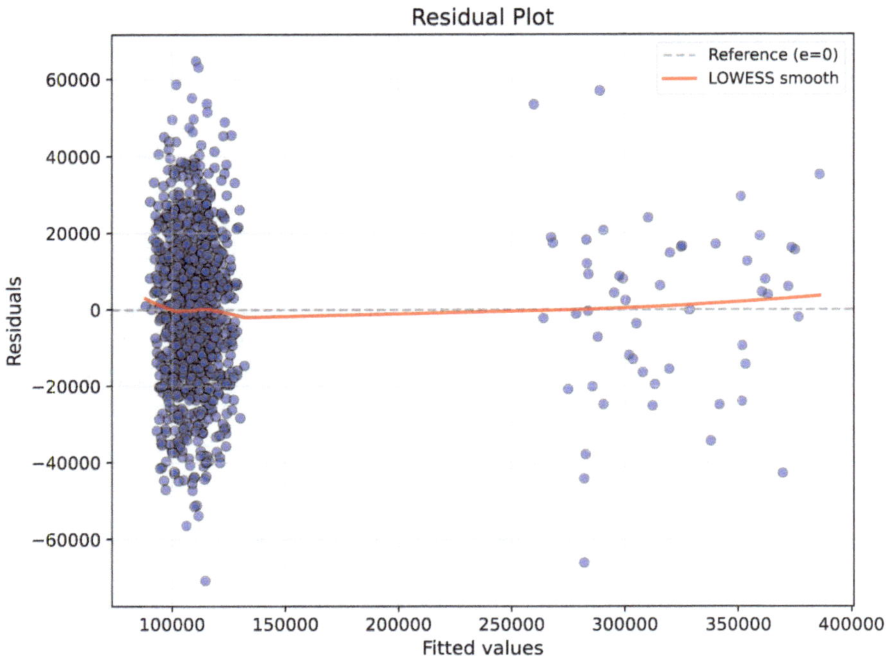

Fig. 9.21 Residual plot for best subsets model

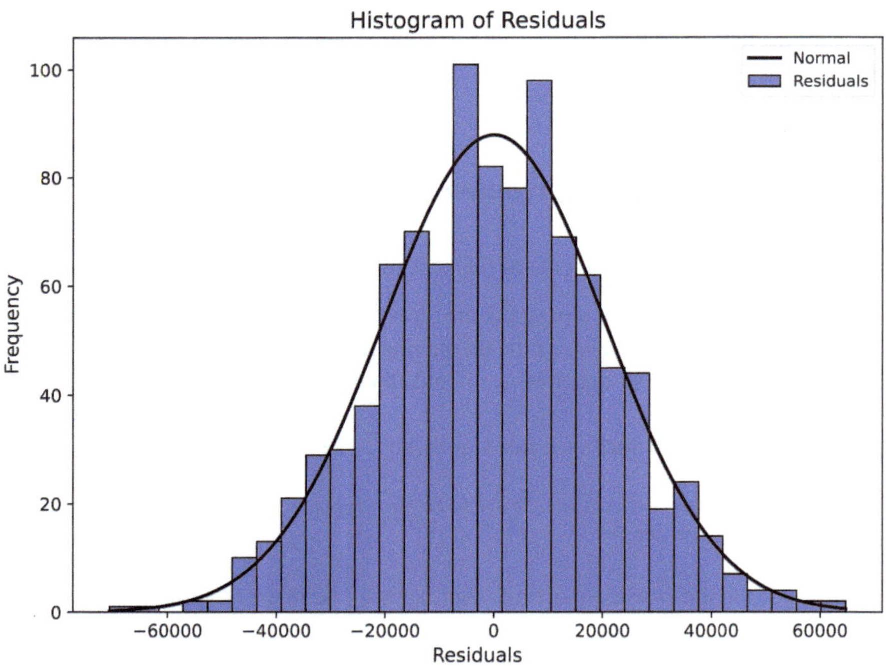

Fig. 9.22 Histogram of residuals for best subsets model

- Plan_3—The p-value indicates this variable is significant. The interpretation is:

The `Charges` *for plan 3 is expected to be $11,650.70 more than the other plans, holding all other variables constant.*

- Age:Obese:Smoke_Y—The p-value indicates this variable is significant. The interpretation is:

The `Charges` *is expected to increase by $3163.97 more for every year increase in* `Age` *for obese smokers as compared to the rest, holding all other variables constant.*

9.17.10 Model Comparison

Table 9.2 summarizes the various models fit throughout the analysis. The preliminary and early stepwise models yield similar adjusted R^2 values just under 0.30, and neither addresses the violations observed in the residual diagnostics. Simple best subsets and stepwise methods also fail to capture the underlying structural relationships. In contrast, the full interaction model substantially improves model fit, raising the adjusted R^2 to 0.8456, though its complexity makes interpretation more difficult. To address this, both best subsets and stepwise selection using BIC were applied to the expanded feature set, yielding the same parsimonious model. This agreement reinforces confidence in the selected predictors and demonstrates that model complexity can be reduced significantly without sacrificing interpretability.

9.17.11 Charges Across BMI Categories

We would now like to test for differences among `Charges` across the various BMI categories for smokers and nonsmokers. Using the `cut` command, we can segment the `BMI` variable according to the cutoffs as specified in the `bins` list. In this case,

Table 9.2 Comparison of model performance metrics

Model	Adj. R^2	AIC	BIC	Notes
Preliminary	0.2957	24221.6	24260.9	Includes ID and numeric plan; no transformations/interactions
Stepwise (simple)	0.2956	24219.7	24249.1	No transformations/interactions
Best subsets (simple)	0.2956	24219.7	24249.1	No transformations/interactions
Full interaction	0.8456	22711.5	22790.0	All main effects + interactions
Best subsets (BIC)	0.8451	22704.1	22728.6	Parsimonious interaction model
Stepwise (BIC)	0.8451	22704.1	22728.6	Same model as best subsets (BIC)

an individual with a BMI of below 18.5 is placed in the "Under" category, while 18.5 to below 25 is categorized as "Normal," while 25 to under 30 is "Over," and 30 or above as "Obese." These categories are specified in the `labels` list.

```
from ravix import hist
hist(BSR)
```

The smokers are specified in the dataframe by indexing the values where `Smoke` is "Y".

```
import pandas as pd
```

Using the `Categories` column, we use the `boxplot` function to get a visual comparison of the smokers by `Categories`. Ravix handles numeric vectors and creates boxplots from them using the `boxplot` function.

```
bins = [-np.inf, 18.5, 25, 30, np.inf]
labels = ['Under', 'Normal', 'Over', 'Obese']
df['Categories'] = pd.cut(df['BMI'], bins = bins,
                          labels = labels)
```

Notice from the boxplot in Fig. 9.23 that the `Obese` category significantly differs from the remaining categories. Therefore, there is a notable difference in `Charges` among categories. The result can be formalized with a statistical test. Here, μ_1, μ_2, μ_3, and μ_4 represent the mean `Charges` of the respective `Categories` for smokers.

Null Hypothesis: Charges is the same across different BMI categories (for smokers).

$$H_0 : \mu_1 = \mu_2 = \mu_3 = \mu_4$$

Alternative Hypothesis: At least one mean is different.

The alternative hypothesis is equivalent to saying that there is a difference in `Charges` among the categories. Using the `ols` and `summary` commands, we calculate an ANOVA test.

```
smokers_df = df[df['Smoke'] == "Y"]
```

```
#                df      Sum Sq     Mean Sq          F  p-value
# Regression      3   1.741e+12   5.805e+11   717.9403  < 2e-16
# Residual      210   1.698e+11   8.085e+08
# Total         213   1.911e+12
```

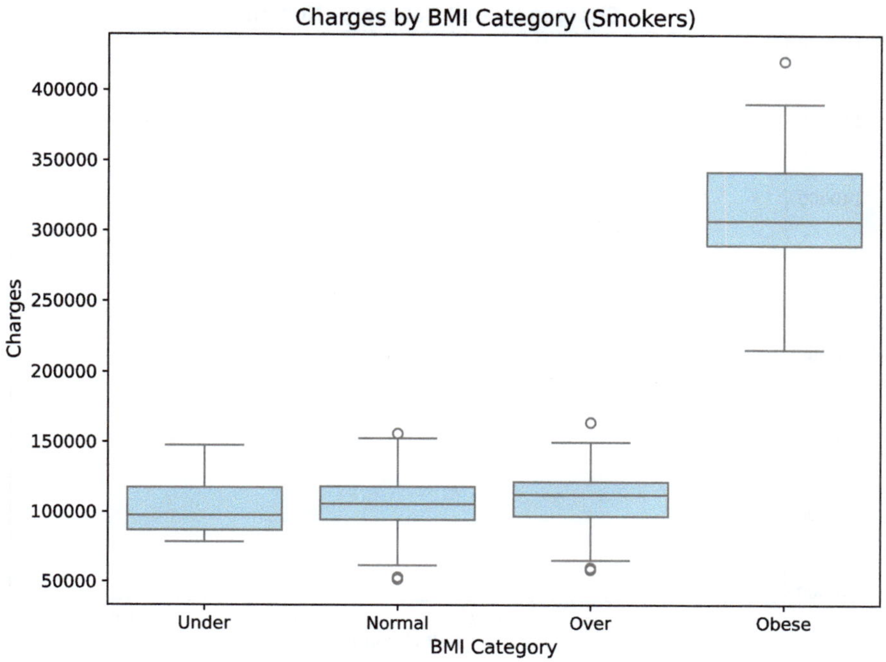

Fig. 9.23 Boxplot of charges by BMI category (Smokers)

The p-value of the F-test is extremely small in this case, indicating statistical significance. Therefore, the ANOVA test is in agreement with our assessment made by looking at the boxplots.

9.17.12 *Charges Across BMI Categories for Nonsmokers*

Next, we analyze the relationship of nonsmokers in a similar way. The nonsmokers are indexed in the dataframe by selecting rows where the `Smoke` variable is set to `"N"`.

```
from ravix import boxplot
boxplot("Charges ~ Categories", data = smokers_df,
        color = "lightblue",
        xlab = "BMI Category",
        title ="Charges by BMI Category (Smokers)")
```

The boxplot of nonsmokers by category can be colored “green” by specifying the `color` argument within the `boxplot` command.

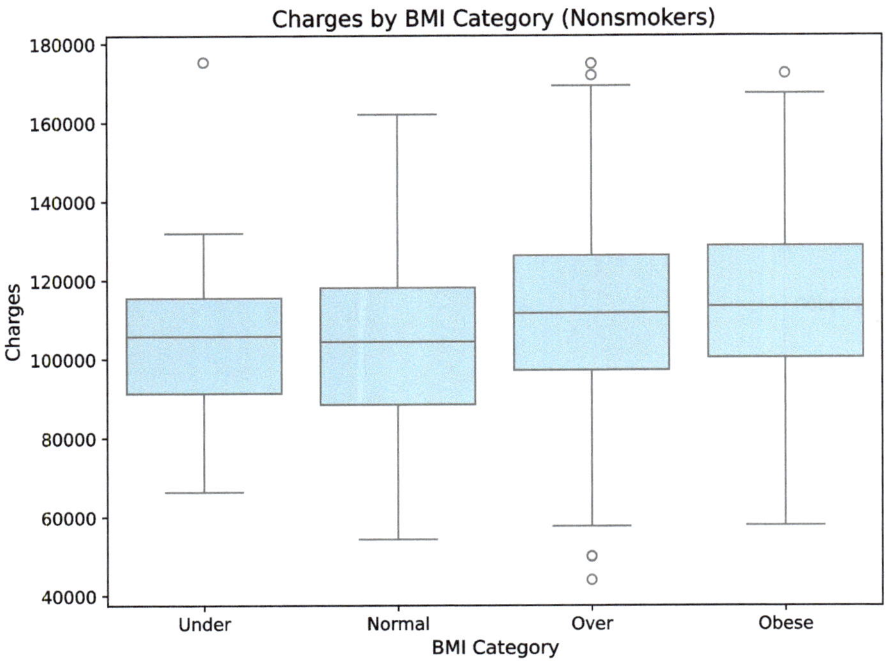

Fig. 9.24 Boxplot of charges by BMI category (nonsmokers)

```
from ravix import ols
aov1 = ols("Charges ~ Categories", data = smokers_df)
aov1.summary(out = "anova")
```

From Fig. 9.24, we notice that the median does differ for each of the `Categories`, but not as drastically as they do in the smoker boxplot shown in Fig. 9.23. While the ANOVA test looks for differences in the means, the mean and the median are both measures of the center of the distribution. Here, we formalize the result with a statistical test. For this analysis, we let μ_1, μ_2, μ_3, and μ_4 represent the mean `Charges` of the respective `Categories` for nonsmokers.

Null Hypothesis: Charges are the same across different BMI categories (for nonsmokers).

$$H_0 : \mu_1 = \mu_2 = \mu_3 = \mu_4$$

Alternative Hypothesis: At least one mean is different.

The test is concluded using the newly created dataframe `nonsmokers_df`.

```
nonsmokers_df = df[df['Smoke'] == "N"]
```

```
#                df      Sum Sq     Mean Sq        F    p-value
# Regression      3   1.143e+10   3.809e+09   7.9686   3.08e-05
# Residual      782   3.738e+11   4.780e+08
# Total         785   3.852e+11
```

The p-value of the F-test indicates that there is a difference among the categories for nonsmokers. Notice here that the p-value is significantly larger than the p-value for the smoker F-test, which indicates that the difference among smokers is more significant than nonsmokers. This result is consistent with our conclusion from observing the boxplots.

9.17.13 Income Across Plans

Our last task will consist of analyzing the three plans by `Income`, since we consider this factor to differ across the three plans.

While the mean income of individuals with plans 1 and 2 is relatively similar, the mean income of individuals with plan 3 is very different. Here, we produce a boxplot to visually show the difference of `Income` among `Plans`.

```
from ravix import boxplot
boxplot("Charges ~ Categories", data = nonsmokers_df,
        color = "lightblue",
        xlab = "BMI Category",
        title = "Charges by BMI Category (Nonsmokers)")
```

In Fig. 9.25, the plans correspond to various income values. Lastly, we will test for a difference using a formal F-test. We let μ_1, μ_2, and μ_3 represent the mean income corresponding to plans 1, 2, and 3.

Null Hypothesis: Income is not statistically significant across different plans.

$$H_0 : \mu_1 = \mu_2 = \mu_3$$

Alternative Hypothesis: At least one mean is different among the plans.

As seen in Fig. 9.25, income varies by plan. Next, we have to determine if income is statistically significant across the separate plans. To do so, we need to create a regression model between `Income` and `Plan`, and examine the significance of the model and individual variables.

```
from ravix import ols
aov2 = ols("Charges ~ Categories", data = nonsmokers_df)
aov2.summary(out = "anova")
```

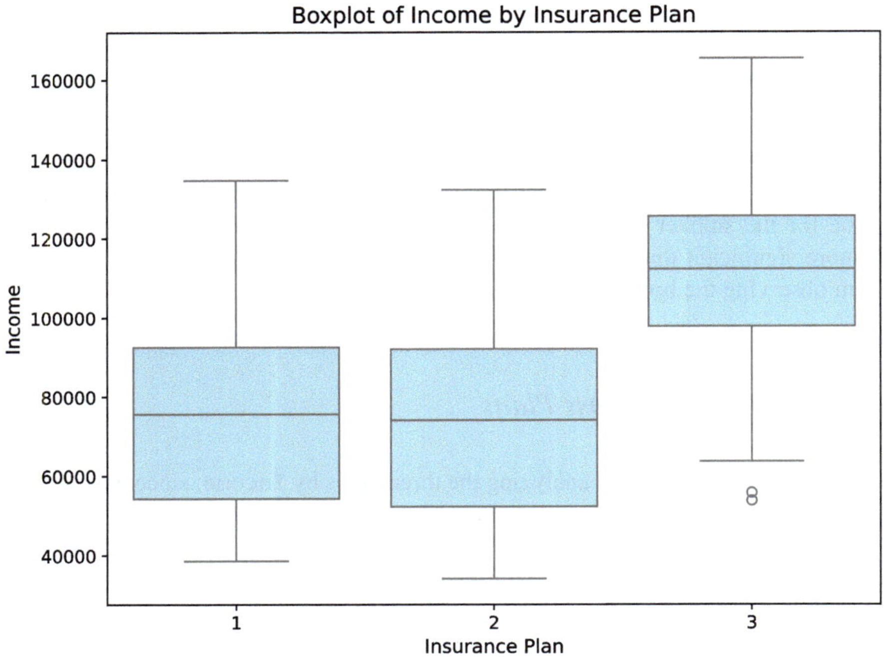

Fig. 9.25 Boxplots by plan

```
#                 df       Sum Sq     Mean Sq          F  p-value
# Regression       2   3.009e+11   1.504e+11   328.1247  < 2e-16
# Residual       997   4.571e+11   4.585e+08
# Total          999   7.580e+11
```

This ANOVA table shows that plans are different based on the p-value of the F-statistic. In particular, this indicates that the model is statistically significant when compared to an alpha of 5%, meaning that not all plans are the same and at least one of the plans differs. For more detail, a summary of the regression model is produced.

```
from ravix import boxplot
boxplot("Income ~ Plan", data = df, color = "lightblue",
        xlab = 'Insurance Plan',
        title ='Boxplot of Income by Insurance Plan')
```

```
# Summary of OLS Regression Analysis:
# ========================================================
#
# Coefficients:
# --------------------------------------------------------
#                    Estimate  Std. Error  t-value    p-value
# Intercept           75300.4      1354.2   55.603   < 2e-16 ***
# Plan_2              -1168.2      1706.0   -0.685    0.4936
# Plan_3              36314.9      1802.5   20.147   < 2e-16 ***
#
# Model Statistics:
# --------------------------------------------------------
# Residual Std. Error: 21412.54
# R-squared:        0.3969          AIC: 22784.34
# Adj. R-squared: 0.3957            BIC: 22799.06
# F-statistic: 328.1247 on 2 and 997 DF, p-value: < 2e-16
# ========================================================
```

Note that the F-statistic is the same in the ANOVA output and summary. However, the summary shows that Plans 1 and 2 do not differ significantly based on the individual significance t-test of `Plan_2`. `Plan_3`, on the other hand, differs significantly from `Plan_1`, which is the underlying reason for the significance of the F-statistic noted by the low p-value.

9.17.14 Recommendations and Findings

We find the following with respect to the tasks that we were given:

1. Assess the regression model fit from the preliminary analysis.

   A review of the preliminary analysis showed that the model not only violated the linear regression assumptions, but it also used invalid variables and neglected some important interactions.

2. Find a parsimonious model that predicts insurance charges and interpret the coefficients.

   After applying best subsets and finding the minimum BIC, a model with only three predictor variables and one interaction was able to achieve an adjusted R^2 of 0.8451. The coefficients are interpreted in the Coefficient Interpretation section above.

3. Test for differences in charges across the various BMI categories for smokers and nonsmokers.

Both tests across the different categories for smokers and nonsmokers indicated a significant difference among categories, with the most extreme difference being that of smokers in the obese category.

4. Test for income differences across the three insurance plans.

The test revealed that there were differences in the income for insurance plans. In particular, insurance plan 3 differs from other plans.

9.17.15 Case Conclusion

This case demonstrates how modern health care is being transformed by analytics, and more specifically, how analytics can provide insights from healthcare data using a detailed example of the process to determine healthcare charges. Since greater insight is known about what determines healthcare charges, QI's decision-makers can adjust the cost of premiums to offset these charges. With this information, insurance coverage can be more accurately priced to account for high-cost claims.

The analysis in this case study utilized a variety of modeling and diagnostic tools available in the Ravix Python package. These include `ols` for model estimation, `summary` for coefficient interpretation and ANOVA results, and plotting functions such as `plot`, `hist`, and `boxplot` to visually assess assumptions and relationships. We also used `bsr` for best subsets regression and `step` for stepwise selection. These tools allowed us to build and evaluate multiple regression models, examine interactions, and assess differences across groups—providing a comprehensive approach to healthcare cost prediction using Python.

Problems

1. Healthcare Costs Iterative Methods
Load in the data:

```
from ravix import ols
aov3 = ols("Income ~ Plan", data = df)
aov3.summary(out = "anova")
```

Using the insurance charges dataset, build a model to predict the `charges` considering all other variables from the dataset.

a. Perform backward elimination and produce a summary.
b. Perform forward selection and produce a summary.
c. Perform stepwise regression and produce a summary.
d. Are the models the same? If not, which one is different?

e. How well does the backward elimination model fit the data?
f. Construct a residual plot and a histogram of the residuals of the backward elimination model and discuss. By observing the plots, is it evident if any of the linear regression model assumptions are violated?

2. Country GDP Iterative Methods
Load in the data:

```
aov3.summary()
```

Using the countries dataset, build a model to predict the GDP, considering all other relevant variables from the dataset.

a. Perform backward elimination and produce a summary.
b. Perform forward selection and produce a summary.
c. Perform stepwise regression and produce a summary.
d. Are the models the same? If not, which one is different?
e. How well does the backward elimination model fit the data?
f. Construct a residual plot and a histogram of the residuals of the backward elimination model and discuss. By observing the plots, is it evident if any of the linear regression model assumptions are violated?

3. Country GDP Best Subsets Regression
Load in the data:

```
from ravix import get_data
df = get_data("charges.csv")
```

Using the countries dataset, build a model to predict the GDP, considering all other relevant variables from the dataset.

a. Fit a best subsets model to predict GDP, considering all other relevant variables from the dataset.
b. Use the plot function on the best subsets object from part a.
c. Which variables are in the best subset?
d. Plot the best subsets object again, but use the adjusted R-squared as the metric.
e. Which variables are in the best subset using adjusted R-squared?
f. Print out the best adjusted R-squared to at least 4 decimal places.
g. Construct a residual plot and a histogram of the residuals of the best subsets model indicated by BIC, and discuss if the linear regression model assumptions are met.

4. Country GDP Best Subsets Regression 2
Load in the data:

```
from ravix import get_data
df = get_data("countries.csv")
```

Using the countries dataset, build a model to predict the GDP, considering all other relevant variables from the dataset.

a. Fit a best subsets model to predict the GDP, considering all other relevant variables and the interactions of all predictor variables. **Hint:** After removing all of the nonrelevant predictor variables from the data, use the formula "`GDP ~ .**2`" to include all of the two-variable interactions.
b. Use the plot function on the best subsets object from part a.
c. Which variables are in the best subset?
d. Plot the best subsets object again, but use the adjusted R-squared as the metric.
e. Which variables are in the best subset using adjusted R-squared?
f. How does the model fit compare with the best model from the previous problem?

5. Airbnb Best Subsets Regression
Load in the data:

```
from ravix import get_data
df = get_data("countries.csv")
```

Using the Airbnb dataset, build a model to predict the `Price` considering all other relevant variables from the dataset.

a. Fit a best subsets model to predict `Price`, considering all other relevant variables and the interactions of all predictor variables. **Hint:** After removing all of the irrelevant predictor variables from the data, use the formula "`Price ~ .**2`" to include all of the two-variable interactions.
b. Use the `plot` function on the best subsets object from part a.
c. Which variables are in the best subset?
d. Plot the best subsets object again, but use the adjusted R-squared as the metric.
e. Which variables are in the best subset using adjusted R-squared?
f. How does the model fit compare with the best model from the previous problem?

Appendix A
Ravix Modeling Architecture

A.1 Purpose of the Modeling Module

The modeling functionality presented in this appendix reflects the scope of regression techniques covered in this text, which focuses on ordinary least squares (OLS) regression. Ravix provides additional modeling capabilities beyond OLS; however, these are not discussed here, as the emphasis of this book is on linear regression theory, inference, diagnostics, and model selection.

Ravix leverages established statistical libraries for model estimation and inference while providing a streamlined interface designed for clarity and applied analytics workflows. By abstracting lower-level computational details, the package enables users to concentrate on model specification, interpretation, and decision-making rather than implementation mechanics.

In addition to its modeling capabilities, Ravix provides a cohesive visualization framework for exploratory analysis, diagnostic evaluation, and presentation-quality graphics. Modeling and plotting are designed to operate as a unified system, allowing users to transition seamlessly from estimation to visualization without restructuring data or rewriting code.

Although Ravix was originally developed to support regression instruction, its functionality extends beyond the classroom. The framework is equally suited for professional analytics, reproducible research, and publication-quality visualization, offering standardized modeling and plotting workflows that scale across applied settings.

At its core, the modeling module is guided by four principles:

- **Designed for clarity and reproducibility**—syntax and output are structured to support both instructional use and professional analytical workflows.
- **Built on established statistical libraries**—leveraging proven estimation engines.

D. P. McGibney, *Applied Linear Regression for Business Analytics with Python*,
International Series in Operations Research & Management Science 378,
https://doi.org/10.1007/978-3-032-23806-1

- **Standardizes regression and visualization workflows**—promoting consistency across analyses and graphical outputs.
- **Emphasizes interpretability over automation**—prioritizing explanation and transparency rather than black-box modeling.

These principles align directly with the applied teaching focus of this text, where regression is presented as both a statistical framework and a decision-support tool, supported by an integrated modeling and visualization environment.

Supplemental documentation, modeling examples, and instructional resources for Ravix are available at ravixanalytics.com.

A.2 Modeling Workflow in Ravix

Regression modeling in Ravix follows a structured analytical pipeline. While individual components of this workflow are introduced throughout the chapters, they are not presented in a single unified sequence. This appendix consolidates the modeling process into a canonical workflow that reflects how analyses are conducted in practice.

The standard Ravix modeling pipeline is:

- **Data acquisition:** `get_data`
- **Visualization:** `plot`, `hist`, `plot_cor`, `boxplot`, `barplot`
- **Formula specification:** `"Y ~ X1 + X2"`
- **Model estimation:** `ols`
- **Output interpretation:** `summary`
- **Diagnostics:** `plot`, `hist`, `ncv`, `shapiro`

This workflow mirrors the progression of regression analysis developed across the text:

- Data preparation and visualization precede modeling.
- Estimation produces interpretable coefficients.
- Visualization supports both exploratory insight and model communication.
- Diagnostics assess model validity.
- Inference supports statistical conclusions.

Because visualization is tightly integrated with estimation outputs, Ravix functions as both a modeling framework and a plotting environment within a single analytical workflow. By centralizing this pipeline, the appendix serves as a unified analytical map connecting conceptual instruction with computational implementation.

A.3 Core Modeling Functions (Grouped by Role)

Ravix modeling functions are organized according to their role within the analytical workflow rather than alphabetically. This structure reflects how regression analyses are conducted in practice and reinforces the staged progression introduced throughout the text. Functions are therefore grouped into estimation, inference, and model selection components that align directly with the modeling pipeline.

A.3.1 Estimation Functions

Estimation functions are responsible for fitting regression models using formula-based specification. These functions generate structured model objects that serve as the foundation for inference, visualization, and diagnostic evaluation.

- **`ols`**

Estimates ordinary least squares regression models. This is the primary modeling function used throughout the text and supports full formula syntax, including interactions and transformed predictors.

Although Ravix includes additional estimation capabilities beyond OLS, they fall outside the scope of this textbook and are not documented here. The appendix therefore focuses exclusively on the modeling structures necessary to support linear regression analysis.

All estimation functions return standardized model objects designed for compatibility with Ravix summary, diagnostic, and visualization tools.

A.3.2 Inference and Reporting Functions

Inference functionality in Ravix is centralized within the summary framework. Rather than distributing inferential outputs across multiple functions, Ravix consolidates statistical reporting into a unified model reporting system controlled through the `out` argument.

- **`summary`**

Generates structured statistical reports for fitted regression models. Supported output modes include:

- Coefficient tables (`out = "simple"`)
- Analysis of variance tables (`out = "anova"`)
- Type I sums of squares ANOVA (`out = "anova1"`)
- Confidence intervals for coefficient estimates (`out = "confint"`)

By integrating these elements into a single reporting structure, Ravix standardizes inferential interpretation while maintaining consistency across analytical workflows.

A.3.3 Model Selection Functions

Model evaluation and selection extend the estimation workflow by enabling structured comparison across competing model specifications. While detailed instructional coverage of selection procedures is provided elsewhere in the text, the functions themselves are noted here for architectural completeness.

- **`stepwise`**

Implements stepwise regression procedures for automated predictor selection based on specified criteria.

- **`bsr`**

Performs best subsets regression to evaluate competing predictor combinations and identify optimal model structures.

These selection tools support systematic model refinement while maintaining compatibility with Ravix visualization and summary outputs.

By grouping functions according to analytical role, Ravix reinforces a coherent modeling workflow in which estimation, inference, and selection operate as interconnected components of a unified regression and visualization environment.

A.4 Model Object Structure

Models estimated in Ravix return fitted model objects that store estimation results and support downstream analysis. These objects serve as the central interface connecting modeling, visualization, inference, and diagnostic procedures within the analytical workflow.

Fitted model objects contain core regression outputs consistent with standard statistical estimation frameworks. Key stored components include:

These components are accessed internally by summary, visualization, and diagnostic functions, allowing users to evaluate model performance without manually extracting intermediate results.

By encapsulating estimation outputs within a unified object structure, Ravix supports reproducible analytical workflows while maintaining compatibility across modeling, plotting, and inference procedures presented throughout the text. The Ravix source code, version history, and development updates are publicly available at github.com/danielmcgibney/ravix.[1]

[1] For the Ravix source code, version history, and development updates, see McGibney (2026b).

Attribute	Description
params	Estimated regression coefficients
bse	Standard errors of coefficient estimates
tvalues	t-Statistics for individual coefficients
pvalues	Significance levels for hypothesis tests
rsquared	Coefficient of determination
rsquared_adj	Adjusted coefficient of determination
fittedvalues	Model-predicted response values
resid	Model residuals

A.5 Formula Processing Engine

Regression models in Ravix are specified using formula syntax introduced earlier in the text. While formula construction is taught conceptually in Chapter 2***, this section formalizes the parsing rules governing model specification within the modeling framework.

Supported syntax elements include:

Syntax	Meaning
+	Add predictor
-	Remove predictor
:	Interaction only
*	Main effects plus interaction
.	Include all predictors
-1	Remove intercept
**	Exponent transformation

Additional formula processing conventions include the following:

- Interactions are expanded according to hierarchical modeling principles.
- Transformed predictors may be included when specified explicitly.
- Intercepts are included by default unless removed.
- Categorical predictors must be encoded prior to estimation where applicable.

Formalizing these rules ensures consistent model specification and reduces ambiguity when constructing complex regression structures.

A.6 Estimation Mechanics

To preserve statistical transparency, Ravix documents the computational foundations underlying model estimation while avoiding unnecessary mathematical derivation. This section summarizes the estimation mechanics relevant to the modeling procedures covered in the text.

Key implementation elements include:

- Ordinary least squares estimation for linear regression models
- Parameter estimation based on minimization of sum of squared error
- Variance estimation derived from residual standard error
- Degrees of freedom determined by sample size and number of estimated parameters

These mechanics align directly with the statistical theory developed throughout the book. By maintaining transparency in estimation procedures, Ravix balances computational abstraction with methodological rigor.

A.7 Integration with Visualization and Diagnostics

Ravix is designed as an integrated analytical environment in which modeling, visualization, and diagnostic evaluation operate within a unified workflow. Rather than requiring users to manually extract fitted values, residuals, or predictor structures, Ravix maintains compatibility across analytical functions through a shared interface.

This integration is facilitated through three design features:

- Formula compatibility across modeling and visualization functions
- Direct accessibility of residuals and fitted values from model objects
- Consistent data inheritance across estimation, plotting, and diagnostics

As a result, users can transition seamlessly from estimation to graphical evaluation and diagnostic assessment without restructuring data or generating intermediate objects. Visualization and diagnostic procedures can operate on raw data, formula specifications, or fitted model outputs, depending on the analytical objective.

This shared analytical interface enables Ravix to function not only as a regression modeling framework but also as a cohesive visualization and diagnostic environment within a single workflow.

A.8 Reproducibility and Analytical Consistency

All modeling procedures presented in this text are implemented using the Ravix analytical framework described throughout the appendix. The modeling and visualization workflows are designed to produce consistent, interpretable outputs that align directly with the statistical methods developed in each chapter.

Datasets used in case studies and demonstrations are accessible through the Ravix data interface, allowing readers to replicate analyses and explore model variations within the same computational environment.

By standardizing model specification, reporting structures, and graphical outputs, Ravix supports reproducible regression analysis while maintaining a workflow that remains stable across instructional and applied settings. The version of Ravix used in this text is given below.

References

Anderson, D.R., D.J. Sweeney, T.A. Williams, J.D. Camm, and J.J. Cochran. 2016. *Statistics for Business & Economics*. Cengage Learning.

Joshua D. Angrist, and Jörn-Steffen Pischke. 2008. *Mostly Harmless Econometrics: An Empiricist's Companion*. Princeton: Princeton University Press.

Borle, S. 2017. Course notes for linear regression for business statistics. https://www.coursera.org/learn/linear-regression-business-statistics.

Brase, C.H., and C.P. Brase. 2016. *Understanding Basic Statistics, Enhanced*. Cengage Learning.

Central Intelligence Agency. 2021. The world factbook 2021. https://www.cia.gov/the-world-factbook/.

Chatterjee, S., and A.S. Hadi. 2006. *Regression Analysis by Example*. Wiley Series in Probability and Statistics. Hoboken: Wiley.

John Fox, Sanford Weisberg, Brad Price, Daniel Adler, Douglas Bates, Gabriel Baud-Bovy, Ben Bolker, Steve Ellison, David Firth, Michael Friendly, Gregor Gorjanc, Spencer Graves, Richard Heiberger, Pavel Krivitsky, Rafael Laboissiere, Martin Maechler, Georges Monette, Duncan Murdoch, Henric Nilsson, Derek Ogle, Brian Ripley, Tom Short, William Venables, Steve Walker, David Winsemius, Achim Zeileis, and R-Core. 2022. *R package 'car': Companion to Applied Regression*. (Version 3.1-1).

Francis Galton. 1989. Kinship and correlation. *Statistical Science* 4(2): 81–86.

Charles R. Harris, K. Jarrod Millman, Stéfan J. van der Walt, Ralf Gommers, Pauli Virtanen, David Cournapeau, Eric Wieser, Julian Taylor, Sebastian Berg, Nathaniel J. Smith, et al. 2020. Array programming with numpy. *Nature* 585: 357–362.

John D. Hunter. 2007. Matplotlib: A 2d graphics environment. *Computing in Science & Engineering* 9(3): 90–95.

James, G., D. Witten, T. Hastie, and R. Tibshirani. 2014. *An Introduction to Statistical Learning: with Applications in R*. Springer Texts in Statistics. New York: Springer.

Gareth James, Daniela Witten, Trevor Hastie, and Robert Tibshirani. 2021. *ISLR: Data for An Introduction to Statistical Learning with Applications in R*. R package version 1.4.

Gareth James, Daniela Witten, Trevor Hastie, and Robert Tibshirani. 2023. *An Introduction to Statistical Learning: with Applications in Python*. Berlin: Springer International Publishing.

Kutner, M.H., C.J. Nachtsheim, and J. Neter. 2003. *Applied Linear Regression Models*. The McGraw-Hill/Irwin Series Operations and Decision Sciences. Columbus: McGraw-Hill Higher Education.

D. P. McGibney, *Applied Linear Regression for Business Analytics with Python*, International Series in Operations Research & Management Science 378,
https://doi.org/10.1007/978-3-032-23806-1

Legendre, A.M. 1805. *Nouvelles méthodes pour la détermination des orbites des comètes*. Nineteenth Century Collections Online (NCCO): Science, Technology, and Medicine: 1780–1925. Paris: F. Didot.

Daniel P. McGibney. 2023. *Applied Linear Regression for Business Analytics with R*. International Series in Operations Research & Management Science. Berlin: Springer International Publishing.

Daniel P. McGibney. 2024. Business Regression Companion Resources. https://businessregression.com. Accessed 16 Feb 2026.

Daniel P. McGibney. 2026a. Ravix Analytics: Regression Modeling and Visualization Framework. https://ravixanalytics.com. Accessed 16 Feb 2026.

Daniel P. McGibney. 2026b. Ravix: Regression modeling and visualization framework for Python. https://github.com/danielmcgibney/ravix. GitHub repository.

Wes McKinney. 2010. Data structures for statistical computing in Python. In *Proceedings of the 9th Python in Science Conference*, 56–61.

Penn State Department of Statistics. 2022. Course notes for stat 501: Regression methods. https://online.stat.psu.edu/stat501/.

Skipper Seabold and Josef Perktold. 2010. Statsmodels: Econometric and statistical modeling with Python. In *Proceedings of the 9th Python in Science Conference*, 92–96.

Sullivan, M. 2010. *Statistics: Informed Decisions Using Data*. Sullivan Statistics Series. London: Pearson.

Venables, W.N., and B. D. Ripley. 2002. *Modern Applied Statistics with S*, Fourth edition. New York: Springer. ISBN 0-387-95457-0.

Pauli Virtanen, Ralf Gommers, Travis E. Oliphant, Matt Haberland, Tyler Reddy, David Cournapeau, Evgeni Burovski, Pearu Peterson, Warren Weckesser, Jonathan Bright, et al. 2020. Scipy 1.0: Fundamental algorithms for scientific computing in Python. *Nature Methods* 17: 261–272.

Walpole, R.E., R.H. Myers, S.L. Myers, and K. Ye. 2012. *Probability and Statistics for Engineers and Scientists*. Upper Saddle River: Prentice Hall.

Michael L. Waskom. 2021. Seaborn: Statistical data visualization. *Journal of Open Source Software* 6(60): 3021.

Wojtkiewicz, R.A. 2016. *Elementary Regression Modeling: A Discrete Approach*. Thousand Oaks: Sage Publications.

Zhao, C. 2022. Edinburgh airbnb data. https://www.kaggle.com/datasets/candicezhao28/edinburgh-airbnb-data .

Index

A
Alternative hypothesis, 120, 155, 306, 308, 309
ANOVA, 2, 150–152, 160, 306–308, 310–312, 317
Assumptions, *see* Model assumptions

B
Backward elimination, 268, 269, 273, 276, 285, 291
Best subsets regression, 276, 277, 284, 285, 312, 318
Box–Cox transformation, 223, 225, 226, 255
Boxplots, 238, 239, 241, 306, 307, 309

C
Categorical variables, 169, 170, 289, 295, 296
CDF, *see* Cumulative distribution function (CDF)
Centering, 188, 194, 305
Coefficient of determination, *see* R-squared
Coefficients, 64, 73, 113, 143, 170, 214, 286, 316
Confidence intervals, 8, 73, 79, 88, 90, 134, 135, 139, 141, 143, 146, 147, 317
Constant variance, 75, 217, 218, 247, 250
Cook's distance, 234, 235
Correlation coefficient, 50, 51, 53, 63, 64, 295
Critical value, 79–82, 89, 92–94, 120, 141, 142, 146
Cumulative distribution function (CDF), 78
Curvilinear, 183, 187, 194, 203, 216

D
Degrees of freedom, 10, 62, 75, 79, 81, 82, 84, 85, 88, 89, 91–94, 117, 119, 120, 122, 141, 146, 150, 151, 156, 320
  denominator, 63, 75, 91–94, 120, 151, 152
  numerator, 62, 91–94, 120, 150–152, 156
Descriptive statistics, 98, 236
Diagnostics, *see* Model diagnostics
Dummy variables, 8, 156, 169, 170, 176–178, 180, 181, 208

E
Error term, 74, 114, 140, 183, 213

F
Fitted values, 215–217, 226, 235, 287, 293, 301, 320
Forward selection, 271–275, 285, 292

H
Heteroscedasticity, *see* Nonconstant variance
Histograms, 6, 9, 36, 40, 46
Homoscedasticity, *see* Constant variance
Hypothesis testing, 2, 79, 80

D. P. McGibney, *Applied Linear Regression for Business Analytics with Python*, International Series in Operations Research & Management Science 378,
https://doi.org/10.1007/978-3-032-23806-1

I

Independence, 231
Interaction, *see* Interaction effects
Intercept, 57, 65, 97, 118, 119, 133, 143, 159, 171, 173–176, 178, 181, 182, 187–189, 191, 195–198, 205–207, 232, 271, 303, 319
Irreducible error, *see* Error term

L

Least squares, 2, 55, 57, 58, 73, 142, 194, 315, 317, 320
Leverage, 233–235
Linearity, 74, 215, 216, 243, 287, 293

M

Matplotlib, 6
Mean squared error (MSE), 62, 92, 120, 267
Model assumptions, 74, 75, 213, 214, 217, 247, 256, 287
Model selection, 266–268, 315, 317
Multicollinearity, *see* Variance Inflation Factor (VIF)

N

Nonconstant variance, 217–220, 226, 228, 252, 253, 255–257
Normal distribution, 79, 84, 214, 228, 229, 243, 288
Normality, 228–231, 243, 256, 287, 288, 301
Null hypothesis, 80, 85, 89, 90, 92, 117, 119, 120, 134, 155, 220, 230, 306, 308, 309
NumPy, 6, 64, 66, 68, 97, 99, 142, 174, 179, 189, 201, 208, 296
- corrcoef, 64, 65
- std, 99
- where, 174, 179, 189, 208, 296

O

Outliers, 35, 46, 103, 231, 301
Overfitting, 121

P

Pandas, 6, 65, 67, 68, 90, 97, 99, 105, 116, 126, 143, 157, 172, 179, 181, 208
- corr, 126
- cut, 305
- DataFrame, 36, 65, 67, 90, 99, 105, 116, 143, 157
- describe, 41, 98, 191, 236, 237
- head, 40, 97, 157, 199, 208, 237, 262, 289
- quantile, 41
- std, 99

Parsimonious model, 284, 286, 291, 303, 305, 311
Percent point function (PPF), 78
PPF, *see* Percent point function (PPF)
Prediction intervals, 139, 141, 146, 147, 215
Predictions, 2, 63, 67, 76, 140, 213, 231, 261
Predictor variables, 8, 65, 91, 113, 141, 169, 213, 261, 299
Python, 1, 2, 4–9, 40, 41, 68, 73, 81–83, 88–90, 92, 93, 98, 102, 113, 115–117, 125, 126, 132, 133, 135, 139, 142, 161, 178, 188, 208, 213, 216, 233, 234, 312

Q

Quantile function, *see* Percent point function (PPF)

R

Ravix package, 1, 7, 8, 42, 46, 64–66, 116, 135, 208, 218, 253, 291
- data
  - get_data, 40, 97, 98, 126, 184, 188, 262, 316
- diagnostics
  - ncv, 218, 219, 226, 253, 257, 259, 316
  - shapiro, 256, 259, 287, 316
  - vif, 125, 128, 135
- modeling
  - bsr, 277, 278, 280, 283, 285, 291, 303, 312, 318
  - intervals, 141, 146
  - ols, 64, 65, 90, 103, 105, 116, 129, 132, 135, 177, 178, 181, 184, 186, 189, 208, 219, 242, 250, 259, 263, 277, 306, 312, 316, 317
  - predict, 67
  - stepwise, 318
  - summary, 65, 88, 90, 116, 117, 157, 160, 172, 181, 184, 186, 208, 242, 250, 263, 270, 272, 278, 292, 306, 312, 316, 317

plotting
abline, 191, 196, 197
barplot, 100, 237, 316
boxplot, 35, 36, 102, 238, 259, 306, 307, 312, 316
hist, 36, 44, 230, 243, 259, 312, 316
plot, 36, 42, 66, 125, 126, 128, 135, 186, 187, 189, 191, 200, 201, 208, 216, 219, 226, 259, 295, 296, 298, 312, 316
plot_bsr, 277, 278, 280, 282, 291
plot_cor, 127, 135, 200, 208, 295, 316
qq, 259
transforms
box_cox, 225, 226, 255, 259
Regression, 1–10, 49, 52, 57, 58, 62–66, 68, 73–76, 79–81, 83, 85, 86, 88–92, 94, 97, 104, 108, 109, 113–118, 120, 122, 125, 130, 133, 135, 136, 140, 141, 143, 147, 151, 156, 158–160, 169–174, 176, 180, 181, 183, 191–194, 199, 203, 207, 208, 213–216, 219, 223, 225, 226, 232–235, 243, 247, 250, 256–259, 261, 262, 266, 267, 273–275, 286–288, 292, 293, 295, 303, 309–311, 315–321
linear, 49, 64, 73–75, 91–93, 108, 121, 129, 131, 132, 135, 139–141, 150–152, 215, 235, 252, 258
multiple, 113–115, 117, 118, 120–123, 131–133, 135, 140, 150, 155, 158, 164, 172, 202, 213, 214
polynomial, 193, 216
simple, 114, 214, 265
Residuals, 55, 76, 215, 287, 319
Residual standard error (RSE), 76, 320
Response variable, 43, 64, 73, 114, 169, 213
Root mean squared error (RMSE), *see* Residual standard error (RSE), *see* Residual standard error (RSE)

S
Scatterplots, 40, 43, 46, 52, 126, 127, 187, 295–297, 299
Scikit-learn, 4, 6, 7
SciPy, 6, 81, 82
cdf, 82, 83, 93, 117, 118
ppf, 81, 83, 84, 89, 92–94
Seaborn, 6, 66
Shapiro-Wilk test, 230, 256, 287, 288
Significance level, 92
Sklearn, *see* Scikit-learn
Slope, 52, 57, 58, 65, 84, 86, 88, 97, 104, 105, 119, 130, 131, 143, 191, 195–198, 221, 232, 250
Standard deviation, 9, 11, 41, 49, 75, 76, 79, 81, 85, 99, 101–103, 107, 215, 234
Standard error, 85, 86, 88, 117, 118, 140, 142, 146, 147, 215
Statsmodels, 6, 88
Sum of squares, 10, 11, 49, 50, 52, 62, 121, 122, 139, 160, 161

T
Test statistic, 80, 82, 83, 86, 88, 92, 94, 117, 120, 155, 218
Transformations, 8, 169, 220, 223, 259, 293, 295, 305
centering, 188, 194, 305

V
Variance, 2, 8–11, 62, 74, 75, 86, 139, 214, 215, 219, 223, 225, 226, 243, 293, 317, 320

MIX
Papier aus verantwortungsvollen Quellen
Paper from responsible sources
FSC® C105338

FSC
www.fsc.org

If you have any concerns about our products,
you can contact us on
ProductSafety@springernature.com

In case Publisher is established outside the EU,
the EU authorized representative is:
Springer Nature Customer Service Center GmbH
Europaplatz 3, 69115 Heidelberg, Germany

Printed by Libri Plureos GmbH
in Hamburg, Germany